W9-BSO-356

WITHDRAWN

BIOLOGY OF THE GENE

575.1
L664b

Biology of the
❧❦ GENE ❧❦

LOUIS LEVINE

Professor of Biology
City College of New York

SECOND EDITION
with 199 illustrations

WITHDRAWN

THE C. V. MOSBY COMPANY

SAINT LOUIS   1973

SECOND EDITION

Copyright © 1973 by The C. V. Mosby Company

All rights reserved. No part of this book may be reproduced
in any manner without written permission of the publisher.

Previous edition copyrighted 1969

Printed in the United States of America

International Standard Book Number 0-8016-2987-X

Library of Congress Catalog Card Number 72-89495

Distributed in Great Britain by Henry Kimpton, London

CB/CB/B   9   8   7   6   5   4   3   2

*To my professors of genetics who
inspired as they taught*

**Theodosius Dobzhansky
L. C. Dunn
Howard Levene
Francis J. Ryan**

29529

58252

Preface

This book is written for a single-semester undergraduate course in genetics. Two factors influence its content and organization. The first factor is the central and unifying position of genetics in the biological sciences, which requires considering such diverse areas as biochemistry, physiology, cytology, development, behavior, and evolution. The second factor is a desire to present, wherever possible, the experimental procedures and data that have led to our present concepts in genetics. This manner of presentation is an outgrowth of my firm conviction that education should not result solely in the accumulation of a body of information but must also include an understanding of the methodology used in the discipline studied and of the limitations in our knowledge about the subject.

It has been traditional to begin textbooks in genetics with a consideration of the work of Mendel and the principles of inheritance that his findings elucidated. However, recent discoveries on the nature of the genetic material and the manner in which it operates have shed new light on the chemical bases of many previously known genetic phenomena. In order that the student may benefit from this more recently acquired information in his study of inheritance patterns, the first topics considered are the nature and functions of hereditary material, the genetic code, and the physical basis of inheritance. Discussions then follow of gene interactions, multiple-factor inheritance, sex linkage and sex determination, chromosome numbers, chromosome mapping, chromosomal rearrangements, and extrachromosomal inheritance. In all these discussions, the insight provided by recent findings on the nature and functions of the genetic material is stressed. Next follows an analysis of concepts of the gene as a mutable unit, a recombinational unit, and a functional unit. This consideration of the different aspects of the gene leads into a study of the regulation of gene action and the control that genes have over metabolism, development, and behavior. The final topics discussed are the genetic composition of populations of organisms and the fate of genes in succeeding generations of a given population. Other subjects included in these latter discussions are the roles of mutation, selection, migration, and genetic drift in determining the genetic composition of different populations of a species and the permanent establishment of genetically diverse populations through species formation.

Wherever possible, the relationship of genetic phenomena to man has been stressed. This is done not only to increase the student's interest in the material being discussed but also because human genetics has become one of the very active fields in modern research. The present explosive period of genetic research and publication has added a tremendous body of information to an already considerable amount available from the older literature. Because of the necessity of limiting the various discussions in the book to workable dimensions, a list of further readings is provided at the end of each chapter. Some of these references are designed to give the student historical background for topics covered within the chapter, while others are provided to afford the student a wider acquaintance with modern research efforts in the field.

It has been four years since the first edition of this book was published. In that time, the field of genetics has continued its explosive contribution to virtually every aspect of biology and medicine. The current edition has been revised with this recent material in mind. New topics have been added, and a number of original discussions have been rewritten in the light of new information. In addition, there have been a number of substantial changes in the organization of the book's contents. It is hoped that these additions and rearrangements will help present the material of genetics as the exciting body of information it is.

Since its publication, this book has been used by many genetics classes both in the United States and abroad. I am deeply grateful to those instructors and students who have written either to the publisher or directly to me and have made suggestions for the book's improvement. Their recommendations have received careful study during the preparation of the present edition. I am especially indebted to two of my own students, Miss Harriet Rubenstein and Miss Helena Stuler, who spent many hours giving me the advantage of the student's point of view. The present effort has been guided by and has benefited a great deal from all of the above. I sincerely hope that those who use this book will continue to advise me of their reactions to it.

This book has benefited from the efforts of many persons. I am especially indebted to the following associates, each of whom graciously consented to review parts of the manuscript and also offered valuable suggestions for its improvement: Dr. Leonard C. Norkin, Dr. Rose R. Feiner, Dr. George C. Carmody, Dr. Betty C. Moore, Dr. Donald J. Komma, Dr. Norman M. Schwartz, Dr. Muriel Lederman, Dr. William N. Tavolga, and Dr. Frederick E. Warburton. I also wish to express my gratitude to the reviewer for the thoroughness with which he read the entire manuscript and for the comments he made. Any shortcomings of the book, however, are my responsibility alone. Finally, I wish to express my appreciation to Mr. Joseph T. Fevoli for preparing all the illustrations, Mrs. Rita Berkowitz and Mrs. Anne W. McCartney for performing the arduous task of typing the manuscript, and Mrs. Joan Sobel and Miss Helena Stuler for their aid in preparing the questions and problems. Credits for tables and figures from other publications are given in the legends according to the wishes of the author or publisher. I am indebted to the literary executor of the late Sir Ronald A. Fisher, F.R.S.; to Dr. Frank Yates, F.R.S.; and to Oliver & Boyd, Ltd., Edinburgh, for permission to reprint Tables 3 and 4 from their book *Statistical Tables for Biological, Agricultural and Medical Research*.

Louis Levine

Contents

13   Genes in populations, 283

14   Race and species formation, 312

BIOLOGY OF THE GENE

1 Nature and functions of hereditary material

Genetics is the branch of biology that deals with heredity and variation. This definition would appear to limit the area of study to the transmission of characteristics from parents to offspring and to the study of the variability in traits that may occur from one generation to the next. However, the field of genetics is in reality very broad, for within its scope lie such topics as (1) identification of the hereditary material and the nature of its chemical and structural properties; (2) study of the organization of the genes into chromosomes and the transmission of the chromosomes from parents to progeny either in asexual or sexual reproduction; (3) analysis of the interactions of the different genes and the role of environment in producing the characteristics of the individual; (4) study of the different types of genetic diversity that can occur and the consequences of this diversity to the individual and to the population. Genetic studies may attempt to gather information on such differing subjects as the origin of living material from nonliving matter and the future of mankind.

It is obvious that our knowledge concerning many of these topics is incomplete. In our discussions the limits of our information on a particular subject will be stated, and whenever possible, the areas in which future research might bring meaningful answers to questions will be indicated. Let us begin our study of genetics with a review of our knowledge of the nature and functions of the hereditary material.

IDENTIFICATION OF GENETIC MATERIAL

A prime goal in the study of genetics has been the identification and analysis of the actual genetic material. As it turned out, we had acquired much information about the modes of inheritance and about the relationships of genes to one another before we were able to demonstrate which of the many chemical compounds of the cell contains the genetic material. Furthermore, much of the chemistry of the genetic material was known long before its significance in genetics achieved wide understanding and acceptance.

As long ago as 1807 the distinction between inorganic and organic compounds was made. By 1820 it had become customary to think of the organic compounds as falling into one or another of three broad groups: the carbohydrates, the lipids, and the proteins. By the mid-nineteenth century it seemed clear that, of the three organic compounds, *proteins* were the most complicated in structure and the most important in function. However, in 1871 a chemist named F. Miescher reported that he had isolated from pus cells a substance that turned out not to be carbohydrate, lipid, or protein. Since he had obtained the new substance from nuclei, Miescher named it *nuclein*. Later the substance was discovered to have acid properties, and it was renamed *nucleic acid.*

Transformation in bacteria

The identification of the genetic material became a point of dispute between investigators who thought the material resided in the protein of the nucleus and those who believed it to be in the nucleic acids. A resolution of the problem did not come until 1944 when Avery, MacLeod, and McCarty reported their work on *transformation* in pneumococci. In man, pneumonia is sometimes caused by the bacterium *Diplococcus pneumoniae,* commonly known as *pneumococcus.* There are two types of pneumococcal cells. In one type of cell a considerable amount of polysaccharide material is secreted by the cell, and a large capsule

forms around the cell. The colony produced by these cells has a glistening appearance and is called "smooth" (S). In the other type of cell no polysaccharide slime layer is secreted by the cell. The colony formed by such cells has an irregular appearance and is termed "rough" (R). Smooth (S) cells are virulent and can cause pneumonia, but rough (R) cells are nonvirulent.

Investigations of the S form of *pneumococcus* revealed the existence of many kinds of capsules, each distinguishable on the basis of differences in the chemical composition of its polysaccharide. The S pneumococci were classified as type I-S, type II-S, type III-S, etc. Each type, when it divides, produces cells of the same type as the parental cell. Occasionally an S bacterium will change to an R bacterium (1 per 10^6 or

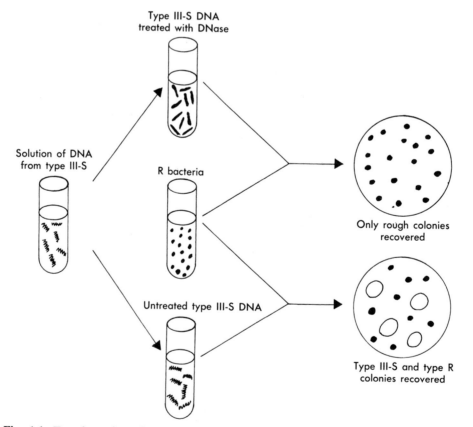

Fig. 1-1. Transformation of pneumococcus, using DNA extracted from heat-killed smooth cells. DNase is an enzyme that breaks down DNA.

10^7 cells). The reverse change, R to S, almost never occurs. When the R cells divide, they always give rise to more R cells. Avery and co-workers disrupted encapsulated cells of type III-S. They then fractionated the debris from the disrupted cells into its various chemical components (carbohydrates, lipids, etc.). After this they took R cells that had been derived from type II-S cells and separately mixed different samples of them with each of the cell components from the type III-S cells. Only the DNA fraction of the type III-S cells was found capable of transforming some of the unencapsulated (R) cells to encapsulated (S) cells. The transformed encapsulated cells were type III-S, the same type as the cell from which the DNA was obtained. This experiment, shown in Fig. 1-1, clearly demonstrated that the genetic property for pneumococcus capsule formation resides only in the DNA of the cell.

In recent years transformation of a large number and variety of hereditary characters has been demonstrated among different species of bacteria. In many cases the character transformed represents the ability to make an enzyme. One such example, in the bacterium *Bacillus subtilis,* is the enzyme β-galactosidase that catalyzes the hydrolytic splitting of the disaccharide lactose into galactose and glucose. Bacteria that either fail to make this protein or make an inactive form of it cannot ferment lactose. They are nutritionally defective and are called *auxotrophic,* whereas those that can utilize lactose are nutritionally competent and are characterized as *prototrophic.* When the DNA from prototrophic *B. subtilis* is mixed in the medium with auxotrophic cells, some of the auxotrophs are transformed into prototrophs.

The role of DNA in bacterial transformation was put to a further test. In a separate experiment the DNA was first exposed to the enzyme deoxyribonuclease (DNase), which selectively breaks down DNA. This treatment resulted in the abolition of the transforming activity. Treatment of the DNA with enzymes that degrade proteins was without effect. Thus the possibility that a protein contaminant might be responsible for the transforming activity can be excluded. The mechanism by which the DNA accomplishes transformation has been shown to be an incorporation of the donor DNA into the chromosome of the recipient cell. This mechanism will be considered further in the discussion of the phenomenon of crossing-over in Chapter 7.

• • •

Returning now to the problem of the identification of the hereditary material, we find that the identification of DNA as the genetic material was further substantiated by experiments involving viruses that multiply inside bacteria (bacteriophage).

Bacteriophage T2

Viruses are small but well-organized entities. They can be seen and photographed through the electron microscope. Their shapes are quite characteristic, and each kind of virus exhibits a specific geometric pattern in its external morphology. Viruses are also specific as to the type of organism and cell that each kind will normally attack. One type of virus attacks the leaves of the tobacco plant (tobacco mosaic virus); another attacks the motor cells in the spinal cord of man (poliomyelitis virus); still others invade bacteria (T2, T4, ΦX174, etc., collectively called *bacteriophage*). All viruses contain at least protein and nucleic acid. The nucleic acid is of either the RNA or the DNA type. Differences in the two types of nucleic acid will be discussed later in this chapter.

There are a number of viruses that will invade the colon bacillus *Escherichia coli,* multiply within it, split (lyse) the bacterium, thereby killing it, and at the same time release more bacteriophage. T2 is such a virus. It is composed of only protein and DNA. The protein is organized into a "coat" that covers a DNA "core." Hershey and Chase in 1952 reported on their studies of the T2 life cycle, made by using radioactive isotopes of sulfur (^{35}S) and phosphorus (^{32}P). Protein molecules almost invariably contain sulfur atoms but very few if any phosphorus atoms, whereas nucleic acids always contain phosphorus atoms but never any sulfur atoms. The viruses were allowed to go through a number of life cycles in bacterial hosts in a food medium containing ^{35}S and ^{32}P. As a result the coats of the viruses were labeled with ^{35}S and the DNA with ^{32}P. Hershey and Chase then infected unlabeled bacteria with labeled T2. They found that the ^{35}S remained outside the bacterial cell, in the protein coat. The coat could even be shaken off the bacterial cell shortly after contact without interfering with the subsequent production of new bacteriophage. The ^{32}P, on the other hand, entered the bacterial cell. On lysis, which occurs about 20 to 25 minutes after initial infection, one finds some 100 to 150 new

bacterial viruses complete with DNA cores and protein coats. The conclusion, as in the case of bacterial transformation, was simple and clear: the DNA contained the hereditary material (genetic code) that controlled the production of the entire virus. The experimental procedure is shown in Fig. 1-2.

Tobacco mosaic virus

A life cycle similar to that found in T2 has been discovered in the tobacco mosaic virus, which kills the cells of the tobacco leaf and results in a yellow mottling of the leaf (tobacco mosaic disease). This virus also consists of an outer protein coat and an inner core of nucleic acid. The nucleic acid in this case is RNA. In 1956 H. Fraenkel-Conrat reported that preparations of the viral RNA that had been completely

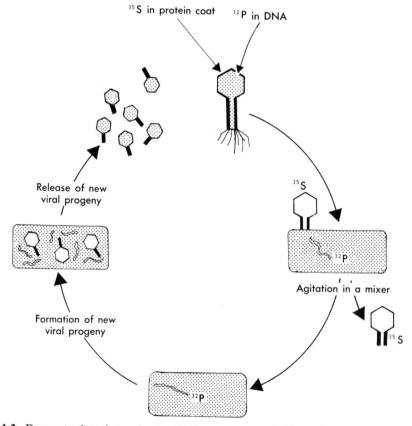

Fig. 1-2. Demonstration that only the DNA component of T2 carries genetic information. (From Watson, J. D. 1965. Molecular biology of the gene. W. A. Benjamin, Inc., New York.)

freed from protein could cause the tobacco mosaic disease when introduced into healthy tobacco plants. From the sap of the diseased plants, tobacco mosaic virus indistinguishable in all respects from the original strain could be isolated. Inoculation of a healthy plant with the protein of the virus did not produce the disease or the virus. This experiment showed that viral RNA alone can direct the formation of normal virus whose protein component is identical to that of the original virus. The experiment is diagrammed in Fig. 1-3.

• • •

The investigations outlined above have identified the nucleic acids as the hereditary material. They have also shown that, depending on the organism, DNA or RNA can be the genetic material. It is important at this time to establish what the functions are of the hereditary material of an organism.

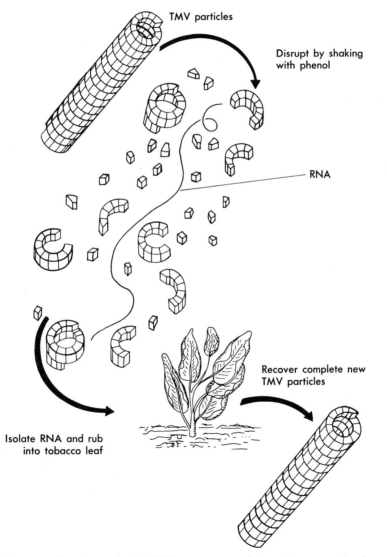

TMV particles

Disrupt by shaking with phenol

RNA

Recover complete new TMV particles

Isolate RNA and rub into tobacco leaf

Fig. 1-3. Demonstration that only the RNA component of the tobacco mosaic virus carries genetic information. (From Stahl, F. W. 1964. The mechanics of inheritance. Prentice-Hall, Inc., Englewood Cliffs, N. J.)

First, the hereditary material must be capable of duplicating itself. Second, it must be capable of directing the formation of all the other structures of the organism. This latter function can be achieved if the genetic material can direct the formation of both structural proteins and enzymatic proteins. The enzymes can, in turn, direct the metabolism of the organism. To understand how the genetic material carries on its necessary functions, we shall now study the organization of DNA and of RNA.

NATURE OF THE HEREDITARY MATERIAL

Nucleic acids, as is also true of proteins, are examples of *polymeric molecules;* that is, they are formed by the combination of smaller building blocks. The building blocks of the nucleic acids are called *nucleotides.* Each nucleotide consists of three subunits: a nitrogen-containing purine or pyrimidine derivative called the nucleotide *base,* a 5-carbon *sugar,* and *phosphoric acid.* There are two types of nucleic acid: *ribonucleic acid* (RNA) and *deoxyribonucleic acid* (DNA). The nucleotides of RNA and DNA are called, respectively, *ribonucleotides* (ribotides) and *deoxyribonucleotides* (deoxyribotides).

One of the ways in which RNA and DNA differ from each other is in the sugar part of their nucleotides. In the nucleotides of RNA, a hydroxyl (OH) group is present at the 2' position. In DNA the 2' oxygen is absent (hence the prefix deoxy); instead of the hydroxyl group, there is a hydrogen atom. The structural formulae of the two pentoses are shown in Fig. 1-4. Another dif-

ference between RNA and DNA lies in the kinds of bases found in each. There are four principal kinds of ribonucleotides. They differ from each other solely in the nature of their bases. The four bases that occur to an important extent in RNA are *adenine, uracil, cytosine,* and *guanine.* Of these, adenine and guanine are purines, whereas uracil and cytosine are pyrimidines. The bases appearing in DNA are the same, except that uracil is replaced by its methylated derivative *thymine.* Other bases also occur and will be considered at the appropriate times. Structural formulae of the common bases are shown in Fig. 1-5.

That part of a nucleotide which consists of its base plus its sugar is called a *nucleoside.* It is named according to its base as follows, for the ribose series (riboside):

Base	Nucleoside	Abbreviation
Adenine	Adenosine	A
Uracil	Uridine	U
Cytosine	Cytidine	C
Guanine	Guanosine	G

For the deoxyribose series (deoxyriboside) each name is preceded by deoxy, as

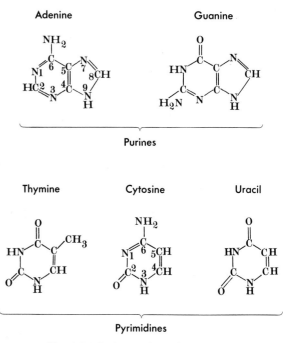

Fig. 1-5. The bases of DNA and RNA.

Fig. 1-4. Pentoses found in nucleic acids.

follows: deoxyadenosine, deoxycytidine, de-oxyguanosine. An exception is the nucleo-side that contains thymine, which is called thymidine. It is unnecessary to specify the prefix deoxy for the nucleoside containing thymine, as this base is found only in DNA. The abbreviations for the four deoxyribo-nucleosides are dA, dC, dG, and T.

The complete nucleotides (base plus sugar plus phosphoric acid) of the ribose series are named as follows:

Base	Nucleotide	Abbreviation
Adenine	Adenylic acid	AMP
Uracil	Uridylic acid	UMP
Cytosine	Cytidylic acid	CMP
Guanine	Guanylic acid	GMP

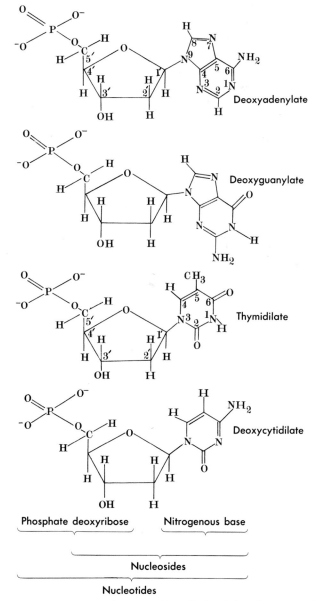

Fig. 1-6. The four common deoxyribonucleotides. Each of the nitrogenous bases is attached to deoxyribose by a bond that connects a ring nitrogen atom of the base to the number 1 atom of the deoxyribose. The suffix "ate" equals "acid."

The nucleotides can also be named adenosine monophosphate, etc. As before, the names of the deoxyribose series have the prefix deoxy, as in deoxyadenylic acid (deoxyadenosine monophosphate), abbreviated dAMP, etc., with the exception of thymidylic acid (thymidine monophosphate), abbreviated TMP. Examples of the four common deoxyribonucleotides are shown in Fig. 1-6.

One does not always find the nucleotides of the cell in the form of nucleoside monophosphates. They may be found to contain diphosphate or triphosphate groups. These, likewise, are named from the corresponding nucleoside—for example, adenosine diphosphate (ADP) and adenosine triphosphate (ATP). The importance of the nucleoside triphosphates lies in the fact that the synthetic machinery of the living cell is not capable of assembling either DNA or RNA directly from nucleotides containing a monophosphate or a diphosphate. The cellular enzymes that catalyze the formation of the nucleic acids can act only on nucleotides containing a triphosphate group. The synthetic process results in the splitting off of two of the three phosphates of each nucleotide and the involvement of the third in a phosphodiester bond. The precursors of DNA are the deoxyribonucleoside triphosphates: dATP, dCTP, dGTP, and TTP. RNA is synthesized from the ribonucleoside triphosphates: ATP, CTP, GTP, and UTP. The ribonucleoside triphosphates have, quite apart from their role as precursors of RNA, the indispensable role of energy carriers. The energy-liberating process involves the removal of one or two phosphate groups. The energy released is utilized in such energy-consuming reactions as muscle contraction, emission of light by fireflies, photosynthetic formation of sugars by green plants, and formation of nucleic acids and proteins.

As stated earlier, nucleic acids are polymeric molecules. The linkage between adjacent nucleotides is of the ester type, in which the 3′ and 5′ hydroxyls of two different sugars form a double ester with phosphoric acid. This is known as a phos-

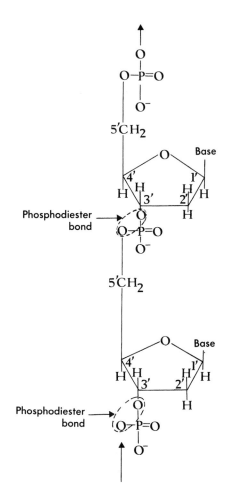

Fig. 1-7. Polydeoxyribonucleotide. "Base" is either a purine or a pyrimidine.

phodiester bond and is shown in Fig. 1-7. The nucleotides, linked in this fashion, form a *polynucleotide chain,* in which the sugar of one nucleotide is bound to the phosphate group of another nucleotide and so on.

Geometric organization of DNA

A knowledge of the component parts of DNA does not, by itself, tell us how the molecule is geometrically organized. Any model of the physical structure of the molecule must also suggest hypotheses as to how DNA carries on its two essential functions: self-replication and protein synthesis.

The first step in understanding the organization of DNA came from the work of

Table 1-1. Base composition of DNA from various organisms

	Adenine %	Thymine %	Guanine %	Cytosine %
Man (sperm)	31.0	31.5	19.1	18.4
Salmon	29.7	29.1	20.8	20.4
Sea urchin	32.8	32.1	17.7	17.7
Yeast	31.7	32.6	18.8	17.4
Mycobacterium tuberculosis	15.1	14.6	34.9	35.4
Escherichia coli	26.1	23.9	24.9	25.1
Vaccinia virus	29.5	29.9	20.6	20.3
E. coli bacteriophage T2	32.6	32.6	18.2	16.6*

*5-Hydroxymethyl cytosine.

E. Chargaff, which was reported in 1950. The chemical analyses of DNA obtained from many different organisms showed that there exists a specific quantitative relationship between the different nucleotides. For each nucleotide of adenine, there is one of thymine; and for each nucleotide of guanine, there is one of cytosine. But, although the A:T and G:C ratios are one to one in all DNA, the ratio of A + T to G + C varies in different organisms. Examples of the percents of the various nucleotides obtained from different organisms are shown in Table 1-1.

As stated earlier, the DNA of some organisms contains a base that is not adenine, thymine, guanine, or cytosine. An example of this is given in Table 1-1, which shows that the DNA of the bacterial virus T2 does not contain cytosine but instead contains 5-hydroxymethyl cytosine. Note that the amount of 5-hydroxymethyl cytosine in T2 roughly equals the amount of guanine found in the virus. This indicates that in T2, 5-hydroxymethyl cytosine takes the place that cytosine has in the DNA of other organisms. Other cases of base substitution are known. However, these are rare, and their role in the evolution of the organisms involved is not understood.

Although chemical analysis provided information regarding the composition of DNA, it did not give any indication of the physical structure of DNA. That information came from a study by M. H. F. Wilkins in 1953 of the *x-ray diffraction pattern* of DNA fibers. In such studies a beam of x rays is passed through a substance and allowed to fall on a photographic plate. If the substance has no regular arrangement of its subunits, the x rays will produce a central spot on the photographic plate, marking the position of the main beam. About this central spot is a light fog caused by the effect of diverted x rays. This fog fades off continuously with increasing distance from the central spot. However, if the subunits of the material have an orderly arrangement, the x rays will be diverted in some directions more than in others. If the arrangement of the subunits is in some regularly repeated pattern, there will be a reinforcement of some of the x-ray bands, resulting in the production of very sharp bands on the photographic plate. Between these bands there will be completely clear areas. Wilkins found that the DNA from several different organisms produced very definite and almost identical diffraction patterns. This meant that the nucleotides within the different DNA macromolecules were oriented spatially in the same fashion. The diffraction patterns further revealed that the DNA was helical in structure. An example

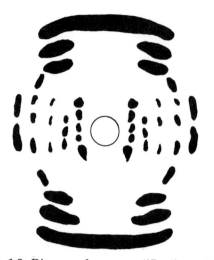

Fig. 1-8. Diagram of an x-ray diffraction pattern of DNA. (From Wilkins, M. H. F. 1963. Science **140:**941-950.)

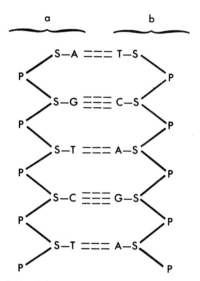

Fig. 1-10. Diagram of section of double helix: **P,** phosphate; **S,** sugar; **A,** adenine; **T,** thymine; **G,** guanine; and **C,** cytosine; **a** shows one strand and **b** the other strand. Double and triple broken lines in center represent hydrogen bonds holding the two strands together.

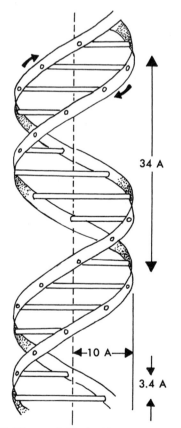

Fig. 1-9. Watson-Crick double-stranded helix configuration of DNA.

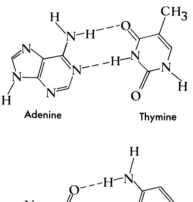

Adenine   Thymine

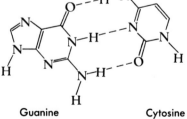

Guanine   Cytosine

Fig. 1-11. Examples of base pairs formed between single DNA strands.

or an x-ray diffraction pattern of DNA is shown in Fig. 1-8.

The information obtained from the x-ray diffraction studies and from the chemical analyses led J. D. Watson and F. H. C. Crick in 1953 to propose the model for DNA structure shown in Fig. 1-9. They hypothesized that DNA consists of two helically coiled polynucleotide chains *(double helix)*. The ribbon or backbone of each chain is made up of phosphates and pentose sugars, while the nucleotide bases project between them as bars and hold the two strands of the double helix together through chemical bonds. This is diagrammed in Fig. 1-10. The coiling of the double helix is right-handed, and a complete turn occurs every 34 A. (Angström units). Since each nucleotide occupies 3.4 A. along the length of a strand, ten nucleotides occur per complete turn.

As just mentioned, the two strands of the double helix are held together by chemical bonds between the bases on the adjacent strands. The simplest hypothesis, based on the work of Chargaff, was that adenine is always linked to thymine and guanine is always linked to cytosine. The forces responsible for this linkage are the *hydrogen bonds* (H bonds) between the paired bases. It is true that hydrogen can form only one true covalent bond, but under certain conditions a hydrogen that is chemically bound to one atom can form a weak linkage of a non-covalent type to a second atom. Hydrogen bonds are much weaker than ordinary covalent bonds and may be ruptured easily by an increase in temperature and then reversibly formed again when the temperature is lowered. Of great importance to the Watson-Crick model of DNA is the fact that H bonding between different purines and pyrimidines is highly specific. A nucleotide of adenine forms two H bonds with thymine, whereas one of guanine forms three H bonds with cytosine. A diagram of the different types of H bonds is shown in Fig. 1-11. Any "abnormal" combinations of purines and pyrimidines are very difficult to effect and are extremely unstable. Yet, as we shall discuss in a later chapter, such abnormal combinations do occur and they provide one of the possible mechanisms of producing mutations.

The arrangement whereby a purine is always linked to a pyrimidine results in a double helix with a uniform diameter of 20 A. This satisfies estimates of the width of the DNA made through electron microscope studies. It follows from the Watson-Crick model of DNA that the base sequences of the two strands are complementary. That is, the sequence of either strand may be converted to that of its partner by replacing adenine by thymine, and vice versa, and guanine by cytosine, and vice versa. Schematically the sequence of bases in a segment of a hypothetical DNA molecule might be as follows:

$$- A - T - T - A - C - A - G - G - C -$$
$$\cdot \quad \cdot \quad \cdot \quad \cdot \quad \cdot \quad \cdot \quad \cdot \quad \cdot \quad \cdot$$
$$\cdot \quad \cdot \quad \cdot \quad \cdot \quad \cdot \quad \cdot \quad \cdot \quad \cdot \quad \cdot$$
$$- T - A - A - T - G - T - C - C - G -$$

A diagram of an extremely short section of a DNA molecule is shown in Fig. 1-12.

We are going to find throughout our study of genetics that there are exceptions to every one of our general statements. This is also true of our concept of the double-helix nature of DNA. There are a number of virus particles that consist of protein and a single strand of DNA. The most studied of these is a virus named ΦX174 that attacks the colon bacillus *Escherichia coli*. Its single-stranded nature is reflected in the ratios of its different nucleotides. It contains adenine, thymine, guanine, and cytosine in the ratios of 1 to 1.33 to 0.98 to 0.75, respectively. In addition to the evidence obtained from chemical analysis, we find that x-ray diffraction photographs do not yield the typical DNA picture. We shall return to ΦX174 in later discussions. For the present, however, let us consider how the Watson-Crick model of DNA allows for the required functions of the hereditary material.

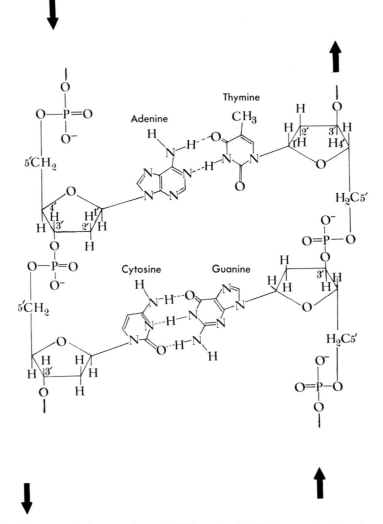

Fig. 1-12. Diagram of short section of DNA molecule. Note opposite direction of sugar-phosphate linkages in the two strands.

DNA replication

One of the requirements of any model of the hereditary material is that the model provide a reasonable explanation of how the hereditary material replicates itself. In discussions of the geometric organization of DNA, it was pointed out that the base sequences of the two strands are complementary. Watson and Crick hypothesized that, prior to duplication, the hydrogen bonds are broken and the two chains unwind and separate. Each chain then acts as a template for the formation onto itself of a new complementary companion chain so that eventually there are two pairs of chains where before there was only one. Under these conditions the sequence of the pairs of bases would have been duplicated exactly. A diagrammatic representation of the hypothesis is shown in Fig. 1-13. Under this model of replication, each newly formed double helix of DNA contains one "old" and one "new" polynucleotide strand.

The mechanism of DNA replication that

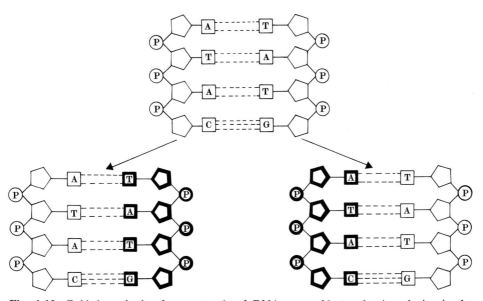

Fig. 1-13. Guided synthesis of new strands of DNA upon old strands. An adenine in the original strand is replaced by a thymine in the new strand, a guanine is replaced by a cytosine, and so forth.

Watson and Crick proposed has received support through a series of experiments with bacteria, algae, and other organisms. The now classic experiment in this field was reported by Meselson and Stahl in 1958. They made use of a *density gradient* technique that permits the detection of the different isotopes of an element. DNA can be labeled with a number of different isotopes, including those of nitrogen. This element has an isotope known as ^{15}N that has a greater mass than does ordinary nitrogen, ^{14}N. DNA molecules consisting of solely ^{15}N can be distinguished, by this technique, from those containing mixtures of ^{15}N and ^{14}N, which in turn can be distinguished from molecules containing only ^{14}N. When using the density gradient technique, one places a sample containing all three types of nitrogen molecules in a concentrated solution of a heavy salt, such as *cesium chloride,* and spins the mixture at high speed in an ultracentrifuge. The cesium chloride is more dense than the water and tends to sediment to the outside of the tube. This tendency to sediment is opposed by diffusion; after several hours the concentration distribution of the CsCl in the cell becomes essen-

tially stable, with the highest salt concentration at the outer, or centrifugal, end of the tube and the lowest concentration at the inner, or centripetal, end. The density of the resulting solution increases smoothly from the inner to the outer end of the cell, forming a density gradient. Each of the three types of nitrogen-containing DNA molecules in the sample will come to equilibrium in the gradient at the point that corresponds to its own effective density. The bands into which the DNA collects are of finite width, since the DNA is also subject to diffusion and its bands result from an equilibrium between the sedimentation and diffusion forces acting on it. After the equilibrium density gradient has been formed and while the material is still spinning in the ultracentrifuge, a beam of ultraviolet (UV) light with a wavelength of 260 nm. (nanometer, 10^{-9} m.) is directed at the centrifuge tube. Each of the three layers of DNA concentration will absorb the UV rays and produce a black band. A photograph may be taken of the centrifuge tube, yielding a permanent record of the banding patterns and their locations in the tube.

A further test for this layering effect of

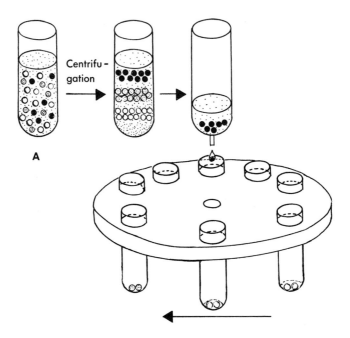

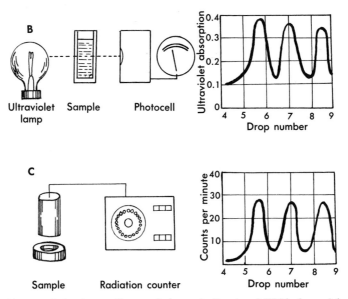

Fig. 1-14. Diagram of density gradient technique. **A,** Bands of DNA formed by centrifugation. **B,** UV light absorption curves, indicating which drops of solution are rich in DNA. **C,** Radiation curves, indicating which drops are rich in DNA that has been tagged with radioactive isotopes. (From "Single-Stranded DNA" by R. L. Sinsheimer. Copyright © 1962 by Scientific American, Inc. All rights reserved.)

the different isotopes of nitrogen can be made. After the centrifugation is completed, a small hole is bored in the bottom of the plastic centrifuge tube and successive drops of the solution are collected separately. The first drops represent the denser end of the solution and the last drops represent the lighter end, with a regular progression in between. Each sample is then placed in a spectrophotometer equipped with an ultraviolet lamp, and the absorption of UV light by each sample is measured by a photocell. Only the DNA in the solution absorbs the UV rays. The amount of light absorbed is then plotted against the particular drop of solution. When the drops that come from a layer of DNA are measured, a sharp rise in the amount of UV light absorbed is indicated. This procedure is shown in Fig. 1-14, *A* and *B*. With our sample, there are three sharp peaks on the graph.

Meselson and Stahl grew *E. coli* for several generations on a medium that contained a nitrogen source consisting solely of ^{15}N. They then transferred these cells, whose DNA was fully labeled with ^{15}N, to a food medium that contained only ^{14}N as a nitrogen source. After waiting for the cells to divide, they sampled the culture and found that the DNA in the cells corres-

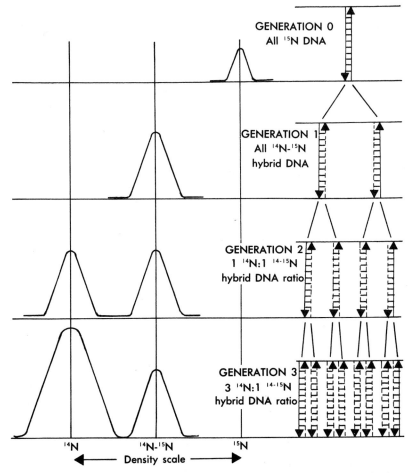

Fig. 1-15. Schematic comparison of the Watson-Crick proposal for DNA replication and the Meselson-Stahl experiment. (From Johnson, W. H., Laubengayer, R. A., Delanney, L. E., and Cole, T. A. 1966. Biology. 3rd ed. Holt, Rinehart & Winston, Inc., New York.)

ponded to the $^{14}N^{15}N$ (hybrid) variety. After the second doubling of the cells, and hence of the DNA in the ^{14}N medium, they found two kinds of DNA in equal amounts: $^{14}N^{15}N$ and $^{14}N^{14}N$. After the third cell division, they found the same amount of hybrid DNA as in the previous sample, but three times as much of the $^{14}N^{14}N$ DNA was present. Their results are shown in Fig. 1-15. These findings are interpreted to lend support to the scheme proposed by Watson and Crick in which every newly formed double helix of DNA contains one old and one new polynucleotide strand. This type of DNA replication, in which one strand of the double helix acts as a template for the formation onto itself of a new complementary companion strand, is called *semi-conservative* replication.

Other experiments, which will not be considered in detail, have demonstrated that DNA replication requires the action of a specific enzyme. Initially this enzyme was called simply *DNA polymerase* because its function is to form a polymer of DNA. For reasons that will become apparent later in this chapter, it is necessary to specify that the enzyme uses a previously existing molecule of DNA as a template. As a result of this requirement in specifying the polymerase, the more appropriate name of the

enzyme is *DNA-directed DNA polymerase* (also called DNA-dependent DNA polymerase).

Having considered the nature of the hereditary material and its replication, we shall now turn our attention to the problem of protein synthesis. Here, too, we shall consider whether the Watson-Crick model permits a reasonable explanation of the process.

Proteins and RNA

Proteins, as stated earlier, are polymeric molecules. All natural proteins are, in effect, linear combinations of smaller molecules, namely, *amino acids*. The amino acids are joined through the elimination of a molecule of water between the carboxyl and the amino groups of adjacent amino acids, forming a linkage of the amide (—CO-NH—) type. This kind of linkage is called a *peptide bond,* and the linear chains of amino acids built up thereby are known as polypeptides. An illustration of the formation of a peptide bond is shown in Fig. 1-16. The formation of a peptide bond requires the action of the enzyme *peptide polymerase* and the energy-rich molecule *GTP* (analogous to ATP, with adenine replaced by guanine). There are about twenty amino acids of common occurrence. The

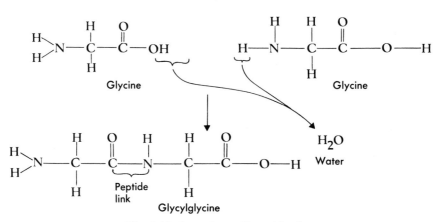

Fig. 1-16. An amino acid combination.

chemical and biological individuality of each protein is a function of the amino acid sequence of that protein.

Studying the various activities of a cell, one finds a close correlation between the ability of a cell to carry out protein synthesis and the cell's content of RNA. Cells that are physiologically very active but synthesize very little protein (e.g., heart, voluntary muscle, kidney) are poor in RNA. On the other hand, large amounts of RNA are always found in the nucleoli and cytoplasm of cells that produce large amounts of protein (e.g., pancreas, liver). An analysis of the RNA in cells shows that there are three distinct types of RNA: ribosomal, transfer (soluble, adaptor), and messenger.

Ribosomal RNA (rRNA) is found in association with protein. This type of RNA constitutes some 85% to 90% of the total cellular RNA. The combination of ribosomal RNA and protein forms a cellular body called the *ribosome*. Ribosomes, found mainly in the cytoplasm, are classified by their sedimentation constants (using the Svedberg unit, s), which are determined by their rates of sedimentation in an ultracentrifuge. The ribosomes from *E. coli* consist of two subunits whose sedimentation constants are 30s and 50s. These subunits may unite, forming a larger complex of 70s.

The association and dissociation of the ribosomal subunits in vitro are controlled by the magnesium ion concentration in the suspending medium. Ribosomal RNA of *E. coli* falls into three classes whose sedimentation constants are 5s, 16s, and 23s. All are single-stranded and have unequal amounts of adenine and uracil and of guanine and cytosine. The characteristics of ribosomes taken from *E. coli* are shown in Fig. 1-17. The *E. coli* ribosomes contain a ratio of 65% RNA and 35% protein. In mammalian cells the ribosomes consist of 40s and 60s subunits, and the ratio of RNA to protein is one to one. Ribosomal RNA from different organisms will have different sedimentation constants (s values).

Ribosomes are the sites of protein synthesis. This has been shown by the use of radioactive amino acids in the food medium. Shortly after these amino acids enter a cell, they become associated with the ribosomes, which then exhibit radioactivity. The radioactivity of the ribosomes gradually disappears, and the newly formed protein containing the "labeled" amino acids exhibits radioactivity. Although the function of the ribosomes is clear, it is not at all obvious why the ribosomes should have such complex structures nor why they should contain RNA, since the ribosomes appear to be

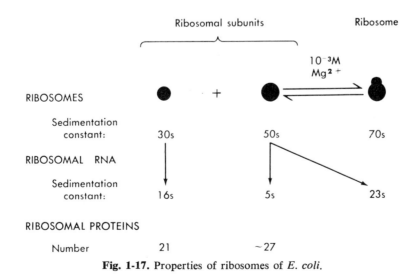

Fig. 1-17. Properties of ribosomes of *E. coli*.

nonspecific workbenches on which any number of proteins can be made.

Transfer RNA (tRNA) is also called soluble or adaptor RNA. The term is applied to a group of small molecules, each one of which has a specific attraction for one of the amino acids. All tRNA molecules have a C-C-A sequence at one end and usually have a G at the other, as shown here:

G . . . . RNA . . . . C-C-A

The molecule, consisting of about eighty nucleotides, is thought to be composed of

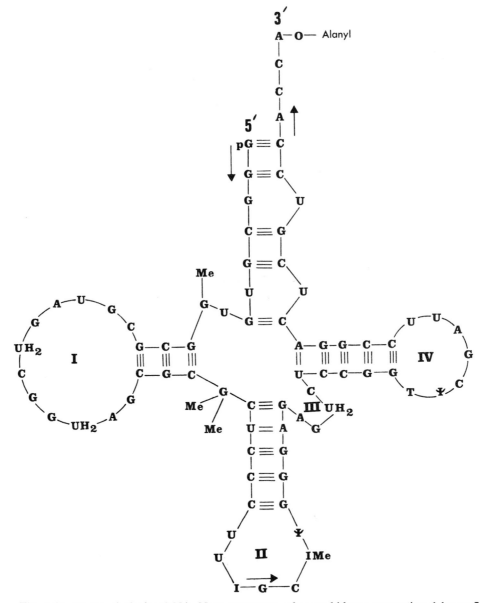

Fig. 1-18. Model of alanine-tRNA. Note occurrence of some hitherto unmentioned bases: **I,** inosine, a purine base that closely resembles guanine; **IMe,** methylinosin; **Me-G,** methylguanosine; Ψ, pseudouridine; **UH₂,** dihydrouridine. (From Jukes, T. H. 1966. Biochem. Biophys. Res. Commun. **24:**744-749.)

four helical regions in the form of a "cloverleaf." This is illustrated in Fig. 1-18, which shows the complete nucleotide sequence of the tRNA that is specific for the amino acid *alanine,* as reported by R. W. Holley and co-workers in 1965. A study of the sedimentation constants of the various tRNAs shows them to be about 4s. The tRNAs comprise about 5% of the cell's total RNA. By the use of radioactively labeled amino acids, the amino acids were found to arrive at the ribosomes individually, with each amino acid attached to the C-A terminal group of its specific tRNA.

Messenger RNA (mRNA) is the third type of RNA involved in protein synthesis. Its existence was hypothesized as a necessary intermediary between the genes, located in the nucleus, and the ribosomes, located in the cytoplasm where the proteins are actually formed. As its name implies, mRNA carries the genetic code contained in the structure of the DNA. Since the DNAs from various organisms differ only in the sequence of their bases, mRNA must reflect this difference in base sequence. The simplest model to hypothesize for mRNA formation is to assume that it is complementary to one of the strands of the DNA double helix. If such is the case, one should find that the base compositions of the mRNA and the DNA from the same organism are equal (allowing for the substitution of uracil in mRNA for thymine in DNA). This fact was demonstrated by Volkin and Astrachan in 1956 for *E. coli* cells infected with phage

T2. They showed that the RNA synthesized in the bacterial cell after infection had the same base composition as the DNA of T2 and not that of *E. coli,* whereas the RNA synthesized in the bacterial cell before infection reflected the base composition of *E. coli* DNA. The type of data obtained is shown in Table 1-2. Neither the ribosomal nor the transfer RNA of the *E. coli* cell changes after infection with phage T2.

The proof that mRNA is a complementary copy of DNA was demonstrated by the use of another technique. In 1960 Marmur and Lane showed that heating DNA molecules to temperatures just below 100° C. breaks the hydrogen bonds holding the two strands together, and the strands quickly separate from each other (DNA *denaturation*). If the temperature is lowered, the complementary strands again form the correct hydrogen bonds, and the double helical form is regained (*renaturation* of DNA). These investigators showed further that reconstitution of the double-strand molecule occurs only between strands that originate from the same or from closely related organisms. This suggested that double-strand hybrid structures could be formed from mixtures of single-strand DNA and RNA and that the appearance of such hybrids could be accepted as evidence of the complementarity of their base sequences. The isolation and identification of DNA/RNA hybrids, in a solution that also contains pure DNA and pure RNA, can be made by the density gradient technique described earlier, pro-

Table 1-2. Comparison of the base compositions of mRNA in *E. coli* cells, before and after T2 infection, with the base compositions of T2 DNA and *E. coli* DNA

Source of DNA	$\dfrac{A + T}{G + C}$	$\dfrac{A + U}{G + C}$	
		Before infection	After infection
T2	1.84	—	1.86
E. coli	0.97	0.92	—

vided the DNA and RNA are labeled with different radioactive isotopes. The procedure is shown in Fig. 1-14, *A* and *C*.

In 1961 Hall and Spiegelman arranged experimentally to have T2 RNA labeled with radioactive phosphorus (^{32}P) and T2 DNA labeled with radioactive hydrogen (^{3}H). The beta particles emitted by ^{32}P have a characteristic energy different from those emitted by ^{3}H; therefore the isotopes can be assayed in each other's presence. The existence of the DNA/RNA hybrid in the centrifuged fraction would be indicated by the appearance of a layer containing both the ^{32}P label of the RNA and the ^{3}H label of the DNA. They found that the mRNA present in the *E. coli* cell after T2 infection hybridized with T2 DNA but not with *E. coli* DNA. Conversely, the mRNA present in the *E. coli* cell before T2 infection hybridized with *E. coli* DNA but not with T2

DNA. An example of the type of results obtained is shown in Fig. 1-19.

Although not shown in Fig. 1-19 as a separate curve, a variable number of DNA/DNA molecules are formed in this type of experiment. They represent the result of the coming together of some of the single, complementary DNA strands of the virus T2 to form double helixes. The DNA/DNA molecules, as seen by the very sharp rise of ^{3}H in Fig. 1-19, are located in that fraction of the liquid between 15 and 20 ml. above the bottom of the tube. This is the same fraction of the liquid that contains the DNA/RNA hybrid molecules. Since the DNA/RNA and DNA/DNA molecules are found together, the labeling of DNA with radioactive isotopes can be omitted. The location of any DNA/DNA molecules can be ascertained by using UV radiation and checking for its absorption pattern as shown

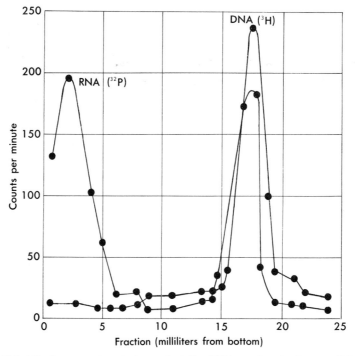

Fig. 1-19. Hybridization experiment shows that the RNA produced after a cell has been infected with the T2 virus is genetically related to the DNA of the virus. Although some of the RNA has been driven to the bottom of the sample tube, much of it has hybridized with the lighter DNA fraction and thus appears between 15 and 20 ml. above the bottom. (From Hall, B. D., and Spiegelman, S. 1961. Proc. Nat. Acad. Sci. U. S. A. **47:**137-163.)

in Fig. 1-14, *A* and *B*. If there is a corresponding, overlapping radioactivity curve of RNA (similar to that shown in Fig. 1-19), the formation of DNA/RNA hybrid molecules has been demonstrated. The omission of DNA labeling has become routine in this type of experiment.

Demonstration that there is a type of RNA which acts as a messenger of the gene and does this by being complementary to the base sequence of the DNA was very clear in the foregoing experiments. This discovery limited the role of DNA in protein synthesis to that of *transcription,* namely, the production of mRNA that would reflect the genetic code contained in the base sequence of the DNA. The sizes of the different mRNAs are very variable. They constitute about 5% of the total cellular RNA.

Transcription requires the action of a specific enzyme, which was initially called *RNA polymerase* because mRNA is in fact a polymer of RNA. However, as was the case with DNA replication, here, too, it is necessary to specify that the enzyme uses a preexisting molecule of DNA as a template. Therefore a more appropriate name for the enzyme is *DNA-directed RNA polymerase* (also called DNA-dependent RNA polymerase).

One strand or two? The proof that messenger RNA was formed by complementarity from DNA raised a number of questions. One question was whether, in the formation of mRNA, one strand or both strands of the DNA double helix were transcribed. This problem was solved through the use of ΦX174, a small DNA virus consisting of only a single strand. When this virus infects an *E. coli* cell, the single strand serves as a template for the synthesis of a complementary strand, resulting in a normal double-stranded DNA molecule. This molecule is called the *replicating form* of the virus and can be isolated for experimental purposes. In the experiment, ΦX174 infected *E. coli* in the presence of ^{32}P, and labeled mRNA was extracted from the bacterial cells. The labeled mRNA was then brought together with an unlabeled single-strand DNA of ΦX174 and, separately, with a denatured sample of an unlabeled double-strand replicative form (RF). As indicated previously, labeling the DNA with radioactive elements as was done in earlier experiments had been found to be unnecessary. The results of the experiments with ΦX174 were clear-cut. No hybrids were formed with the single-strand DNA, but excellent hybrids were produced with the half of the separated strands of DNA from the double-strand replicative form. This implied that the mRNA was not complementary to the single strand normally found in the ΦX174 particle but rather was complementary to the newly formed strand of the replicative form. The implication was confirmed by an analysis of the base composition of the mRNA, which showed that it was complementary to only one of the two strands of the RF of ΦX174 DNA. These findings are identical to those obtained from other organisms. There seems little doubt that in all organisms only one strand of the DNA double helix serves as a template for RNA synthesis.

RNA viruses. Our discussion of nucleic acid replication and transcription thus far has been concerned with DNA. The stress on DNA has been determined by the fact that most organisms, man included, have their genetic information encoded in DNA. In fact there are no known organisms (other than the RNA viruses) in which the genetic material is solely in the form of RNA. RNA viruses, as is true of DNA viruses, are found to be either in double-strand or single-strand form. Double-stranded RNA viruses have been found in both plant and animal cells. Those viruses attacking plant cells cause diseases in a great variety of plant species and tumors in some (e.g., wound tumor virus). Those attacking animal cells include, among others, the *reoviruses.* They occur widely in the respiratory and digestive tracts of man and other animals. However, they have yet to be associated with any specific disease. The pattern of replication and transcription in the double-stranded RNA viruses appears to be similar to that found in double-stranded DNA viruses (e.g., T2). A different terminology has, however, developed for the enzymes involved in these activities. The enzyme that controls RNA virus replication is called *replicase,* whereas the enzyme that controls RNA virus transcription is called *transcriptase.*

Single-stranded RNA viruses fall into two classes. The first class includes, among others, the tobacco mosaic virus discussed earlier in this chapter. These viruses follow the replication and transcription pattern outlined above for ΦX174, which you will recall is a single-stranded DNA virus. The *poliovirus* that attacks the nerve cells of man and other primates is another example of this type of single-stranded RNA virus. The second class of single-stranded RNA viruses contains a number that cause tumors in animals (e.g., Rous sarcoma virus, mouse leukemia virus, mouse mammary tumor virus). Upon entering a cell, this type of virus makes a complementary strand of itself which is composed of DNA. This single-strand of DNA in turn makes a complementary DNA strand of itself, forming a DNA double helix. The newly formed DNA double helix becomes incorporated into one of the chromosomes of the host cell where it is replicated along with the host DNA. While in the host cell, the RNA-derived viral DNA will produce single-stranded RNA viruses that will leave the host cell without destroying it and will enter other cells.

The hypothesis that an RNA virus replicates through a DNA intermediate was first advanced by H. M. Temin in 1963, based on his work with the Rous sarcoma virus that causes malignant connective tissue tumors in chickens, mice, rats, hamsters, monkeys, and many other animal species. Unequivocal confirmation of his findings occurred in 1970 when it was demonstrated, in two different laboratories, that Rous sarcoma virus particles contain a DNA polymerase that uses the RNA strand of the virus as a template. This newly discovered type of enzyme was assigned the name *RNA-directed DNA polymerase* (also referred to as RNA-dependent DNA poly-

merase). One other term has been applied to this enzyme, based on its action of seeming to reverse the transcription process, which, as described previously, involves the making of RNA from DNA. Since this enzyme forms DNA from RNA, it has also been called *reverse transcriptase.* The importance of this type of virus in changing our concept of species-specific DNA will be considered in Chapter 8 and its importance in studying the cancer problem, in Chapter 11.

Relationship of DNA to rRNA and tRNA

After it had been shown that mRNA was complementary to DNA, the question was raised as to whether rRNA and tRNA were also complementary to DNA. In one experiment 16s rRNA was taken from a bacterial culture whose RNA had been labeled with ^{3}H, and 23s rRNA was taken from a culture whose RNA had been labeled with ^{32}P. Initially, one of the two types of rRNA was hybridized to homologous DNA (from the same species) until no more hybridization occurred. The other type of rRNA was then added, and the amounts of secondary hybridization were determined. The two kinds of RNA were found to hybridize without interference from each other, and the conclusion was that they are derived from different sequences of the DNA molecule. Similar experiments utilizing tRNA gave results indicating that tRNA is derived from an entirely separate sequence of the DNA molecule. An extension of these studies determined how much of the DNA molecule is involved in turning out ribosomal and transfer RNA. The experiment involved adding increasing amounts of 4s, 16s, or 23s to a fixed amount of DNA and determining the ratio of RNA to DNA in the hybrid at saturation. The results indicated that, in the bacterium *Bacillus megatherium,* approximately 0.18% of the total DNA molecule is complementary to 23s RNA, 0.14% to 16s RNA, and 0.025% to 4s RNA.

The DNA/RNA hybridization experiments have been extended to higher organisms. In 1965 Ritossa and Spiegelman reported on their findings of DNA/RNA hybrids in the fruit fly *Drosophila melanogaster*. The nucleoli of *Drosophila* are associated with a particular region of the X chromosomes known as the "nucleolar organizer region." Through appropriate matings, stocks of flies were obtained that had 1 to 4 doses of the nucleolar organizer region. Hybridization experiments between the DNA derived from these stocks and radioactively labeled ribosomal RNA were then conducted. As shown in Fig. 1-20, the amount of RNA hybridizable per unit of DNA was directly proportional to the doses of nucleolar organizer regions. It was concluded that the nucleolar organizer region of the X chromosome contains the DNA templates for the ribosomal RNA components. Presumably, the location of the DNA template for tRNA will be found at some future time.

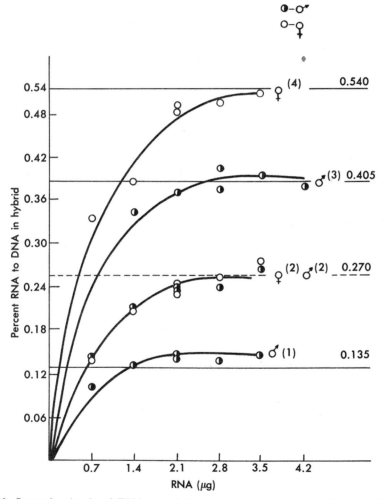

Fig. 1-20. Saturation levels of DNA containing various dosages of nucleolar organizer (NO) region. Dosage of NO is indicated by the number in parentheses. Numerical values of the plateaus are given on the right. Replicate determinations are indicated by multiple circles at the same input levels. (From Ritossa, F. M., and Spiegelman, S. 1965. Proc. Nat. Acad. Sci. U. S. A. **53:**737-745.)

We have reviewed the various types of RNA and their relationships to DNA. In the next section of this chapter we shall examine the interrelationship of RNA and DNA in protein synthesis.

Protein synthesis

The pattern for protein synthesis that emerges from our knowledge of DNA, RNA, and proteins is as follows. There is a partial unwinding of the DNA double helix, permitting the formation of a strand of mRNA that is complementary to one of the two strands of the DNA. After its formation, mRNA passes out of the nucleus and into the cytoplasm. In the cytoplasm, mRNA acts as an attachment site, binding temporarily to a 30s ribosomal subunit.

Elsewhere in the cytoplasm, and apparently independent of mRNA formation, amino acids are selected from an intracellular amino acid pool, for *activation*. Activation involves the reaction of an amino acid with adenosine triphosphate (ATP); the reaction is catalyzed by a specific *amino-acyl synthetase* for each of the twenty amino acids. The product formed by the amino acid and ATP is an amino acid adenylate containing the energy required for the next reaction, which is the attachment of the amino acid to a specific tRNA molecule. The adenylate remains bound to the enzyme until it is transferred to the tRNA, a transfer that is catalyzed by the same amino-acyl synthetase enzyme. Thus the enzyme is specific both for the amino acid and for the tRNA, implying that the enzyme has two different combining sites: one that recognizes the side group of an amino acid and another that recognizes the tRNA specific for that amino acid. The two reactions are diagrammed in Fig. 1-21, with glycine as the amino acid. Note that, for the sake of convenience, the tRNA is pictured in rectangular form rather than as a cloverleaf in this and all subsequent diagrams.

The fate of the amino acid is fixed, once it is attached to its tRNA. The amino acid is carried to the mRNA, which is attached to a 30s ribosomal subunit. If the amino acid carried by the tRNA corresponds to the one indicated at the beginning of the "message," a complex is formed between the 30s ribosomal subunit, the aa $\sim$ tRNA, and mRNA. At this time a 50s ribosomal subunit joins the complex, and a complete 70s ribosome is formed. Each 70s ribosome contains two cavities which are formed by parts of both the 30s and 50s subunits. These cavities act as insertion sites for tRNA molecules. One cavity, called the *A* (amino-acyl) *site,* acts as the initial "binding site" for the aa $\sim$ tRNA molecules. The other cavity, called the *P* (peptidyl) *site,* acts as the "growing site" for the developing polypeptide chain. After the 70s ribosome has been formed, it moves along on the messenger, by some as yet unknown mechanism (probably involving the 30s subunit of the ribosome) to the next amino acid–specifying message. At this next message-point of the mRNA, another tRNA with its appropriate amino acid becomes attached to the ribosome at the A (binding) site. The two amino acids are then joined by a peptide bond, and the first tRNA is released. There are always two tRNA molecules attached to a particular ribosome at any one time: one in the cavity called the P (growing) site and the other adjacent to it at the A site. The second tRNA moves onto the P site when the first tRNA is released, and another tRNA then joins the process. To carry out its functions, tRNA must have two recognition sites: one for the activating enzyme that binds the tRNA to its amino acid, as discussed earlier, and the other for a specific group of template nucleotides of the mRNA. This latter recognition site of the tRNA is presumed to be located at a bend of the molecule and to consist of three nucleotides that are complementary to the template nucleotides of the mRNA that specify a particular amino acid. The production of a polypeptide chain according to the specification of mRNA is called *translation*. The process is diagrammed in Fig. 1-22.

At one end of every polypeptide chain

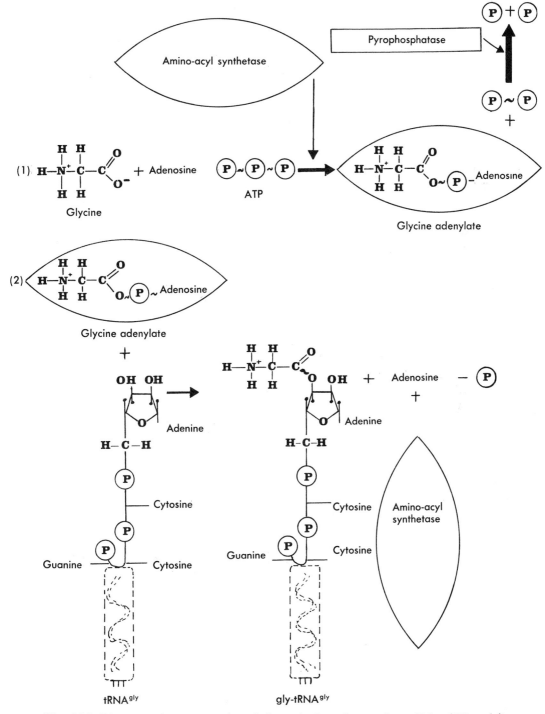

Fig. 1-21. Diagrammatic representation of the activation of an amino acid by ATP and its transfer to the C-C-A end of its specific tRNA adaptor. (From Watson, J. D. 1965. Molecular biology of the gene. W. A. Benjamin, Inc., New York.)

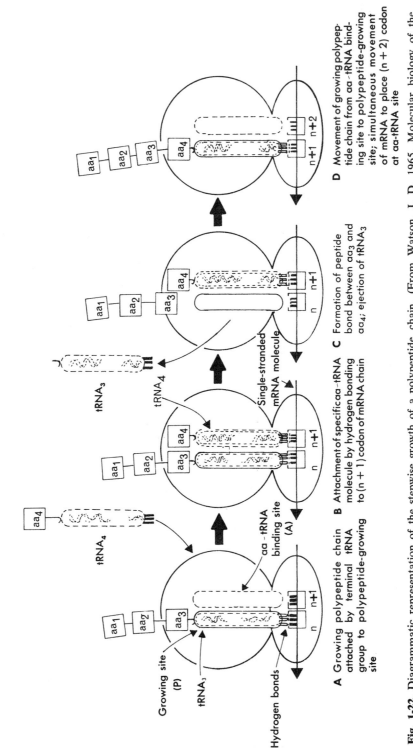

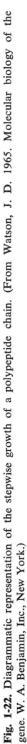

A Growing polypeptide chain attached by terminal tRNA group to polypeptide-growing site

B Attachment of specific aa-tRNA molecule by hydrogen bonding to (n + 1) codon of mRNA chain

C Formation of peptide bond between aa_3 and aa_4; ejection of $tRNA_3$

D Movement of growing polypeptide chain from aa-tRNA binding site to polypeptide-growing site; simultaneous movement of mRNA to place (n + 2) codon at aa-tRNA site

Fig. 1-22. Diagrammatic representation of the stepwise growth of a polypeptide chain. (From Watson, J. D. 1965. Molecular biology of the gene. W. A. Benjamin, Inc., New York.)

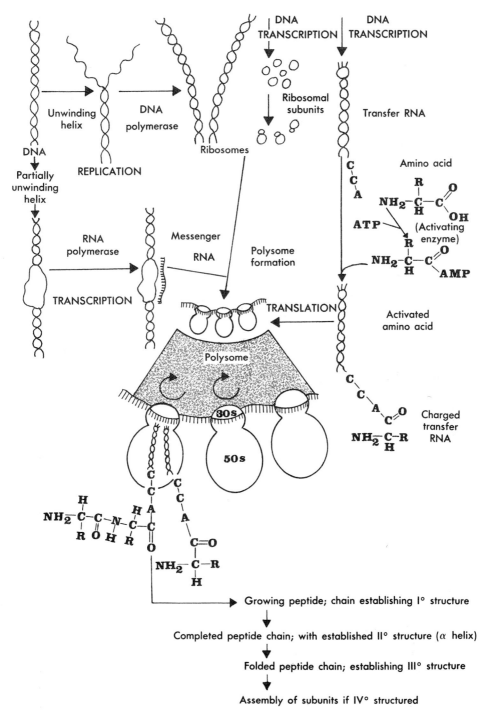

Fig. 1-23. Overall scheme of replication, transcription, and protein synthesis. (From Johnson, W. H., Laubengayer, R. A., Delanney, L. E., and Cole, T. A. 1966. Biology. 3rd ed. Holt, Rinehart & Winston, Inc., New York.)

there is an amino acid bearing a free amino group, and at the other end is one bearing a free carboxyl group. Chains always grow by stepwise addition of single amino acids, starting with the amino-terminal amino acid and ending with the carboxyl-terminal amino acid. The growing carboxyl end of a chain is always terminated by a tRNA molecule. The terminal tRNA molecule provides the mechanism for holding the growing polypeptide chain to the ribosome. After the polypeptide chain has grown to full length, the terminal tRNA must be split off, thereby creating a free terminal carboxyl group and releasing the chain from the ribosome. The mechanism of this release phenomenon is at present unknown. After the completed polypeptide chain is released, the ribosome dissociates into its 30s and 50s subunits. The 30s subunit may then attach to the starting point of the same or some other mRNA molecule. The 50s subunit may subsequently become attached to the same or some other 30s subunit.

The portion of an mRNA molecule that is in contact with a single ribosome is relatively short. This permits several ribosomes, collectively called a *polysome,* to work simultaneously on a single mRNA and form a number of polypeptide chains. At any given time the length of the chains attached to successive ribosomes will vary in direct proportion to the fraction of the message to which the ribosome has been exposed. Messenger RNA molecules vary greatly in length, reflecting the large spread in the length of polypeptide chain products. In *E. coli* the average size of mRNA is 900 to 1500 nucleotides, corresponding to the fact that the average *E. coli* polypeptide chain contains from 300 to 500 amino acids.

Depending on the organism, mRNA may be utilized for relatively short or long periods of time. The duration (half-life) of mRNA in a cell is studied by exposing the cell to *actinomycin D.* This antibiotic binds with the guanosine (G) containing sites in DNA, thus selectively suppressing the synthesis of mRNA by the enzyme DNA-directed RNA polymerase. The persistence of protein synthesis reflects the continued utilization of preexisting mRNA. In bacteria, mRNA is relatively short-lived (half life = 2 minutes), and fresh mRNA is made continuously. In chick embryos, the eye lens proteins are coded by two types of mRNA: one with a half-life of 3 hours and the other with a half-life of more than 30 hours. In mammalian reticulocytes, the mRNA for hemoglobin formation appears to persist indefinitely. It is assumed that the presence or absence of long-lived mRNA is determined by the continuous or transient need of the cell for a particular protein.

The overall pattern of *replication* (synthesis of DNA on a DNA template), *transcription* (synthesis of RNA on a DNA template), and *translation* (synthesis of protein as specified by messenger RNA) is shown in Fig. 1-23. Note that mRNA specifies only the primary structure of the polypeptide. Current thinking and experimentation indicate that no other information is needed to specify the complete structure of the protein molecule.

GENETIC CODE

The discovery that both proteins and nucleic acids are linear arrays of their respective building blocks (i.e., amino acids and nucleotides) led to the hypothesis that the linear sequence of the amino acids in a protein is specified by the linear sequence of nucleotides in a gene. The *colinearity* of protein structure and gene structure was demonstrated by showing that changes in the primary structure of a protein (e.g., A-protein of the enzyme tryptophan synthetase of *E. coli*) always corresponded to changes in the gene (e.g., A-gene of *E. coli*) controlling the production of the enzyme. Experimental evidence for the colinearity of the protein and its gene was reviewed and extended by C. Yanofsky and co-workers in 1964.

With the establishment of the concept of colinearity of protein structure and gene structure, it was necessary to hypothesize a *genetic code* such that the sequence of amino acids in proteins would be found in the sequence of nucleotides of DNA and expressed in the nucleotide sequence of mRNA. The basic problem of such a genetic code is to indicate how information written in a four-letter language (four nucleotides of DNA) can be translated into a twenty-letter language (twenty amino acids of proteins). The group of nucleotides that specifies one amino acid is a code word or *codon*. The simplest possible code is a singlet code, in which one nucleotide codes for one amino acid. Such a code is inadequate, for only four amino acids could be specified. A doublet code, also inadequate, could specify sixteen (4 × 4) amino acids, whereas a triplet code could specify sixty-four (4 × 4 × 4) amino acids. Because the triplet code is the simplest code that can specify the twenty common amino acids found in proteins, it has received the most attention. Since there are more codons (64) than amino acids (20), investigators hypothesized that the genetic code would be *degenerate;* that is, some amino acids would

have more than one DNA (hence, RNA) codon. This means that a messenger RNA may provide for the incorporation of a particular amino acid by more than one triplet. In addition, one would expect to find more than one transfer RNA for an amino acid that is degenerately coded. All the foregoing expectations have, in fact, been found to be true.

Experimental evidence supporting the concept of a *triplet code* was provided by F. H. C. Crick and co-workers in 1961. They worked with the bacteriophage T4 which, like the T2 virus described earlier in this chapter, attacks the colon bacillus *E. coli.* Crick and his associates were able to obtain a large number of T4 viruses that differed from the normal virus in the number of nucleotides contained in a particular region of the DNA. It was found that viruses with either one additional nucleotide in this region, or with two, could not function normally. However, phage with three additional nucleotides would function as usual. Al-

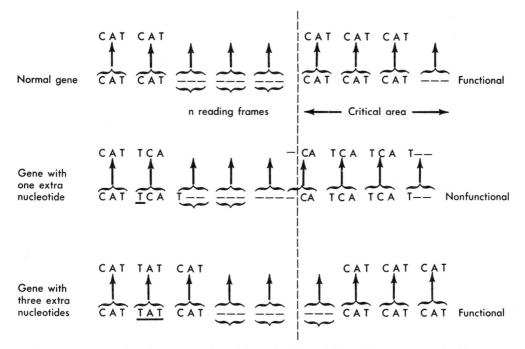

Fig. 1-24. Nucleotide changes and the triplet code. The addition of one extra nucleotide puts the reading out of order. The addition of three extra nucleotides puts the reading back in phase.

though the protein involved in this experiment has not yet been isolated, the normal functioning of the virus clearly must involve the production of some specific protein. This protein was not formed when one or two nucleotides were added to the DNA of the virus. The conclusion reached from the above experiment was that the genetic code is in *triplet* form and that the addition of one or two nucleotides had put the reading of the code out of order. The addition of the third nucleotide resulted in a return to the proper reading of the message. It should be anticipated that the addition of three nucleotides to the DNA will spoil the genetic message over that stretch of the gene that includes the three nucleotides. This is called "misreading." However, the remainder of the message is in normal order and will result in the production of a functional protein, provided the misreading has occurred in some region of the protein that is not critical for its proper functioning. The experimental findings are interpreted in Fig. 1-24.

Initially, some geneticists found it disturbing to think of degeneracy in connection with the genetic code. It appeared wasteful for sixty-four triplets to stand for only twenty amino acids. Attempts were therefore made to hypothesize a triplet code in which the code was overlapping. With an overlapping triplet model, the number of codons could be reduced to about twenty. However, this proposal has serious shortcomings. Let us examine the case of lysine, whose codons appear to be AAG and AAA, and that of phenylalanine, whose codons have been designated as UUU and UUC. Since the code letters for lysine and phenylalanine are mutually exclusive, lysine could not follow phenylalanine, or vice versa, in any protein; this would constitute a forbidden combination. Yet, we know that there are no forbidden combinations of amino acids in proteins. Any two–amino acid combination is possible, any three–amino acid combination, and so on. The conclusion to be drawn from the above is that the genetic code is *nonoverlapping.* A diagrammatic arrangement

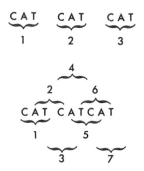

Fig. 1-25. Nonoverlapping and overlapping triplet codes.

of nonoverlapping and overlapping triplet codes is shown in Fig. 1-25.

Another argument against the concept of an overlapping code stems from the fact that a change in a single end nucleotide of a triplet would bring about changes in three amino acids, since the change would simultaneously affect three triplets. When chemical analysis has been possible in various forms of the same protein, as in the case of the different human hemoglobins (A vs. S vs. C), it has been found that the proteins differ from one another only in single amino acids.

The ability to break the code and spell out the triplets that correspond to the various amino acids required synthetic messenger RNAs of known composition. They became available as a result of the work of S. Ochoa who, in 1955, isolated from bacteria the enzyme *polynucleotide phosphorylase,* which catalyzes the conversion of ribonucleoside diphosphates to synthetic polynucleotides similar to natural RNA. This enzyme is not to be confused with DNA-directed RNA polymerase, mentioned earlier in this chapter. Polynucleotide phosphorylase normally acts in the cell to break down RNA; but when tremendous amounts of the ribonucleoside diphosphates are added, it can be used in a test tube to synthesize RNA. The discovery of this enzyme permitted the synthesis of RNA molecules of any base composition desired.

In 1961 Nirenberg and Matthaei, using

the above technique, manufactured some polyuridylic acid. They found that the addition of polyuridylic acid (poly-U) and a mixture of amino acids to a cell-free, protein-synthesizing system obtained from *E. coli* resulted in the formation of a polypeptide that contained only the single amino acid phenylalanine. According to the triplet model, UUU must then be the code for phenylalanine. Very rapidly investigators in different laboratories showed that AAA coded for lysine, CCC for proline, and GGG for glycine. Since RNA molecules of any base composition can be made, mixtures of nucleotides were soon used to synthesize other artificial messages. An example of this was the synthesizing of a polyribonucleotide from a mixture containing 5U:1A. The distribution of the nucleotides in the polynucleotide is at random, so that the above-mentioned mixture should furnish the following proportions of triplets: UUU = 125, UUA = UAU = AUU = 25, AAU = AUA = UAA = 5, AAA = 1. A ratio of phenylalanine:tyrosine:asparagine:lysine of 125:25:5:1 in the polypeptide obtained by using 5U:1A should therefore indicate that the code for tyrosine could be UUA, or UAU, or AUU whereas the code for asparagine could be AAU, or AUA, or UAA. This experimental approach gave much information on the composition of the genetic coding units but could not specify the order of the nucleotides within the triplets. The ability to do this had to await the discovery of still another experimental technique.

The necessary procedure, developed in the laboratory of M. W. Nirenberg in 1964, takes advantage of the fact that specific tRNA molecules bind to ribosome-mRNA complexes, even when the mRNA consists of a single trinucleotide. Single trinucleotides of known linear arrangements are relatively simple to synthesize. The experiment is aided by the use of special filters that permit the passage of tRNA molecules but retain ribosomes. In separate test tubes, each type of tRNA molecule is first charged with its respective amino acid that has previously been made radioactive. The charged

Table 1-3. The twenty amino acids found in proteins

Amino acid	Abbreviation
Alanine	Ala
Arginine	Arg
Asparagine	AspN
Aspartic acid	Asp
Cysteine	Cys
Glutamic acid	Glu
Glutamine	GluN
Glycine	Gly
Histidine	His
Isoleucine	Ileu
Leucine	Leu
Lysine	Lys
Methionine	Met
Phenylalanine	Phe
Proline	Pro
Serine	Ser
Threonine	Thr
Tryptophan	Tryp
Tyrosine	Tyr
Valine	Val

tRNAs are then incubated with ribosomes and known trinucleotides and placed on the filter. If the radioactive amino acid goes through the filter, the trinucleotide in question is not active in binding the tRNA to the ribosome and so is not acting as a messenger. If the radioactive amino acid is retained by the filter, the trinucleotide sequence used is the proper code for the particular tRNA and presumably for the amino acid in question. Using this method separately for each of the twenty amino acids and for each of the sixty-four triplets makes it possible to assign codons for all the amino acids.

If the above experiment is performed with mononucleotides or dinucleotides, the tRNAs are not bound to the ribosomes. This finding offers further experimental proof of the triplet nature of the code. The twenty amino acids and their abbreviations are shown in Table 1-3, and the sixty-four codons and their assignments are shown in Table 1-4.

Table 1-4. The genetic code, consisting of sixty-four triplet combinations and their corresponding amino acids, shown in its most likely version

		Second letter				
		U	C	A	G	
First letter	U	UUU ⎤ Phe UUC ⎦ UUA ⎤ Leu UUG ⎦	UCU ⎤ UCC Ser UCA UCG ⎦	UAU ⎤ Tyr UAC ⎦ UAA Ochre (terminator) UAG Amber (terminator)	UGU ⎤ Cys UGC ⎦ UGA (terminator) UGG Tryp	U C A G
	C	CUU ⎤ CUC Leu CUA CUG ⎦	CCU ⎤ CCC Pro CCA CCG ⎦	CAU ⎤ His CAC ⎦ CAA ⎤ GluN CAG ⎦	CGU ⎤ CGC Arg CGA CGG ⎦	U C A G
	A	AUU ⎤ AUC Ileu AUA ⎦ AUG Met (initiator)	ACU ⎤ ACC Thr ACA ACG ⎦	AAU ⎤ AspN AAC ⎦ AAA ⎤ Lys AAG ⎦	AGU ⎤ Ser AGC ⎦ AGA ⎤ Arg AGG ⎦	U C A G
	G	GUU ⎤ GUC Val GUA GUG ⎦	GCU ⎤ GCC Ala GCA GCG ⎦	GAU ⎤ Asp GAC ⎦ GAA ⎤ Glu GAG ⎦	GGU ⎤ GGC Gly GGA GGG ⎦	U C A G

Third letter

An examination of Table 1-4 shows that there appear to be three codons, UAA, UAG, and UGA, that do not code for any amino acid. These triplets are called *nonsense* codons or *terminators*.

The name "terminators" for the nonsense codons stems from the fact that their inclusion in any mRNA results in the abrupt termination of the message at the point of their location even though the polypeptide chain has not been completed. Table 1-4 further shows that almost all the amino acids have at least two codons while some have as many as six (arginine and leucine). This variable amount of degeneracy has raised questions as to the role of degeneracy and its variability. It is generally agreed that the more degenerate the code for a particular amino acid, the better protected is that amino acid from substitution by another amino acid in a protein. This follows from the reasoning that, should a change in the mRNA occur, it could more easily be to a different triplet that specifies the same amino acid, if the genetic code is highly degenerate for that particular amino acid. As it turns out, the most protected amino acids are also, on the average, the most frequent amino acids in proteins. One codon, AUG for methionine, appears to be an *initiator* of messages as well as the specifier of an amino acid.

With the elucidation of the genetic code, questions were raised as to the *universality* of the code. Does the sequence of nucleotides that codes for a particular amino acid in a bacterial protein also code for the same amino acid in a human protein? The answer appears to be in the affirmative. As we have discussed previously, bacterial ribosomes and tRNAs will participate in polyphenylalanine synthesis if poly-U is used. Rabbit reticulocyte ribosomes and tRNAs will also make polyphenylalanine if poly-U is used. Similar findings have been obtained for poly-C and proline and for poly-A and lysine. Another question is whether mRNA from one organism can be used with the protein-producing machinery from another. It has been found that when amino acid–charged tRNAs from the bacterium *E. coli* were placed with rabbit reticulocyte mRNA and ribosomes, the product was rabbit hemoglobin. By chemical analysis, the hemoglobin was shown to be identical in its primary structure to the hemoglobin made with reticulocyte tRNA. The bacterial tRNA is able to recognize perfectly the mammalian mRNA for hemoglobin and also to work harmoniously with mammalian ribosomes. Thus the code would appear to be largely, if not entirely, universal.

In our discussion of the genetic code, we have stressed the role and importance of the RNA triplet and mRNA. It is not to be forgotten that the mRNA, in turn, reflects the nucleotide sequence of the DNA and that the genetic code is the DNA of the cell and organism.

Codon-anticodon pairing

Our discussion of nucleic acids, protein synthesis, and the genetic code did not indicate the nature of the temporary attraction between the codons of an mRNA and the various tRNAs in the translation process. The basis of this attraction appears to rest on *complementarity*. As was true of DNA replication and DNA-to-RNA transcription, there is a complementary pairing of purine with pyrimidine, in this case A with U and G with C. This means that for every *codon* found in an mRNA, except for the terminator triplets, a corresponding, complementary *anticodon* must be present on the appropriate tRNA. At this point let us look once more at Fig. 1-18 and describe the probable functions of the component parts of a tRNA molecule. Loop I is believed to function in the binding of the specific activating enzyme (amino-acyl synthetase) to the tRNA. Loop II contains the anticodon, which in this case is IGC. Loop III has, at the present time, no known function. Loop IV is believed to function in the binding of tRNA to the ribosomal surface.

In considering codon-anticodon pairing, let us first look at the initiator triplet, AUG, which in an mRNA strand indicates the point at which translation begins. An analysis of the tRNA for methionine has shown that its anticodon is CAU, which is the RNA complement of the initiator codon when read in reverse order. Since nucleotides are read from the 5′ to the 3′ position, the antiparallel juxtaposition of these two triplets on a ribosome would occur as follows:

mRNA . . . . . 5′ AUG 3′ (codon)

methionine tRNA . . . . . 3′ UAC 5′ (anticodon)

When, because of the degeneracy of the genetic code, more than one codon exists for a given amino acid, each corresponding type of tRNA is usually found to have an appropriate complementary anticodon. As an example, we can consider two of the six codons for leucine:

mRNA . . . . . 5′ CUU 3′ (codon)

leucine tRNA^{I} . . . . . 3′ GAA 5′ (anticodon)

and

mRNA . . . . . 5′ UUG 3′ (codon)

leucine tRNA^{II} . . . . . 3′ AAC 5′ (anticodon)

Table 1-5. Possible pairing combinations of bases according to the wobble hypothesis

Initial base in anticodon	Terminal base in codon
A	U
C	G
G	U or C
U	A or G
I	A, U, or C

Unfortunately, the strict regularity of codon-anticodon pairing indicated above does not always occur. If we look at Fig. 1-18, we find that the anticodon for alanine is IGC. The possible advantage of having I (inosine), a purine base that closely resembles guanine, in this anticodon is unknown. However, it is known that this particular tRNA molecule can pair with three of the four codons that specify alanine— GCC, GCU, and GCA. In the case of the tRNA for tyrosine, the anticodon is GΨA (the pseudouridine acting as would uridine). Here the tRNA molecule can pair with both codons that specify the amino acid—UAU and UAC. Note that it is always the terminal nucleotide of the codon that is substitutable. Other examples of this situation have been found. To explain this phenomenon, in 1966 F. H. C. Crick proposed his *wobble hypothesis*. He postulated that the first two positions of any codon (or last two of any anticodon) were fixed as to their complementing nucleotides. However, the third base of at least some codons (or first of some anticodons) have a certain amount of play, or wobble, so that more than just one possible type of pairing can occur. The possible pairing combinations are shown in Table 1-5.

It is important to ask whether there is any advantage to this flexibility of pairing combinations for the terminal bases of at least some codons. A possible answer can be

reached from an examination of Table 1-4. The multiple codons for virtually every amino acid are seen to be identical in their first two bases and to differ only in the third. The freedom to "wobble" may be linked to the degeneracy of the genetic code. With a degenerate code, the ability of at least some tRNAs to pair with alternative codons for the same amino acid would certainly appear to be an efficient manner of carrying out protein synthesis. In contrast to this flexibility, it is interesting to note that an initial C base in an anticodon can only pair with a terminal G base in a codon (Table 1-5). This fact takes on great importance when we consider that the initiator triplet is AUG and the only codon for tryptophan is UGG. Quite apparently, the C-G requirement provides rigidity in the translation process where it is needed. The successful evolution of a genetic system requires that it provide, as necessary, for both flexibility and rigidity.

SUMMARY

A discussion of the breadth and scope of modern genetics opened this chapter. We then considered the nature of the hereditary material and found that it can be in the form of DNA or RNA, depending on the organism. This was followed by an analysis of the geometric organization of DNA as found in most organisms. An important requirement of the model of the DNA molecule is that it provide for replication and transcription.

After it was established that the model met these requirements, there was a review of the different types of RNA (ribosomal, transfer, and messenger) and the roles of each in protein synthesis. An analysis of the genetic code, including its triplet nature, degeneracy, nonoverlapping character, and universality, followed. The chapter closed with a discussion of codon-anticodon pairing.

Questions and problems

1. Discuss the lines of evidence for the belief that nucleic acids are the genetic material.

2. Define the following terms with respect to nucleic acid structure:
   a. Nucleotide
   b. Nucleoside
   c. Polynucleotide chain
   d. Double helix
   e. Hydrogen bond
   f. Purine
   g. Pyrimidine
   h. Complementary strand

3. What are the chemical and structural differences and similarities between the genetic material of tobacco mosaic virus and that of the bacteriophage T2?

4. How does the double-stranded nature of the genetic material lend itself to the prerequisite that genetic material must be able to replicate?

5. In light of your answer to question 4, how can ΦX174 replicate, since this virus's DNA is known to be single-stranded?

6. What are the salient features of the genetic code?

7. A missense mutation is one in which one base of a particular triplet is replaced by a different base, resulting in the coding of a different amino acid. A nonsense mutation is one in which the replacement of one base by another results in a triplet for which no amino acid is coded.
   a. What effect do you think a missense mutation has on protein synthesis?
   b. What effect do you think a nonsense mutation has on protein synthesis?

8. Discuss the function of each of the following in protein synthesis:
   a. DNA
   b. Messenger RNA
   c. Ribosomes
   d. Ribosomal RNA
   e. Transfer (soluble) RNA
   f. ATP

9. A segment of one of the complementary strands of T2 DNA is T-T-A-G-C-G.
   a. What is the sequence of the corresponding region on the complementary strand of DNA?
   b. What is the sequence of the corresponding region on the mRNA synthesized from this strand?
   c. What is the sequence of the corresponding region of the two tRNAs used in translating this segment?

10. How has nature provided a control over whether a particular DNA strand will replicate itself or will serve as a template for transcription of mRNA?

11. Modern techniques in molecular biology make it possible for investigators to isolate a particular class of tRNA with its corresponding amino acid attached. In one particular experiment, cysteine tRNA with cysteine attached at the C-C-A end was isolated, the amino acid chemically treated so that it was changed to alanine, and the altered complex intro-duced into a cell-free, protein-synthesizing system. By radioactive labeling it was possible to determine where the altered amino acid was deposited on a growing polypeptide chain.
    a. Will the labeled amino acid be found at a site on the chain corresponding to a cysteine triplet or to an alanine triplet in the mRNA strand?
    b. What hypothesis is being investigated in this type of experiment?

12. Distinguish between auxotrophic and prototrophic bacteria. How are these nutritional states taken advantage of in the study of heredity?

13. Studies conducted by S. Benzer on the rII region (the portion of the genome controlling plaque morphology) of the bacteriophage T4 have revealed that this region is about 2100 nucleotide pairs long. How many amino acids are there in the polypeptides whose synthesis is directed by the rII region?

14. Discuss the role of each of the following in protein synthesis:
    a. Peptide polymerase
    b. GTP
    c. 30s ribosomal subunit
    d. Amino-acyl synthetase
    e. 50s ribosomal subunit
    f. Polysome

15. Discuss the genetic ramifications of Yanofsky's discovery of colinearity between gene and protein.

16. Refer to Table 1-4.
    a. Which amino acid is less likely to be substituted for in protein synthesis, isoleucine (Ileu) or threonine (Thr)? Why?
    b. What is the role in protein synthesis of the codons UAG, UGA, and UAA?

17. What evidence exists for the universality of the genetic code?

18. How have the following experimental tools been used in aiding our understanding of the physical basis of heredity:
    a. X-ray diffraction analysis
    b. Density gradient technique
    c. Hybridization
    d. Polynucleotide phosphorylase

19. Refer to Fig. 1-12.
    a. How is hydrogen bonding reflected in the diagram?
    b. How is the complementary nature of the genetic material reflected in the diagram?
    c. How is the polarity of each strand reflected in the diagram?
    d. Which feature in the diagram reflects the backbone of the molecule?
    e. Which feature in the diagram allows one to say the molecule depicted is not RNA?

20. Using a dark line for labeled chromosomes and a light line for unlabeled chromosomes, draw the products of DNA replication in the following instances (keep in mind the

results obtained in the Meselson and Stahl experiment regarding the manner of replication in *E. coli*):

a. After continued replication in heavy (^{15}N) medium

b. After one round of replication in ^{14}N, following transfer from heavy (^{15}N) medium

c. After two rounds of replication in ^{14}N, following transfer from heavy medium

d. After three rounds of replication in ^{14}N, following transfer from heavy medium

References

Avery, O. T., MacLeod, C. M., and McCarty, M. 1944. Studies on the chemical nature of the substance inducing transformation of pneumococcal types. I. Induction of transformation by a desoxyribonucleic acid fraction isolated from pneumococcus type III. J. Exp. Med. **79:** 137-158. (Evidence from transformation studies for DNA as the genetic material.)

Brenner, S., Jacob, F., and Meselson, M. 1961. An unstable intermediate carrying information from genes to ribosomes for protein synthesis. Nature **190:**576-581. (Evidence for the role of DNA in transcription.)

Chargaff, E. 1950. Chemical specificity of nucleic acids and the mechanisms of their enzymatic degradation. Experientia **6:**201-209. (Evidence for the species specificity of nucleic acid composition.)

Crick, F. H. C., Barnett, L., Brenner, S., and Watts-Tobin, R. J. 1961. General nature of the genetic code for proteins. Nature **192:**1227-1232. (Features of the genetic code, elucidated from experiments on mutations.)

Goodman, H. M., and Rich, A. 1962. Formation of a DNA-soluble RNA hybrid and its relation to the origin, evolution, and degeneracy of soluble RNA. Proc. Nat. Acad. Sci. U. S. A. **48:**2101-2109. (Evidence from hybridization studies of DNA's template role for tRNA.)

Hall, B. D., and Spiegelman, S. 1961. Sequence complementarity of T2-DNA and T2-specific RNA. Proc. Nat. Acad. Sci. U. S. A. **47:**137-146. (Evidence from hybridization studies of DNA's template role for mRNA.)

Hershey, A. D., and Chase, M. 1952. Independent functions of viral protein and nucleic acid in growth of bacteriophage. J. Gen. Physiol. **36:** 39-56. (Evidence from T2 studies for DNA as the genetic material.)

Hoagland, M. B., Stephenson, M. L., Scott, J. F., Hecht, L. I., and Zamecnik, P. C. 1958. A soluble ribonucleic acid intermediate in protein synthesis. J. Biol. Chem. **231:**241-257. (Direct evidence for the existence of a soluble RNA fraction and for its role in protein synthesis.)

Kornberg, A. 1960. Biological synthesis of deoxyribonucleic acid. Science **131:**1503-1508. (A report on the in vitro synthesis of DNA.)

Kurland, C. G. 1970. Ribosome structure and function emergent. Science **169:**1171-1177. (A review of current work on the organization and functions of ribosomes.)

Meselson, M., and Stahl, F. W. 1958. The replication of DNA in *Escherichia coli*. Proc. Nat. Acad. Sci. U. S. A. **44:**671-682. (Evidence for the semiconservative manner of DNA replication.)

Nirenberg, M. W., and Matthaei, J. H. 1961. The dependence of cell-free protein synthesis in *E. coli* upon naturally occurring or synthetic polyribonucleotides. Proc. Nat. Acad. Sci. U. S. A. **47:**1588-1602. (The original description of the use of synthetic polyribonucleotides in amino acid-incorporating systems, including a discussion of their applications to the study of the genetic code.)

Speyer, J. F., Lengyel, P., Basilio, C., and Ochoa, S. 1962. Synthetic polynucleotides and the amino acid code. IV. Proc. Nat. Acad. Sci. U. S. A. **48:**441-448. (An example of the types of experiments used to assign a particular triplet to a particular amino acid.)

The genetic code. 1966. Cold Spring Harbor Symp. Quant. Biol. **31:**1-762. (An excellent collection of papers dealing with various aspects of the code.)

Watson, J. D., and Crick, F. H. C. 1953. Molecular structure of nucleic acids; a structure for desoxyribose nucleic acid. Nature **171:**737-738. (Formulation of the double helix model and a discussion of its genetic implications.)

Yanofsky, C., and Spiegelman, S. 1962. The identification of the ribosomal RNA cistron by sequence complementarity. II. Saturation of and competitive interaction at the RNA cistron. Proc. Nat. Acad. Sci. U. S. A. **48:**1466-1472. (Evidence from hybridization studies of DNA's template role for rRNA.)

Ycas, M. 1969. The biological code. John Wiley & Sons, Inc. (Interscience), New York. (A complete and up-to-date monograph on the genetic code.)

2 Physical basis of inheritance

In the preceding chapter we discussed the nature and functions of the hereditary material. Although we have reviewed the method by which DNA, or RNA in certain viruses, replicates itself, we have not as yet examined either the physical arrangement of the hereditary material within the organism or the methods by which this hereditary material is transmitted from one generation to the next. In considering these two topics, we shall first discuss viruses, then bacteria, and finally those organisms whose cell or cells contain a distinct nucleus and cytoplasm. As will be apparent, we know most about the physical arrangement and transmission of the hereditary material in the bacteria and less about these phenomena in viruses and higher organisms.

VIRUSES

We shall take as our example the bacteriophage T4 that, like T2, infects and lyses *Escherichia coli*. Our discussion will, however, be generally applicable to most other viruses. As described in Chapter 1, a typical virus consists of a protein coat and a nucleic acid core. From a study of electron micrographs, the outer protein layer of the T4 phage is estimated to be about 50 A. thick. When the coat is disrupted, the entire DNA contents of the virus can be obtained. The electron microscope shows the nucleic acid to consist of a single thin circular structure that we may call the *viral chromosome*. It has a width of 20 A., which corresponds to the width of a DNA molecule. The length of the T4 chromosome has been estimated at about 60 μ. Only the chromosome of the

virus enters the bacterial cell. While inside the bacterial cell, the phage chromosome is considered to be in a *vegetative state* during which it undergoes replication, directs protein synthesis, and may engage in an exchange of genetic material with the chromosome of some other phage particle (recombination—a process that will be discussed in Chapter 7). DNA replication precedes protein coat formation. After a new DNA molecule has been formed, subunits of the coat are synthesized and laid down around the DNA, forming the mature phage particle.

In Chapter 1 it was pointed out that at approximately 20 to 25 minutes *(latent period)* after initial infection, lysis of the bacterial cell occurs and 100 to 150 virus particles *(burst size)* are released. A question arose as to what was the pattern of multiplication of the phage within the bacterial cell. The technique that yielded the answer was developed mainly by A. H. Doermann and reported in 1952. The method involved an interruption of phage growth at intervals during the latent period by means of cell poisons or low temperature, followed by the release of the intracellular phage by sonic vibration or other means. Using this technique, one finds that there is no detectable T4 phage present for the first half of the latent period. At the end of that time one phage particle per bacterium is found. The number of intracellular phage particles that can be liberated per infected cell then increases at an approximately linear rate until the normal burst size is reached. This observation indicates that the accumulation of infectious phage particles within the bacterial cells is determined by the rate of synthesis of some phage component that is limiting in the system. Which component is causing the observed linear rate in phage production is as yet unknown.

It will be noticed that in the above description of the virus life cycle, there was no discussion of "growth" as there would have been with any type of cell. The normal life cycle of a cell includes a period of increase in size, which is followed by a splitting process that divides the cellular material into two parts. The viruses, however, do not increase in number as a result of the division of a larger preexisting unit. Normally, they are all of constant size and chemical composition. Their lack of growth and of reproduction-through-division sets them apart from all other forms of life. An additional aspect of the viral life cycle is worth stressing. Viruses are the extreme example of obligate parasites. They are completely dependent on their hosts for all the chemical precursors required for their reproduction and, in addition, are completely dependent on their host's protein-synthesizing machinery.

Viruses can transfer some of their genetic material to one another. This has been viewed as a form of sexual reproduction although it is quite different from the sexual process of higher organisms. The discovery of the recombination of genetic characters in phage was reported by Delbrück and Bailey in 1946. They worked with the viruses T2 and T4. The T2 phage was of the type that formed a small colony *(plaque)* and was designated T2r⁺, whereas the T4 phage was of the type that formed a large plaque and was designated T4r. The experimenters infected *E. coli* strain B with both types of viruses. After the latent period, the bacteria underwent lysis, and the completely formed phage emerged. Among the viruses were found the parental types, T2r⁺ and T4r, and in addition there were two new types, T2r and T4r⁺. These latter types were recombinations of the genetic characters of the parental viruses. It was hypothesized that somewhere in the development of

the phage particles a transfer occurred between some of the parental viruses of the genes controlling the type of plaque that would be formed. Thus we see in viruses some of the components of the life cycle found in higher organisms, including mechanisms both for self-replication and gene transfer. Let us now consider the bacteria, about which we have a good deal of information concerning their hereditary material and life cycle.

BACTERIA

Most investigations of the ultrastructure of the bacterial genetic system have involved *E. coli.* However, the findings have been confirmed by similar studies of other bacterial species; presumably, they will apply also to the blue-green algae. When a bacterial cell is ruptured, its entire DNA content is released. An electron micrograph study shows the *bacterial chromosome* to consist of one long circular structure. Its width is 20 A., and its length is about 1000 μ (1 mm.). The bacterial chromosome thus appears to be organized much like the viral chromosome. The fact that the chromosome is some fifteen to twenty times longer than the T4 chromosome reflects the presence of a larger number of genes in the bacterial chromosome. The greater number of genes provides the necessary structural and enzymatic proteins for the bacterium's structure and physiology, which are more complex than those of the virus.

When the circularity of the bacterial chromosome was established, a question arose as to the mechanism of its duplication. It was uncertain as to whether DNA replication began at one point of the chromosome and then continued completely around the circle or whether replication could begin at many points. The answer was provided by a technique called *autoradiography.* This technique takes advantage of the fact that thymidine, a component of DNA, is not incorporated into any other part of the organism. If the thymine contains some radioactive atoms (radioisotopes of hydrogen or carbon), one can follow the replication of DNA. In this technique, as reported by J. Cairns in 1963, bacteria are grown in a medium containing ³H-thymidine for two generations. The bacteria are then lysed by the use of chemicals, and the DNA is separated from the rest of the cell contents. The DNA molecules are suspended in melted photographic emulsion that is poured onto a smooth glass plate. Once the emulsion has hardened, one has the equivalent of a photographic plate that can be examined and photographed through the light microscope. Wherever the radioactive thymidine has been incorporated into DNA, a black dot will be seen in the photographic emulsion, above the chromosome. The black dot is caused by collision of the electrons ejected from the decaying ³H

atoms with the silver grains in the photographic emulsion. Where an entire section of the DNA has replicated, the black dots will form a track. A diagram of an autoradiograph is shown in Fig. 2-1. From a study of a large number of such autoradiographs, it was concluded that the bacterial chromosome is duplicated without a rupture of its ring and that the replication of the DNA proceeds unidirectionally from a single starting point, producing two Y-shaped double-helix forks. Thus, although the two strands of the DNA double helix have opposite polarity, replication in vivo appears to proceed along both strands in the same direction. After the bacterial chromosome has duplicated itself, the cytoplasm divides, and the bacterium will have transmitted its

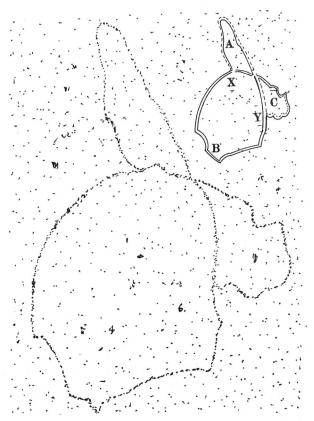

Fig. 2-1. Diagram of an autoradiograph of the chromosome of *E. coli,* labeled with tritiated thymidine for two generations and extracted with lysozyme. Inset: The same structure, divided into three sections (**A, B,** and **C**) that arise at the two forks (**X** and **Y**). (From Cairns, J. 1963. Cold Spring Harbor Symp. Quant. Biol. **28**:43-46.)

hereditary material to the next generation of cells.

In addition to reproducing asexually, bacteria may transfer parts or all of their genetic material to other cells of the same or closely related species. As with viruses, this transfer of genetic material differs from the sexual reproduction of "higher" organisms in that a true fusion cell (zygote) is not formed. Gene transfer in bacteria may take one of three forms: transformation, conjugation, transduction.

Transformation

We have already considered transformation in Chapter 1. In this process, a portion of the bacterial chromosome of one cell enters another cell and replaces the homologous section of the recipient cell's chromosome. The cells of a large number of bacterial species release fragments of DNA into their surrounding food medium. These DNA fragments have been found to be capable of transforming other cells in the culture. In some species the DNA comes from cells that die and undergo self-digestion, or autolysis. In others, cells that are growing and dividing also release DNA fragments. It seems plausible to hypothesize that for some bacterial species transformation may be the natural mechanism for genetic transfer between species members. Transformation may also permit genetic exchange between members of different species. In 1964 B. W. Catlin reported the reciprocal genetic transformation of strepto-mycin-susceptible cells to streptomycin-resistant cells, between the bacteria *Neisseria catarrhalis* (coccus) and *Moraxella non-liquefaciens* (bacillus).

The mechanism involved in the transformation process is not clearly understood. In order to accomplish the process, a number of obstacles must be overcome. First, there is the penetration of the cell wall and the cell membrane by what is a comparatively large molecule. Second, there is the avoidance of chemical degradation of the donor DNA by the enzymes of the recipient cell. Naked DNA does not appear to have any means of its own for accomplishing these tasks. The solutions of these problems must then be sought within the recipient bacterium itself.

A model of how the transformation process is accomplished in the bacterium *Bacillus subtilis* was proposed in 1969 by Akrigg and his co-workers. It involves the following facts. Older bacterial cells are known to be more transformable than younger ones. Associated with aging there is an increased deposit, in its cell wall, of the cell's own autocatalytic enzymes. These enzymes are believed to cause breaks in the cell wall through which the transforming DNA enters and makes contact with the cell membrane. An electron microscope study of *B. subtilis* shows that the cell membrane of this organism invaginates and forms a vesicle, called a *mesosome*, either at one of the poles or at the middle region of the cell. By the use of autoradiography, it

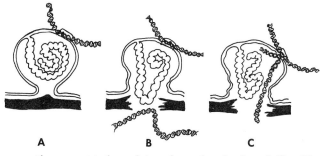

Fig. 2-2. Diagrammatic representation of transformation in *B. subtilis*. (From Akrigg, A., Ayad, S. R., and Blamire, J. 1969. J. Theoret. Biol. **24:**266-272.)

was found that the transforming DNA appears to penetrate the cell wall at the point where the mesosome is located. One last fact has to be stressed. Other research has demonstrated that the chromosome of the recipient cell is attached to the cell's membrane either on the mesosome or very close to it. Thus a direct connection would appear to be possible between the extracellular, transforming DNA and the recipient cell's chromosome. Should such contact be achieved, it would be possible for the transforming DNA to become incorporated into and replace some of the recipient cell's chromosome. Fig. 2-2 illustrates the steps involved in the model of transformation just discussed. How generally applicable this model is to all known cases of transformation is not known.

Conjugation

A second method by which bacteria may transfer genetic material is conjugation. In this process, discovered by Lederberg and Tatum in 1946, there is a transfer of chromosomal material from one cell to another by direct cell-to-cell contact. Most of our knowledge on bacterial conjugation comes from experiments using *E. coli* strain K12— a strain that was found to contain different mating types.

Among these are an F^+ type cell that acts as a genetic donor, or male, and an F^- type cell that acts as a genetic recipient, or female. When a population of F^+ cells is mixed with F^- cells, only about one in 10^4 donor cells transfers chromosomal DNA to its recipient. This transfer is usually attributed to the presence in F^+ populations of rare types called *Hfr,* for "high frequency of recombination." When isolated, Hfr cells produce populations in which virtually all the cells are donors. The relationships of F^+, F^-, and Hfr cells to one another will be discussed further in Chapters 5 and 8.

In their original experiment, Lederberg and Tatum took advantage of the fact that normal *E. coli* cells grow and multiply on a *minimal medium* that contains only minerals and sugar. The sugar is a source of energy and carbon. From these simple materials, the bacterium can manufacture all its needed amino acids, vitamins, purines, and pyrimidines. So that they might demonstrate the transfer of genetic material in *E. coli,* the experimenters chose a donor strain lacking the ability to manufacture biotin and methionine and able to grow only on a food medium that, in addition to minerals and sugar, contains this vitamin and amino acid. For their recipient strain, they chose one that lacks the ability to manufacture threonine and leucine and can grow only when these amino acids are supplied in the food. The two strains were first mixed in a broth and then plated on a food medium lacking the vitamin biotin and the three amino acids. It was found that numerous colonies were formed, which could grow on the minimal medium. There seemed to be little doubt that the portion of the donor chromosome containing the ability to manufacture threonine and leucine had been incorporated into the recipient chromosome. Equally apparent was the fact that the portion of the donor chromosome controlling biotin and methionine production had not been incorporated into the recipient chromosome of the cells that grew on the minimal medium. This type of experiment was repeated, using many different genetic traits, and all confirmed the conclusion that conjugation results in a transfer of genetic material from one bacterium to another.

Later experiments on conjugation revealed that it takes some 110 minutes after a conjugation tube has been formed between the bacteria for the entire chromosome to be transferred. It was also found that one can separate the conjugating bacteria at any time after initial contact, thereby halting the chromosome-transfer process. The results of such interrupted conjugations demonstrated both that the donor's bacterial chromosome is transferred in linear order and that the longer the conjugation time, the more genes are transferred. As will be discussed in Chapter 7, this type of experiment can be used to map the position of the genes on the chromosome. For our present discussion, the linear transfer of the donor's chromosome means that its circular structure must be broken in order to permit its migration. As we shall see below, the linearity of the migratory chromosome is a function of its formation.

After conjugation is terminated, the two bacteria separate and each continues its normal cycle of growth and asexual reproduction. This implies that the donor cell did not transfer its original chromosome but rather made two replicas of it, one of which was transferred while the other was retained. One question asked was whether the migratory chromosome is completely formed in the male before transfer is initiated or whether transfer requires concomitant replication of the male chromosome, with one of the replicas being transferred into the female as it is formed. The answer came from an experiment by Jacob, Brenner, and Cuzin as reported in 1963. They took male cells and grew them in a medium containing both heavy (^{13}C, ^{15}N) and radioactive (^{3}H) isotopes. After a number of cell divisions, all the cells contained the heavy and the radioactive isotopes in their DNA. The experimenters then mixed the male cells with female cells in a medium containing light (^{12}C, ^{14}N) and nonradioactive (^{1}H) isotopes. After a short period of conjugation, the density of the radioactive DNA transferred to the female cells was determined. The transferred DNA was found to contain one light and one heavy strand. Thus it was clearly demonstrated that chromosome transfer is directly geared to chromosome replication. The experiment also indicated that the linearity of the migratory chromosome must be a result of its mode of formation during conjugation.

Conjugation may be another natural mechanism that microorganisms have of transferring genetic material. Conjugation has been reported within such bacterial genera as *Salmonella, Shigella, Pseudomonas,* and *Vibrio*. It has even been possible to transfer *E. coli* chromosomes to *Salmonella* and *Shigella*. However, the yield of such interspecific chromosome transfers is extremely small.

Transduction

In our previous discussions of the interactions of bacteria and phage, we have considered only the situation in which the virus enters the bacterium, multiplies within it, and then lyses the cell, thereby killing the bacterium. Another type of interaction between the organisms is possible: a viral chromosome may enter a bacterial cell, become incorporated into its chromosome, and remain latent within the cell. In this latent state, the bacterial virus is called a *prophage*. The prophage acts like a portion of the bacterium's chromosome. In this role, the viral genes will be reproduced along with the host's own genes for a number of generations and will be found in all the daughter cells of the original bacterium. Eventually, in one of the daughter cells, the prophage may resume its existence as a separate entity, at which time it will replicate itself, produce complete virus particles, and lyse that particular bacterium. Bacteria that contain prophages are called *lysogenic bacteria,* and viruses are called *temperate phages*. The phenomenon of lysogeny will be discussed at greater length in Chapter 8.

The ability of a virus to enter a latent

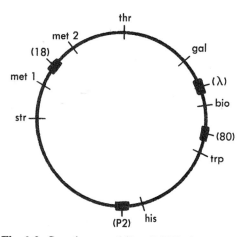

Fig. 2-3. Genetic map of *E. coli* K12 chromosome, showing location of insertion sites of some prophages (in parentheses) and location of some bacterial genes. (From Campbell, A. M. 1969. Episomes. Harper & Row, Publishers, Inc., New York.)

phase within its bacterial host provides the mechanism for a third method by which the genetic material may be transferred from one bacterium to another. In this process, called *transduction,* a phage acts as a vector that transports a portion of a bacterial chromosome from a donor cell to a recipient. The phenomenon was reported by Zinder and Lederberg in 1952. They found that when viruses attacked and lysed the cells of one strain of *Salmonella* and subsequently entered the cells of a lysogenic strain, the lysogenic cells exhibited some of the traits of the first strain. Later investigations showed that there are actually two types of transduction. In one type, called *generalized transduction,* all the donor's genes have a roughly equal chance of being transduced. Such a situation is found to be the case for transduction in *Salmonella* by phage P22. In the other type, called *restricted transduction,* only a restricted region of the bacterial chromosome can be transduced. An example of restricted transduction is found in the temperate phage λ (lambda) and its host *E. coli* K12, in which only a region that includes the cluster of genes controlling galactose fermentation is transducible.

In order to gain a better understanding of the transduction phenomenon, we shall now examine in some detail a model, proposed by A. Campbell in 1963, of λ phage incorporation into, and later detachment from, a bacterial chromosome. As mentioned above, this virus is capable only of restricted transduction. We should therefore not be at all surprised to find that λ always inserts at a particular point on the bacterial chromosome. Fig. 2-3 shows the insertion points of some prophages on the chromosome of *E. coli* strain K12. The prophages are indicated by the numbers or letters within the various parentheses. The other symbols mark the locations of but a few of the many known genes of this bacterium. Phage λ is located between the genes that direct the synthesis of enzymes required for the fermentation of the monosaccharide galactose *(gal)* and for the production of the vitamin biotin *(bio).*

Let us now turn our attention to Fig. 2-4. As indicated in Fig. 2-4, *A,* λ is thought to be in linear form while in its protein coat. The letters *A, J, N,* and *R* symbolize the location of some of the genes of the viral chromosome. When in a host cell, λ is believed to form a ring (Fig. 2-4, *B*) and, as a result of the action of certain enzymes, to become inserted into the bacterial chromosome. Fig. 2-4, *C,* shows a section of the

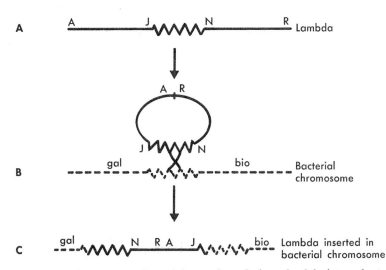

Fig. 2-4. Diagrammatic representation of integration of phage lambda into a bacterial chromosome. (From Campbell, A. M. 1969. Episomes. Harper & Row, Publishers, Inc., New York.)

bacterial chromosome with the inserted virus between the *gal* and *bio* genes. In this condition, the *E. coli* is said to be in a lysogenic condition, and λ is considered to be temperate. Under these circumstances, the viral chromosome is replicated along with the *E. coli* chromosome at each cell cycle, and the division of the bacterial cell results in two lysogenic bacteria.

For reasons not well understood at all, after a number of generations, the λ phage will become detached from the bacterial chromosome in one or more of the lysogenic bacteria. The viral DNA will then replicate itself many times and will also cause the production of many protein coats. After a particular number (burst size) of complete viral particles have been formed, the host cell will be lysed. In the process of detachment of λ from the *E. coli* chromosome, the incorporation of part of the bacterial DNA into the viral chromosome can occur. This process is illustrated in Fig. 2-5. Fig. 2-5, *A*, shows a portion of the bacterial chromosome with λ inserted between the *gal* and *bio* genes. In Fig. 2-5, *B,* we see the reverse formation of a ring by the virus, with the inclusion of the *E. coli gal* gene within the ring. The subsequent detach-

ment of λ, shown in Fig. 2-5, *C,* results in the *gal* gene becoming part of the viral DNA and the concomitant retention of the λ *J* gene region by the *E. coli* chromosome. Since the bacterial cell will die as a result of being lysed by the virus, we cannot determine the possible effect that the loss of the *gal* gene and the addition of the viral *J* gene would have had on the bacterium. However, it has been found that the loss by λ of part of its chromosome, as seen in Fig. 2-5, *D,* prevents the virus from invading by itself any more bacterial cells. The affected viruses are called λ*dg* (defective, galactose-transducing).

As stated above, λ*dg* cannot by itself invade an *E. coli*. However, if a λ*dg* and a normal λ jointly infect a bacterial cell, they may enter the cell and proceed through either the lytic or lysogenic cycle of phage production. In the lytic phase, both types of viruses will be produced in equal numbers. In the lysogenic phase, the chromosomes of both viruses will be inserted into the *E. coli* chromosome, and both will be replicated at each cell cycle. While in the lysogenic state, the *gal* gene in the λ*dg* will function as if it were one of the bacterium's normal genes. The occurrence of transduc-

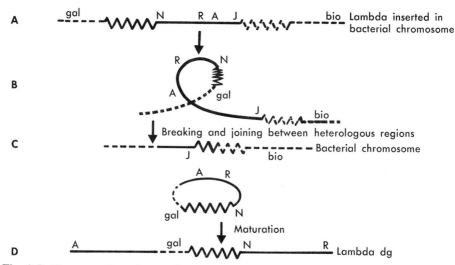

Fig. 2-5. Diagrammatic representation of the origin of lambda *dg*. (From Campbell, A. M. 1969. Episomes. Harper & Row, Publishers, Inc., New York.)

tion will become apparent under the following conditions. Let us assume that the *E. coli,* prior to being lysogenized, did not have the ability to produce the enzymes required for galactose fermentation. Let us further assume that the *gal* gene carried by λ*dg* does have this ability. After lysogeny has occurred, the *E. coli* cell and all its descendants will be found to have the ability to produce the necessary enzymes.

As with transformation and conjugation, transduction also appears to be a widespread phenomenon. It has been found to occur in the following bacterial genera: *Salmonella, Escherichia, Pseudomonas, Vibrio, Staphylococcus,* and *Proteus.* Transduction of genes between *Salmonella typhimurium* and *E. coli* has also been effected. We shall now consider the physical arrangement and transmission of the hereditary material in higher organisms.

HIGHER ORGANISMS

In this category we place Protozoa, Algae (except the blue-green), Fungi, and all the multicellular organisms. In these forms the DNA is associated with protein and is organized into more than a single chromosome. Moreover, the chromosomes are located in a special cellular structure, the nucleus. The occurrence of a nucleus in the cells of higher organisms is so distinctly different from the situation found in bacteria and blue-green algae that special terms have been applied to the organisms in the two groups. Higher organisms are also referred to as *eucaryotes*—those that have a true nucleus. Bacteria and blue-green algae, on the other hand, are called *procaryotes*—those that have their DNA in a prenuclear form. The cell of a higher organism consists typically of a *cell membrane, cytoplasm,* and *nucleus.*

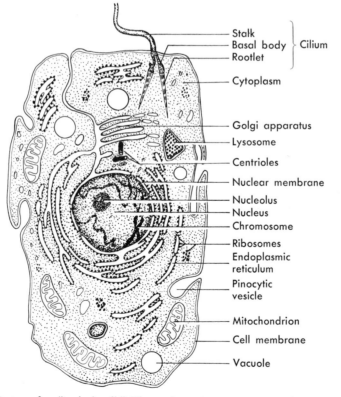

Fig. 2-6. Diagram of a "typical cell." The various structures are not found in all cells but illustrate the differentiations that occur among cell types.

The cell membrane is about 75 A. (Angström units) thick and appears to be composed of three separate layers: two protein layers of 20 A. each on either side of a phospholipid layer of 35 A. The plasma membrane may fold outward to form *microvilli* or inward to form *vesicles*. Some of the invaginations are not delimited by a vesicle but instead lead into the center of the cell. There, channels connect with a complex set of vesicles that interlace the structure of the cell and are called the *endoplasmic reticulum*. The appearance of the endoplasmic reticulum can be "rough," when it carries ribosomes on its cytoplasmic surface, or it can be "smooth," when the ribosomes are absent. The endoplasmic reticulum connects with two other cell structures. It is continuous with the system of tightly packed, smooth-surfaced vesicles called the *Golgi apparatus*, which is thought to function in the production of macromolecules by the cell. The endoplasmic reticulum is also continuous with the outer of the two membranes that surround the nucleus. The nuclear membrane has pores such that its content may be continuous with the cytoplasm of the cell. A diagram of a cell from a multicellular organism is shown in Fig. 2-6.

The cytoplasm and its formed elements carry on most of the metabolic functions of the cell. The cytoplasm may also contain DNA that acts in the pattern outlined in Chapter 1. In fact, the number of cases in which the cytoplasm has been shown to carry hereditary units is constantly increasing. These will be discussed in detail in Chapter 8.

The nucleus is concerned with the control of the metabolic functions carried on in the cytoplasm and with the transmission of the genes to the cells formed as a result of cell division. The genes are organized into a number of structural units, each of which is called a chromosome. Each species has a typical chromosome number. In man the number is 46. Since each human being is formed through sexual reproduction, 23 of the chromosomes are contributed by the mother (through the egg) and 23 by the father (through the sperm). Man, in reality, does not contain 46 unrelated chromosomes but, rather, twenty-three pairs of chromosomes. This association of chromosomes in pairs results from the fact that for every chromosome contributed by the mother there is a corresponding, or *homologous*, chromosome contributed by the father. To say that chromosomes are homologous implies that if a particular chromosome carries a gene that affects eye color, the corresponding, or homologous, chromosome will also contain a gene that affects eye color. The homologous chromosomes need not carry the gene for the same eye color, but they will both carry genes that affect the trait. Furthermore, such pairs of genes are located at exactly corresponding places in the homologous chromosomes. Thus if a gene for eye color is located at the tip of a particular chromosome, the corresponding gene in the homologous chromosome will also be located at the tip of its chromosome. What has been said for eye color genes applies equally well for all other genes.

To understand the modes of inheritance in different organisms, one must study the ways in which chromosomes are transmitted from cell to cell. This involves an analysis of cell division, in which the movements of homologous chromosomes with relation to one another become most important.

ROLES AND TYPES OF CELL DIVISION

As described earlier in this chapter, a procaryote (bacterium or blue-green alga) exhibits a relatively simple type of cell division involving replication of its single chromosome and a splitting of its cytoplasm. With these organisms, division of the cell results in reproduction of the individual and the transition from one generation to the next. When we come to consider the eucaryotes, however, we find that there are two types of cell division, one called *mitosis* and the other *meiosis*. Both types of cell division involve a dissolution of the nuclear membrane with the concomitant disappearance of the nucleus as a cellular structure, a pattern of movement and distribution of the chromosomes of the cell, and at the end of the process, the formation of a new nuclear membrane with the concomitant re-establishment of the nucleus as a cell structure. Completing both types of cell division there is a splitting of the cytoplasm into two halves.

In the case of unicellular eucaryotes, mitosis represents asexual reproduction and a progression of generations. For multicellular eucaryotes, however, mitosis gives rise to cells that remain in the individual's body and contribute to its functioning. With these organisms, mitosis leads to growth. Turning to meiosis, we find that in unicellular eucaryotes meiosis results in the formation of either the gametes or the nuclei that function in sexual reproduction. When considering the multicellular eucaryotes, one must distinguish the end products of meiosis in animals from that in plants. In multicellular animals, meiosis results in the formation of the *gametes* (sperm or eggs) that function in sexual reproduction. In multicellular plants, however, meiosis results in the formation of *spores* that mark the transition from the diploid to the haploid phase of their life cycle.

In both types of cell division, the nucleus and its chromosomes are most important, but their actions differ significantly in the two cases. It is important to compare and contrast mitosis with meiosis and understand the consequences of each for genetics.

Mitosis

For illustration purposes, we shall consider the case in which an organism has a typical chromosome number of 4 in the nuclei of its cells. Let us further assume that this organism is the result of sexual reproduction; hence, its 4 chromosomes are really two pairs of homologous chromosomes.

The process of mitosis can be considered as taking place in four stages, although it should be borne in mind that the stages follow one another in continuous fashion.

The four steps have been called *prophase, metaphase, anaphase,* and *telophase.* The period of time between successive mitoses has been termed *interphase.* It is best to begin our consideration of mitosis with interphase.

Interphase. Cells in interphase have a distinct nucleus, which shows little internal differentiation except for the nucleolus. The chromosomes are very indistinct and stain poorly. During this period, the cell is in an extremely active metabolic state.

Prophase. Cells are considered to be in prophase when the chromosomes become visibly distinct. The chromosomes become increasingly more stainable as prophase proceeds. During this phase of mitosis, the chromosomes coil, thereby becoming short and thick. From the beginning of prophase it is apparent that each chromosome is composed of two filaments, called *chromatids* (or "sister" chromatids), which are closely associated along their entire length. The chromatids are attached to one another at a specific point called the *centromere.* The end of the prophase is marked by the disintegration of the nuclear membrane. Several other phenomena also characterize prophase. These include the disappearance of the nucleoli within the nucleus and, in animal cells, the migration of the *centrioles* around the nucleus, with the accompanying formation of *spindle fibers* and *astral rays* that appear as slender streaks radiating out from the centrioles. The final location of the centrioles determines the position of the poles of the spindle. Most plant cells lack centrioles, but the spindle that forms is quite similar to that formed in animal cells.

Metaphase. The metaphase stage of mitosis is considered to begin with the completed formation of the spindle that looks like a series of parallel delicate fibers stretched across the cell from pole to pole. The fibers are actually protein molecules oriented longitudinally between the poles. The spindle serves to bring the chromosomes to the equatorial plane of the cell. Once there, the chromosomes become attached, in some unknown manner, to the spindle fibers at their centromeres, which during metaphase remain undivided.

Anaphase. Metaphase passes into anaphase when the centromere divides and the chromatids (now called chromosomes) begin to move toward the poles. The members of each pair of homologous chromosomes (actually the sister chromatids of prophase and metaphase) move toward opposite poles of the cell. The chromosomes have, through this process, separated to give two groups of like genetic constitution.

Telophase. In telophase, the final phase of mitosis, there is a regrouping of the chromosomes into a nuclear structure. The nuclear membrane is reconstituted, and the nucleoli reappear within the nucleus. Electron microscope studies have shown that the nuclear membrane reforms at the surface of the chromosomes, suggesting that cytoplasmic substances are excluded from the daughter nuclei at the time of their reconstruction. The chromosomes uncoil and elongate, lose their stainability, and finally take on the appearance of an interphase nucleus. Coincident with the latter stages of mitosis, there is a division of the cytoplasm resulting in two distinct cells. A typical sequence of mitosis in an animal cell is shown in Fig. 2-7.

• • •

A study of the mechanics of mitosis does not complete our consideration of this part of the cell cycle. There remain both the question of just how many copies of each gene are in a chromosome and the question of when in the cell cycle the genes replicate themselves.

The answer as to how many copies of each gene are present in a chromosome varies with the type of cell being studied. There are some cells in which the genes replicate themselves many times but neither the chromosomes nor the cytoplasm divides (e.g., salivary gland cells of *Drosophila*). In other cells one finds that both the genes and the chromosomes undergo periodic duplication and separation (e.g., root tip cells).

Examining the internal structure of a stretched-out prophase chromosome under the electron microscope, one finds that it consists of a two-strand cable 200 A. thick. Each strand (chromatid) of the cable is 100 A. thick and contains one DNA double helix associated with a type of basic protein called *histone*. At the subsequent anaphase the 200 A. cable divides longitudinally, and each resultant chromosome contains one DNA double helix with its associated protein. The anaphase chromosome represents the minimal chromosomal unit as found in higher organisms. Recall that both the viral and the bacterial chromosome initially consisted solely of a single DNA double helix in the form of a ring that replicated itself from a single starting point. It is of interest to ask (1) whether within each chromosome of a higher organism the DNA is also in ring form and (2) whether it replicates from a single starting point. We do not know whether or not the DNA of an interphase chromosome is in the form of a ring. Although genes are arranged linearly along the chromosome, it is possible that sections of the DNA may be in ring form and that the chromosome consists of a linear sequence of these rings. Neither electron microscope studies nor chemical studies have yet given us an unequivocal answer to the first question just asked. We do have an answer to the question of whether replication of the DNA within the chromosome has a single starting point. The experiment that yielded the information was reported by W. Plaut in 1963. He took salivary glands of the fruit fly, *Drosophila melanogaster,* and placed them in a culture medium containing tritiated thymidine for ten minutes. The glands were then put on slides, covered with photographic emulsion, and thus subjected to autoradiography. When the preparations were examined under the phase microscope, it was found that some regions of the chromosomes contained the labeled thymidine but others did not. The discontinuity of labeled regions within a chromosome offers clear evidence for the existence of several replicating points in the DNA complement along the length of the

Drosophila salivary gland chromosome. Conceivably, each replicating point represents a separate ring of DNA. However, as stated above, we have no conclusive information on the arrangement of the DNA in the interphase chromosome of a higher organism.

The question of when in the cell cycle the gene replicates itself remains to be considered. Here a fairly clear picture has been obtained through the use of autoradiography in which cells are permitted to grow and divide in a solution containing radio-active thymidine. Cells are removed after various intervals of time, put on slides, dipped into a photographic emulsion, and eventually made into a photographic plate. A stain is then applied that penetrates the emulsion and shows the outlines of the cells and their structures. Wherever the radioactive thymidine has been incorporated into a chromosome, a black dot will be seen in the photographic emulsion above the cell. When no radioactive thymidine has been incorporated into the cell, no dots will be produced. Such an experimental arrange-

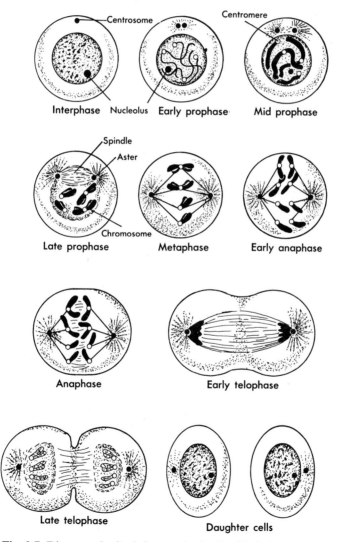

Fig. 2-7. Diagram of mitosis in an animal cell with 4 chromosomes.

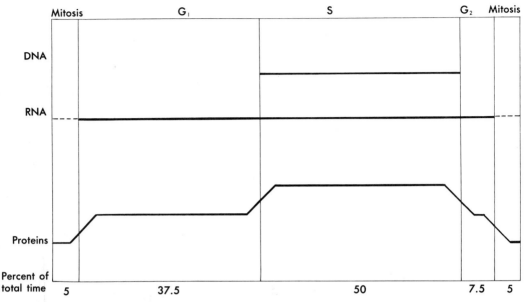

Fig. 2-8. Life cycle of a typical cell. It is divided into three phases between one mitosis, or cell division, and the next. Horizontal lines indicate the levels at which RNA, DNA, and proteins are synthesized during each of the phases and also during mitosis. (From "Autobiographies of Cells" by R. Baserga and W. E. Kisieleski. Copyright © 1963 by Scientific American, Inc. All rights reserved.)

ment enables one to tell when cells are synthesizing DNA. Immediately after cell division (i.e., during interphase) each daughter cell enters a phase called G_1 during which no synthesis of DNA occurs but RNA and proteins are produced. This is followed by an S phase in which the cell synthesizes DNA, RNA, and, at a stepped-up rate, proteins. There then follows a G_2 phase during which the cell stops making DNA and reduces its production of protein. This phase is followed by mitosis during which no DNA, very little protein, and sometimes no RNA is produced. Fig. 2-8 shows the life cycle of a typical epithelial cell from the small intestine of the mouse.

Significance of mitosis

Mitosis results in an exact division and distribution of chromosome material. The longitudinal division of each chromosome into chromatids, with meticulous distribution of the chromatids to the daughter cells, ensures that the daughter cells will have,

quantitatively and qualitatively, the same genetic constitution as the original cell from which they arose. For unicellular organisms, mitosis results in a population whose members are exact replicas of the ancestral cell. For multicellular organisms, mitosis results in every cell of the body's having the same hereditary material as that in the original fertilized egg.

Specialization of tissues (differentiation) must then be the result of the activation of different sets of genes in the cells of the different tissues. This topic will be considered further in Chapter 11. The next type of cell division to be considered, meiosis, is that by which gametes and spores are formed.

Meiosis

Meiosis results in cells that carry half the number of chromosomes that are found in somatic cells. Meiosis consists of two cellular divisions, accompanied by only one duplication of chromosomes. This process results either in the formation of gametes, *gametogenesis,* in animals, or in the formation of spores, *sporogenesis,* in plants.

Gametogenesis. The production of gametes may be further subdivided into *spermatogenesis,* formation of sperm, and *oogenesis,* formation of eggs. Since there are differences between sperm and egg formations, these topics must be treated separately. Since the gonadal cells will have the same hereditary material as the original fertilized egg, an organism with a typical chromosome number of 4 will again be simplest to discuss. We shall first consider the formation of sperm.

SPERMATOGENESIS. The following four stages constitute the process of spermatogenesis.

Spermatogonia are cells in the testis that divide by mitosis. One of the daughter cells usually remains a spermatogonial cell, and the other becomes a primary spermatocyte.

Primary spermatocytes are cells in which, during prophase, the homologous chromosomes pair with one another. The pairing phenomenon is called *synapsis,* and during this process the pairs of chromosomes become twisted around one another. By this time each chromosome is visibly divided into two chromatids, but the centromeres do not separate, so that the sister chromatids remain attached to each other. As a result, the four chromatids of the homologous chromosomes are closely associated in a *tetrad.* In metaphase the tetrads become arranged in the equatorial plane of the cell. Homologous chromosomes, each consisting of two chromatids held together by an undivided centromere, then separate and move toward opposite ends of the cell, and nuclear membranes may be formed. The cytoplasm then divides with each daughter cell, called a *secondary spermatocyte,* receiving one member of each homologous pair of chromosomes. We therefore consider the cell as having but 2 chromosomes (each consisting of two chromatids), and we say that the number of chromosomes has been reduced from *diploid* to *haploid.*

Secondary spermatocytes are cells containing half the number of chromosomes found in normal somatic cells. In the secondary spermatocyte, the chromosomes (pairs of sister chromatids) align themselves on a new spindle (the earlier one disappears), and the centromere of each chromosome divides. The sister chromatids, now called *chromosomes,* separate from each other and move toward opposite ends of the cell. A nuclear membrane is formed around each complement of chromosomes, and the cytoplasm divides, forming two daughter cells that are called *spermatids.*

Spermatids are cells that do not undergo any further divisions. The spermatid undergoes gradual cellular differentiation (flagellum formation, etc.) into active, motile sperm cells. Thus each spermatogonium that undergoes meiosis results in the formation of four sperm.

OOGENESIS. Oogenesis consists of essentially similar chromosomal stages but does differ in its cytoplasmic distribution.

Oogonia are the potential egg cells of the ovary. Periodically one of these becomes a primary oocyte.

Primary oocytes are cells in which synapsis occurs and tetrads are formed. In the separation of chromosomes, each daughter cell receives 1 chromosome from each homologous pair. However, the division of the cytoplasm is very uneven. One cell receives most of the cytoplasm, hence is very large, and becomes the secondary oocyte. The other cell receives very little cytoplasm and is called the *first polar body* (first polocyte).

Secondary oocytes are haploid cells in which the sister chromatids separate from each other and move toward opposite ends of the cell. Here, too, the cytoplasmic division is very uneven. The cell receiving most of the cytoplasm is called the *ootid,* and the other cell is called the *second polar body* (second polocyte). The first polar body may also divide to form two additional polar bodies. The polar bodies normally disintegrate.

Ootids do not undergo any further division. The ootid need not undergo any further cellular differentiation to form a functional egg. Thus each oogonium that undergoes meiosis results in the formation of one egg.

• • •

The union of the haploid sperm and the haploid egg leads to reestablishment of the diploid number of chromosomes. Depending on the species of animal, the egg may be entered by the sperm before, during, or after oogenesis. In many animals, oogenesis either will not occur at all or will not proceed past a particular stage unless a sperm has entered the egg.

Although we have considered meiosis as composed of two cell divisions, it should be recognized that each of the two divisions mentioned is in turn composed of a prophase, metaphase, anaphase, and telophase. We shall now analyze meiosis in terms of its stages of cell division, using the Roman numeral I to indicate the division stages of a primary spermatocyte or oocyte and the Roman numeral II to indicate the division stages of a secondary spermatocyte or oocyte.

PROPHASE I. Prophase in both primary spermatocyte and oocyte is of long duration and is regarded as a series of stages that are characterized by the pairing and intertwining of homologous chromosomes. The first stage is referred to as *leptotene*. During this stage the chromosomes appear as single threads unassociated with one another. However, chemical studies have shown that DNA duplication occurs in the interphase prior to leptotene, although the double-strandedness of the chromosome is not visible under the microscope until later in the first meiotic prophase. The second stage is called *zygotene*. In this stage homologous chromosomes pair with one another, gene by gene, over their entire length. This is the pairing process that has been named *synapsis*. The association of each pair of homologous chromosomes is called a *bivalent*. The third stage is termed *pachytene*. At this stage it is apparent under the microscope that each chromosome consists of two chromatids, which makes it possible to describe the association of the four

chromatids as a *tetrad*. Toward the end of pachytene, by a mechanism not clearly understood, breaks in chromatids may occur. Broken chromatid pieces tend to heal with other broken chromatids through the action of enzymes. Should one or more chromatids break and each heal with its own broken segments, the fact that a break occurred would never become apparent. However, should two chromatids break and healing occur between segments of different chromatids, one would find two chromatids tied to one another. This process is called *crossing-over* and the resulting juncture of the chromatids is called a *chiasma*. Crossing-over has important genetic consequences that will be discussed in Chapter 7. The fourth stage of the first meiotic prophase is called *diplotene*. Here the chromosomes become shorter and thicker. After the homologous chromosomes have shortened and thickened a great deal, they appear to repel one another. However, their separation at this time is limited by the chiasmata that have been formed. The fifth stage is named *diakinesis*. Here the chromosomes complete their shortening and separation and become angular or oval in appearance. As this elongated prophase is completed, the nuclear membrane disappears, and the spindle begins to form.

METAPHASE I. At metaphase, spindle formation is completed. The bivalents (paired homologous chromosomes) become aligned in the center of the spindle with their respective centromeres oriented toward opposite poles.

ANAPHASE I. During anaphase, the chromosomes move toward the poles. In contrast to the anaphase of mitosis, the centromeres do not divide at this time; instead, they remain intact and direct the movement of their respective chromosomes to the pole of their metaphase orientation. The result is that one member of each bivalent goes to each pole.

TELOPHASE I. When the chromosomes reach the poles, a nuclear membrane is formed around each group, and a division of the cytoplasm occurs. It is important to note that at this time the total number of chromosomes in each daughter cell is half the number of the mother cell. However, since the centromeres did not divide and separate in anaphase I, each chromosome consists of two chromatids.

INTERPHASE. Interphase may be long or short, depending on the organism involved.

Note that no DNA replication occurs during this pause between the first and second meiotic divisions. The second meiotic division occurs without the complication of an elongated prophase.

PROPHASE II. The nuclear membrane breaks down, the spindle begins to form, and the chromosomes coil tightly.

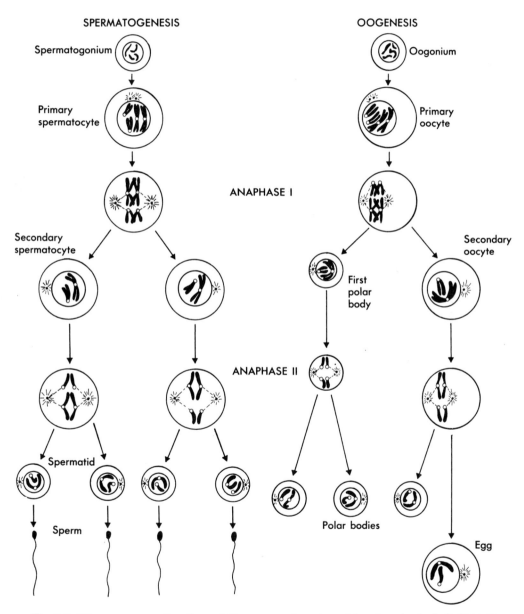

Fig. 2-9. Diagram representing the meiotic sequence in male and female animals. Left: Process of spermatogenesis, resulting in the formation of four sperm. Right: Oogenesis, resulting in the formation of one egg and three polar bodies.

METAPHASE II. The chromosomes, still consisting of two chromatids each, become aligned in the center of the spindle that has been formed.

ANAPHASE II. It is at anaphase II that the centromeres divide and separate, each taking one sister chromatid (chromosome) of each pair to an opposite pole.

TELOPHASE II. During telophase II of the second meiotic division, a nuclear membrane is formed around each group of chromosomes, the chromosomes uncoil, and cytoplasmic division occurs. It should be noted that, as a result of anaphase II, each chromosome, at the end of meiosis, consists of a single chromatid.

• • •

The first meiotic division is called the *reductional division* in that it reduces the number of chromosomes from the diploid to the haploid state. The reductional division separates each maternal from its homologous paternal chromosome. We must realize that most secondary spermatocytes and oocytes will contain some chromosomes of maternal and some of paternal origin. Furthermore, we must remember that,

should paternal and maternal chromatids undergo crossing-over with each other, the resulting cells will contain segments of both parental chromosomes. The second meiotic division is called the *equational division* because it separates the presumably identical chromatids. Figs. 2-9 and 2-10 show the typical stages in meiosis, as well as the elongated and detailed first meiotic prophase.

Sporogenesis. Sporogenesis occurs in plants. In its mechanics, it is most like spermatogenesis in animals. Sporogenesis results in the formation of four haploid spores. These spores give rise to the haploid *(gametophyte)* portion of the plant life cycle, which then produces the gametes without further change of chromosome number. Fusion of the gametes in turn results in the diploid *(sporophyte)* phase of the life cycle.

There is a good deal of variation in size and type of spores produced by the different species of plants. Some plant species have both types of sex organs on the same gametophyte. In these plants only one type of spore is formed. However, most plant species have both male and female gametophytes. In some of these plants the spores are of two

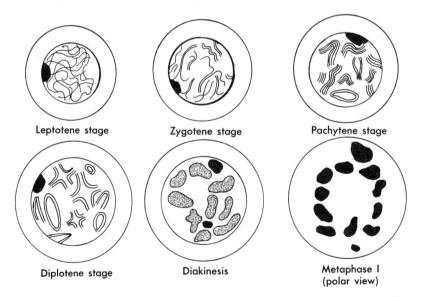

Leptotene stage   Zygotene stage   Pachytene stage

Diplotene stage   Diakinesis   Metaphase I
(polar view)

Fig. 2-10. Diagram of the meiotic prophase I of squash bug, *Anasa tristis,* illustrating chromosome pairing and duplication in the zygotene and pachytene stages, respectively.

types but of equal size. In other species the spore that gives rise to the male gameto-phyte is reduced in size (microspore), whereas the spore that gives rise to the female gametophyte is large (megaspore). In higher plants the microspores and mega-spores are produced within specialized structures of the flower. Some of the higher plants (e.g., lily and tomato) have both mi-crospores and megaspores produced in the same flower. Others (e.g., squash and maize) have the two types of spores pro-duced in separate flowers on the same plant. In still others (e.g., box elder and date palm) the two types of spores are produced on separate plants. As indicated above, the gametophytes produce sperm and egg cells whose union results in the formation of the diploid part of the life cycle. The plant life cycle thus consists of a haploid and a diploid

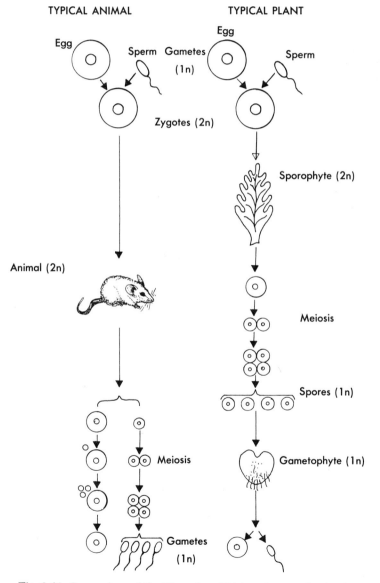

Fig. 2-11. Comparison of the life cycles of higher plants and animals.

phase. The haploid phase contains only one set of genes in each cell and exhibits the characteristics determined by them. The diploid phase contains two sets of genes in each cell, and its characteristics are determined by the interactions between these genes. The various types of gene interactions will be discussed in the next chapter.

The independence of the haploid phase of the plant life cycle is in contrast with that of animals where, with very few exceptions (e.g., male bees), the haploid phase does not carry on an independent existence and, hence, has no opportunity to express the characteristics of most of the genes it contains. A comparison of the life cycles of higher plants and animals is shown in Fig. 2-11.

Significance of meiosis

As we have seen, gamete formation results in a reduction of the chromosome number to one half of its original amount. Gamete formation also requires that one of the chromosomes from each homologous pair be present in each resulting gamete, allowing for the fact that crossing-over can result in a chromosome derived from segments of both homologous chromosomes. Equally important for genetic consideration is the fact that all chromosomes derived from a given parent need not be passed along as a group into a given gamete. Returning to our organism with the diploid number of 4, let us label the two homologous sets of chromosomes as AA^1 and BB^1. We shall assume that chromosomes A and B were derived from one parent while the chromosomes A^1 and B^1 were derived from the other. If large enough numbers of gametes are considered, gametes of the following chromosomal constitutions should be found with equal frequencies: AB, AB^1, A^1B, and A^1B^1. The number of different gametes possible is calculated as $(2)^n$, where n is the number of pairs of homologous chromosomes. Meiosis therefore results in a completely random distribution of chromosomes in the gametes. In the illustration used, with two pairs of homolo-

gous chromosomes, the number of different possible gametes is $(2)^2$, or 4. For man the corresponding figure is $(2)^{23}$, or more than 8 million different possible gametes. The variability of gametes is further increased by the process of crossing-over. Thus gamete formation is the physical basis for the new combinations of hereditary materials that occur in every generation.

SUMMARY

In this chapter we have considered the spatial arrangement of the DNA in the organism and the methods by which the genetic material is transmitted from one generation to another. Our study has indicated that two different spatial arrangements of the DNA can be found. One type characterizes the viruses, bacteria, and blue-green algae. In these forms the DNA is organized into a single structure. In the bacteria experiments have demonstrated that the replication of the chromosome has a single point of initiation and continues around in a ring in one direction. The other type of DNA arrangement is found in the Protozoa, Algae (except the blue-green), Fungi, and all multicellular organisms. In these forms the DNA is associated with protein and is organized into more than a single chromosome. In addition, the chromosomes are found as homologous pairs and are located within a special cellular structure, the nucleus. Whether the DNA in these forms is in ring or linear arrangement is not known. DNA replication is known to occur during the interphase stage of the cell cycle and more than a single point of replication along the chromosome is known to exist.

Viruses, bacteria, and blue-green algae exhibit both asexual and sexual reproduction. Asexual reproduction of viruses consists in the formation of protein coats around DNA cores. Bacteria and blue-green algae reproduce asexually by cell division. The sexual process in these forms does not result in the formation of a true diploid zygote, and normally only part of the genetic material of the donor is transferred to a recipient. Viruses exchange their genetic material during the latent period within the host cell. Bacteria exhibit three methods of transferring genes: transformation, conjugation, and transduction.

In higher organisms, asexual reproduction (mitosis) provides for an exact division and distribution of chromosome material. Sexual reproduction involves an initial reduction in chromosome number from diploid to haploid (meiosis) and a subsequent reestablishment of the diploid state through zygote formation. It was pointed out that meiosis provides the physical basis for new combinations of hereditary materials in every generation. We shall now turn our attention to the different types of interaction that can occur between homologous and nonhomologous genes.

Questions and problems

1. Identify the following: nucleus, cytoplasm, Golgi apparatus, endoplasmic reticulum, plasma membrane.
2. Identify the following and tell what part each plays in the physical basis of heredity:
   a. Homologous chromosomes
   b. Chromatid
   c. Centromere
   d. Histone
   e. Centriole
   f. Spindle fibers
   g. Astral rays
   h. S phase of cell cycle
3. In an organism that has 12 chromosomes in each somatic (diploid) cell, what is the ploidy (the chromosome number) of each of the following cell types?
   a. Kidney tubule cell
   b. Primary oocyte
   c. Spermatogonium
   d. Second polar body
   e. Spermatid
   f. Oogonium
   g. Secondary spermatocyte
   h. Egg
4. Define the following terms:
   a. Oogenesis
   b. Spermatogenesis
   c. Reductional division
   d. Equational division
   e. Bivalent
   f. Tetrad
5. In the genus *Chrysanthemum,* species are known having chromosome numbers of 18, 36, 54, 72, and 90. What does this information suggest regarding the evolutionary history of the different species?
6. What reasons can you suggest for the fact that three of the four products of oogenesis are nonfunctional?

7. State whether the following are true or false. If false, explain why.
   a. Of 24 chromosomes in a mature haploid egg, 12 are always paternal.
   b. Of 20 chromosomes in a primary spermatocyte, 14 may be paternal.
   c. In a plant species in which n = 4, the number of genetically different microspores possible is 64.
   d. With normal spermatogenesis, 50 primary spermatocytes will give rise to 200 sperm.
   e. With normal oogenesis, 50 oogonia will give rise to 100 mature haploid eggs.
8. What are the differences and similarities between the following animal and plant life cycle stages (consider origin and fate)?
   a. Plant sporophyte and animal fertilized egg
   b. Plant spores and animal gametes
   c. Plant gametes and animal gametes
   d. Plant gametophyte and animal fertilized egg
9. Draw all possible gametes resulting from spermatogenesis in a diploid organism having four chromosomes.
10. Draw all possible gametes resulting from oogenesis in a diploid organism having 4 chromosomes.
11. A certain gametophyte, displaying characters *A* and *B*, is crossed to another gametophyte, displaying characters *a* and *b*. With respect to these two traits, what types are expected in the next gametophyte generation?
12. What is the significance of mitosis?
13. What is the significance of meiosis?
14. Which of the following types of organisms would be expected to exhibit the greatest genetic variability and why?
    a. Sexually reproducing
    b. Asexually reproducing
    c. Self-fertilizing
15. Describe the events occurring in the following stages of mitosis:
    a. Prophase
    b. Metaphase
    c. Anaphase
    d. Telophase
16. What are the genetic consequences of crossing-over?
17. Describe the events occurring in the stages of meiosis characterized as follows:
    a. Leptotene
    b. Zygotene
    c. Pachytene
    d. Diplotene
    e. Diakinesis
18. If you saw a single cell at metaphase of meiosis, how could you distinguish whether the cell was undergoing metaphase I or metaphase II?
19. In terms of the mechanisms of cell division discussed in this chapter, explain how the following might arise:
    a. A somatic cell having 24 chromosomes undergoes mitosis to produce two daugh-

ter cells, one having 24 chromosomes and the other 23 chromosomes.

b. A cross between two related species results in a sterile hybrid. When examined cytogenetically, meiotic figures obtained from the hybrid show a mixture of paired and single chromosomes.

c. A somatic cell having 20 chromosomes results directly from a cell having 10 chromosomes.

20. Man has twenty-two pairs of autosomes plus a pair of sex chromosomes (*XX* in females and *XY* in males). With no crossing-over, what percent of a man's sperm will contain all chromosomes derived from the man's father?

References

Akrigg, A., Ayad, S. R., and Blamire, J. 1969. Uptake of DNA by competent bacteria—A possible mechanism. J. Theoret. Biol. **24:**266-272. (A model of the mechanism of transformation in *Bacillus subtilis.*)

Barrer, R. S., Meek, J., and Meek, G. A. 1960. The origin and fate of the nuclear membrane in meiosis. Roy. Soc. (London), Proc., B. **152:**353-366. (A study of nuclear membrane dissolution and reappearance during meiosis.)

Brachet, J. Sept., 1961. The living cell. Sci. Amer. **205:**51-60. (A concise description of the roles of various cellular organelles.)

Brachet, J., and Mirsky, A. E. (eds.). 1961. The cell. Vol. III. Chromosomes, mitosis, and meiosis. Academic Press, Inc., New York. (A review of the knowledge of these processes as of 1961.)

Busch, H., and Smetana, K. 1970. The nucleolus. Academic Press, Inc., New York. (An authoritative and detailed discussion of the role of the nucleolus in the cell.)

Campbell, A. 1963. Segregants from lysogenic heterogenotes carrying recombinant lambda prophages. Virology **20:**344-356. (A model of the mechanism of transduction in *E. coli.*)

Carothers, E. E. 1921. Genetical behavior of heteromorphic homologous chromosomes of *Circotettix* (Orthoptera). J. Morphol. **35:**457-483. (A study of meiosis in species that possess morphologically distinguishable chromosomes.)

Gall, J. G. 1961. Centriole replication; a study of spermatogenesis in the snail, *Viviparus.* J. Biophys. Biochem. Cytol. **10:**163-193. (A study of the maner of centriole duplication and the role of the centriole in cell division.)

Gross, R. P. (ed.). 1960. Second conference on the mechanisms of cell division. Ann. N. Y. Acad. Sci. **90:**345-613. (A conference on the cytological, physical, and chemical features of cell division.)

Huskins, C. L., and Cheng, K. C. 1950. Segregation and reduction in somatic tissues. J. Hered. **41:**13-18. (A study of the meiosis-like processes that occur in some somatic tissues.)

Mazia, D. Sept., 1961. How cells divide. Sci. Amer. **205:**100-120. (An informative article including experimental studies from which events in mitosis have been elucidated.)

Schrader, F. 1953. Mitosis. Columbia University Press, New York. (A review of the movement of chromosomes in cell division.)

Sharp, L. W. 1943. Fundamentals of cytology. McGraw-Hill Book Co., New York. (Pertinent chapters, including Chapter 5 on mitosis and Chapter 8 on meiosis.)

Swanson, C. P., Merz, T., and Young, W. J. 1967. Cytogenetics. Prentice-Hall, Inc., Englewood Cliffs, N. J. (Chapters 1 and 2, discussion of chromosome structure; Chapter 3, meiosis.)

Thomas, C. A., Jr. 1971. The genetic organization of chromosomes. Annu. Rev. Genet. **5:**237-256. (An up-to-date discussion of the chromosomes of both procaryotes and eucaryotes.)

3 Fundamentals of genetics

Since chromosomes are present as homologous pairs in most cells of higher organisms, genes for the various factors must also be present in duplicate. Genes that correspond in location on homologous chromosomes are called *homologous genes*. When homologous genes transcribe different messages, they are called *alleles*.

INTERACTIONS OF ALLELES

Alleles may interact in various ways in determining the characteristics of the organism: they may show simple (complete) dominance of one gene action over the other, they may show partial (incomplete) dominance, or they may show equal expression (codominance) of both alleles. Some genes are located on the sex chromosomes. Sex linkage and sex determination will be discussed separately in Chapter 5. In addition to allelic interactions, independent (nonhomologous) genes located on the same or on different chromosomes may also interact with one another. Their actions may complement each other, supplement each other, produce new phenotypes, or contribute to multiple-factor inheritance.

Simple (complete) dominance

Mendel studied this type of allelic interaction in his original experiments with garden peas. However, we shall illustrate simple dominance by the gene that causes albinism in mammals. For this trait, the gene that permits normal pigmentation of the fur or skin to develop (e.g., in mouse, rat, man) is dominant over the gene that causes albinism (absence of pigment in the skin,

hair, or fur and in the iris of the eye). We shall designate the dominant gene as *C*, and the recessive gene will be designated as *c*. With these two alleles, one can obtain three different genetic combinations: *CC, Cc,* and *cc*. If an individual is *CC* or *Cc*, his appearance *(phenotype)* will be pigmented. If he has the *cc* combination, his appearance will be that of an albino. The genetic constitution *(genotype)* of a *CC* or *cc* individual is said to be *homozygous,* whereas that of the *Cc* individual is said to be *heterozygous*. In simple dominance one gene for the particular characteristic has as much effect as do two genes. The example given in Fig. 3-1 illustrates the cross between a normally pigmented person (Caucasian, Negroid, or Mongoloid) and an albino (Caucasian, Negroid, or Mongoloid).

As was stressed in the discussion of meiosis, homologous chromosomes must separate and enter different gametes. In the albinism case, all the gametes from the pigmented parent will contain *C,* and all the gametes from the albino parent will contain *c*. This separation of homologous genes in the formation of gametes was first enunciated by Mendel in 1865 as the *law of segregation* (also known as the law of purity of gametes). The offspring of the parental cross is called the F_1 (first filial generation). In this cross they are all pigmented, indicating that pigmentation shows simple dominance over albinism. This phenomenon was also described by Mendel in his *"law" of dominance*. This is not a true biological "law," since most contrasting traits do not show simple dominance. However, Mendel was led to this rule by the action of those traits he had studied, most of which showed simple dominance. The Punnett square is used in Fig. 3-1 to ascertain the types and ratios of offspring. It is designed to yield all the possible combinations of gametes that can occur.

Fig. 3-2 shows the results of a cross of two individuals having the genotype of the F_1 of the previous cross. Again homologous chromosomes must segregate in the formation of gametes, and each gamete must carry but one gene for each trait. The offspring of this cross is called an F_2 (second filial generation). We see here the reappearance of albinism, indicating that the gene for this trait was not lost but was only masked in the F_1 by the action of the dominant gene. The ratios of the phenotypes and genotypes are typical for simple dominance.

Fig. 3-3 shows a typical testcross. In such a mating, an F_1 heterozygote is mated to the recessive parental type. Half the offspring show the recessive trait, and the other half exhibit the dominant phenotype.

The testcross is extremely useful in studying an organism showing the dominant phenotype but whose genotype is not known. This type of cross should indicate whether the dominant phenotype is caused by a homozygous or a heterozygous genotype. The appearance of a single recessive phenotype immediately signifies that the organism in question is a heterozygote. The continued appearance of only pigmented offspring would imply that the organism is a homozygote. Inherent in this type of analysis is the assumption that one can deduce genotypes from observing phenotypes. In

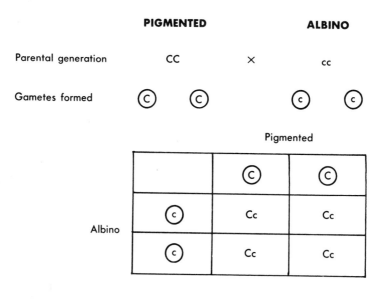

Fig. 3-1. Diagram of a cross between a homozygous pigmented person and an albino.

reality, the only aspect of the organism that we can see is its phenotype. From its phenotype we infer its genotype. When a testcross is performed, the question always arises as to how many pigmented offspring must be observed for the organism in question to be labeled a homozygote. The answer to this question lies in a consideration of the laws of probability.

Laws of probability. Genetic events are governed by the laws of probability. These laws provide a reasonable prediction of the fraction of events that will be of one type or of the other. Two types of probability can be distinguished.

1. *A priori probabilities* are those which can be specified in advance from the nature of the event. An instance is flipping an honest coin, with the probability of obtaining a head being one in two (or ½); another example is the gathering of data on the occurrence of a given event and testing whether the numbers agree with the ratio expected on the basis of some genetic theory (e.g., 1:1, 3:1, etc.).

2. *A posteriori* or *empiric probabilities* are those obtained by counting the number of times a given event occurs in a certain number of cases. An ex-

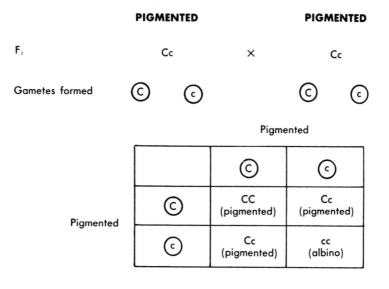

Summary of F₂			
Phenotypes	Phenotypic ratio	Genotypes	Genotypic ratio
Pigmented	3	CC	1
		Cc	2
Albino	1	cc	1

Fig. 3-2. Diagram of a cross between two pigmented persons, both of whom are carriers of the gene for albinism.

ample is the life-expectancy table, in which the number of people who die at age 50, 51, 52, etc. are counted and the data obtained form the basis of predicting the probability of the event occurring for the average member of a given population.

We shall now return to our consideration of the testcross and the problem of how many pigmented offspring must be obtained before we can consider the organism in question to be a homozygote. The problem falls into the category of *a priori* probabilities. Looking at Fig. 3-3, we see that half the offspring are expected to be pigmented. However, one can expect that sometimes each of the first two offspring

may be pigmented even if the pigmented parent is a heterozygote. The same may be true for the first three, four, five, or more offspring. The question is where to stop and consider the pigmented parent a homozygote. *A priori* probability can be used in prediction only if the events under consideration are independent of each other. In our example there is no reason to expect that the production of a pigmented offspring will in any way affect the production of further pigmented progeny. Therefore each pigmented offspring that is produced may be considered an independent event. If two events are independent, the probability of their both occurring is the product of their individual probabilities (e.g., the probability

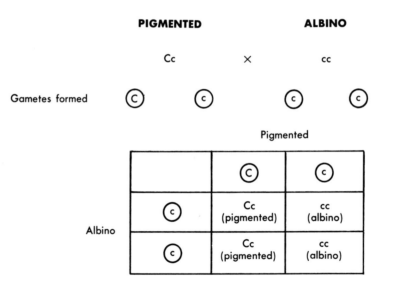

Summary of testcross

Phenotypes	Phenotypic ratio	Genotypes	Genotypic ratio
Pigmented	1	Cc	1
Albino	1	cc	1

Fig. 3-3. Diagram of a cross between a pigmented heterozygote and an albino.

of obtaining two heads in two tosses of an honest coin is ½ × ½ = ¼). In like fashion the probability of obtaining two pigmented offspring in our backcross is one in four (¼). The probability of obtaining three such individuals is ⅛, for four it is ¹⁄₁₆, etc. This progression fits the general expression $(\frac{1}{2})^n$ where n is the number of pigmented offspring. The question of where to stop is still with us. To be fairly certain that we have a firm basis for our decision, we normally collect enough offspring that the chance of obtaining so great a number of pigmented individuals from a heterozygote has less than a 5% chance of occurring. In our example we would have to observe at least five consecutive pigmented offspring. This would yield a probability of $(\frac{1}{2})^5$ which is ¹⁄₃₂ and equals 3.1%. To be more certain, one would collect more progeny before making a decision.

Our consideration of *a priori* probability does not end here. Let us assume that our test organism is a heterozygote. We can then ascertain the probability of obtaining various combinations of pigmented and albino offspring. In our testcross, as we have mentioned, the chance of obtaining a pigmented offspring is ½ and the chance of obtaining an albino offspring is also ½. It will be helpful at this point to use some symbols. Let p equal the probability of obtaining a pigmented offspring and let q equal the probability of obtaining an albino. These are the only two types of offspring that can be produced in our backcross, and it follows that p + q = 1. In our example if p equals ½, then q must equal ½. If we let the number of offspring to be observed equal N and the number of the offspring that are to be pigmented equal X, then the probability of obtaining any particular combination of pigmented and albino offspring can be obtained from the following formula:

$$P(X) = \frac{N!}{X! \ (N-X)!} \ p^X q^{N-X}$$

In this formula, the symbol "N!" (read N factorial) stands for the product of all the integers up to and including N. For exam-

ple, N! = N (N − 1) (N − 2) . . . . 4 · 3 · 2 · 1. Furthermore, we define 0! as equal to one. We can check our earlier reasoning by using this formula in the case of obtaining five pigmented offspring and no albinos. The formula reads:

$$P(5) = \frac{5 \cdot 4 \cdot 3 \cdot 2 \cdot 1}{5 \cdot 4 \cdot 3 \cdot 2 \cdot 1(5-5)!} \ (\tfrac{1}{2})^5(\tfrac{1}{2})^0$$

Since 0! = 1 and anything raised to the zero power also equals 1, the formula quickly reduces to the following:

$$P(5) = (\tfrac{1}{2})^5 = 3.1\%$$

This value for P(X) is the same as we obtained earlier. The formula becomes more involved in other situations. Look at Fig. 3-2 which diagrams a cross of two heterozygous pigmented individuals. Here the p value for a pigmented offspring is ¾, whereas the q value for an albino is ¼. Let us calculate the probability of getting three pigmented and one albino in four offspring. The formula now reads:

$$P(3) = \frac{4 \cdot 3 \cdot 2 \cdot 1}{3 \cdot 2 \cdot 1 \cdot (4-3)!} \ (\tfrac{3}{4})^3(\tfrac{1}{4})^1$$

$$P(3) = 4(\tfrac{3}{4})^3(\tfrac{1}{4}) = (\tfrac{3}{4})^3 = {}^{27}\!/_{64} = 42\%$$

Notice that although a grouping of three pigmented and one albino is exactly what the F_2 generation should be, the probability of obtaining such exact figures in a random sample of four offspring is only 42%. Table 3-1 shows the probabilities of obtaining the various possible combinations of pigmented and albino offspring in an F_2 generation when p = ¾ and q = ¼. Note that all the individual probabilities add up to 1. A probability of 1 indicates that the event is a certainty. In our example it means that any combination of four progeny must fit into one of the five possibilities shown in Table 3-1. This conclusion illustrates another law of probability: when events are mutually exclusive, the probability that one or the other will occur is the sum of their individual probabilities (e.g., probability of obtaining either a head or a tail from the toss of an honest coin is ½ + ½ = 1). While a probability of 1 indicates that the

Table 3-1. Probabilities of obtaining the various possible combinations, X, of pigmented and albino offspring in an F_2 generation of four individuals when $p = \frac{3}{4}$ and $q = \frac{1}{4}$

X		p	Term of binomial expansion
Pigmented	Albino		
4	0	0.317	p^4
3	1	0.422	$4p^3q$
2	2	0.211	$6p^2q^2$
1	3	0.046	$4pq^3$
0	4	0.004	q^4
Total		1.000	1

event is a certainty, a probability of 0 indicates that the event is impossible. All probabilities lie between 0 and 1. Table 3-1 also shows the terms of the binomial expansion of $(p + q)^4$, which correspond to the different possible combinations of pigmented and albino progeny that can be obtained in this case. The generalized formula is $(p + q)^N$, in which, as before, N equals the number of offspring in each observation. The use of the binomial expansion is possible because it represents mathematically Mendelian segregation and recombination.

The usefulness of *a priori* probability lies in the ability it gives the geneticist to estimate the chance of obtaining a specific result in a particular mating. It is extremely helpful in medical counseling in cases where potential parents are known to carry deleterious genes and wish to know the chance of transmitting these deleterious genes to their offspring. We shall consider empiric probability in Chapter 4.

Partial (incomplete) dominance

A gene occurs in man that causes fingers and toes to be abnormally short—a condition known as brachydactyly. Here, too, there are only two alleles—for brachydactyly and for normal-sized fingers. For the gene whose action results in the normal condition of a particular structure the symbol "+" is customarily used. The genes whose effects are dominant are indicated by capital letters, and those whose effects are recessive are designated by lower case letters. In this instance the gene for brachydactyly can be referred to as *Br*. In addition, the two homologous chromosomes can be shown as paired, by the use of two slanted lines "//." Applying these symbols in the case of brachydactyly, we find three possible genotypes: (1) *Br//Br*, (2) *Br//+*, and (3) *+//+*. The genetic combination *Br//+* (heterozygote) results in the abnormally short fingers and toes mentioned above, and the genotype *Br//Br* (homozygote) results in severe crippling with a complete lack of fingers and toes. Although the gene for brachydactyly is dominant, that dominance is incomplete, and two genes for this factor have a much greater effect than a single gene.

Another example of this type of allelic interaction is shown in Fig. 3-4. There we see that the cross between a red-flowered and a white-flowered snapdragon produces an F_1 with pink flowers. In this type of inheritance, the F_1 has a phenotype different from either of the two parents. The F_2 results in ¼ red-, ²⁄₄ pink-, and ¼ white-flowered plants. In crosses involving incomplete dominance both the F_2 genotypes and phenotypes show the 1:2:1 ratio.

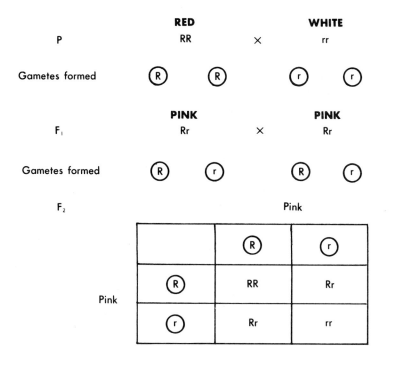

Summary of F₂

Phenotypes	Phenotypic ratio	Genotypes	Genotypic ratio
Red	1	RR	1
Pink	2	Rr	2
White	1	rr	1

Fig. 3-4. Cross between a red-flowered and a white-flowered snapdragon, showing partial (incomplete) dominance.

Partial dominance also includes those genes whose actions in combination with the + (normal) gene are not striking and whose presence in combination with the + gene can, in many cases, be detected only by special tests. Many of these genes were previously thought to be completely recessive. Such a condition is illustrated by the gene whose action in double dose causes thalassemia. This disease develops in childhood and is usually fatal. The gene in single dose

(in combination with the + gene) causes its possessor to suffer a slight anemia with no serious effects. The slight anemia is usually not apparent without clinical tests. Before the advent of such clinical tests, the gene was thought to be a complete recessive.

The importance of the partially dominant gene is that if it is deleterious to its possessor it will handicap him even in single dose, though on a reduced scale. The number of genes found to fit into this category increases as our methods of determining gene action improve.

Codominance (equal expression)

Human blood groups provide an excellent illustration of the codominance type of allelic interaction. We shall consider the ABO series. Here there are three alleles of the same gene. When more than two alleles exist, the series is called *multiple alleles.* There are six different possible combinations of these three alleles: *AA, AO, BB, BO, AB,* and *OO.* Out of these six possibilities, four different blood types are recognized: A, B, AB, and O. Genes *A* and *B* exhibit simple dominance when either of these genes is in combination with the *O* gene. However, when genes *A* and *B* are in combination with each other, each one is expressed as if it were present without the other. This series of alleles therefore exhibits two types of gene interactions: (1) simple dominance between genes *A* and *O* or genes *B* and *O* and (2) *codominance* between genes *A* and *B.*

The blood groups have played an interesting function in cases of disputed paternity. In most instances we can show that a particular male could not have fathered a particular child if in fact he was not the father. However, we cannot prove that a particular male did father the child. As an illustration, let us consider the following case: a male's blood type is O and a fe-male's blood type, O. No child of blood type A or B could have this male as a father. However, we cannot be certain, should the child have blood type O, that this particular male was the father. Hence, one can disprove paternity but not prove it. The courts today need not rely on the ABO series alone for evidence of nonpaternity. There are some five or six other blood factors (e.g., rh, MN, Lewis, Duffy), the combination of which can rule out any improperly accused male.

NONALLELIC GENE INTERACTIONS

Most genes do not act alone in determining the traits of an organism but rather interact with other nonallelic genes in causing their effects. The various modes of interaction are best understood by considering examples of each type.

Combs in fowls (9:3:3:1 ratio)

The analysis of gene interaction in the combs of fowls was one of the classic investigations in genetics and was reported by Bateson and Punnett in a series of papers in the years 1905 to 1908. There are many different breeds of domestic chickens. Each breed is characterized by a type of comb (Fig. 3-5), although different breeds may have the same type. The Wyandotte breed has a comb called "rose," the Brahma breed has a comb called "pea," and the Leghorns have a comb called "single."

A cross of a chicken with a rose comb to

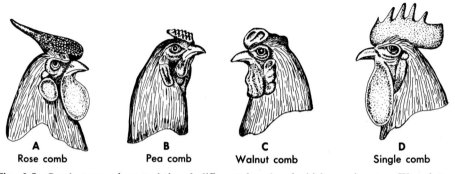

| A | B | C | D |
| Rose comb | Pea comb | Walnut comb | Single comb |

Fig. 3-5. Comb types characteristic of different breeds of chickens: **A,** rose, Wyandottes; **B,** pea, Brahmas; **C,** walnut, hybrid from cross between chickens with rose and pea combs; **D,** single, Leghorns.

one with a pea comb produces offspring with a different type of comb known as "walnut." If the walnut birds are mated to one another, one gets all four types of combs among the offspring in the following ratios: $\frac{9}{16}$ walnut, $\frac{3}{16}$ rose, $\frac{3}{16}$ pea, and $\frac{1}{16}$ single. These matings and the offspring obtained are shown in Fig. 3-6.

The interpretation of the results is as follows. The rose type comb is caused by the combination of a homozygous recessive condition for a gene we shall call p and the presence of at least one dominant gene that we shall call R at a different locus. The pea type comb is due to the combination of a homozygous recessive condition for the gene r and the presence of at least one dominant P gene. The single comb is caused by the double recessive $rrpp$, whereas, the walnut comb is due to the presence of at least one dominant gene at both the R and P loci. If one accepts this model for the inheritance of combs in fowls, one must hypothesize that the Wyandotte breed has

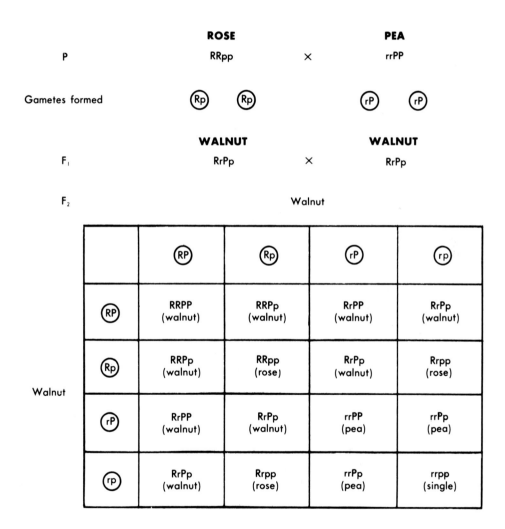

Summary: 9/16 walnut, 3/16 rose,

3/16 pea, 1/16 single

Fig. 3-6. Diagram of a cross between rose comb and pea comb chickens.

Table 3-2. Expected number of gametes, phenotypes, and genotypes to be obtained from various hybrid crosses

Power function	Conditions governed by the power function
2^n	Number of gametes formed by F_1 parents
	Number of phenotypes formed in F_2 in cases of complete dominance
3^n	Number of phenotypes formed in F_2 in cases of incomplete dominance or codominance
	Number of genotypes formed in F_2
4^n	Number of ways to form F_2 offspring

The n here equals the number of allelic pairs for which both F_1 parents are hybrid.

the genotype *RRpp* (rose) and the Brahmas have the genotype *rrPP* (pea). The cross between them yields an F_1 with the genotype *RrPp*. The F_2 then yields the four classes with the ratios indicated above.

The type of inheritance involved in combs in fowls was also studied by Mendel in his garden peas. From his study of various genetic factors that were located on nonhomologous chromosomes, he derived the *"law" of independent assortment* (also known as the "law" of unit characters, which postulated that every characteristic is inherited independently of every other characteristic. This is also not a true biological "law," since it was later shown that different genes on the same chromosome tend to be inherited together.

Checking the F_2 offspring in Fig. 3-6 carefully, we find four kinds of phenotypes, nine kinds of genotypes, and sixteen possible ways to form the F_2. If you look once more at Fig. 3-2, which illustrates the F_2 from a monohybrid F_1, you will see in the F_2 offspring two kinds of phenotypes, three kinds of genotypes, and four possible ways to form the F_2. These numbers fit a pattern which is shown in Table 3-2 for the various types of allelic interactions and which may also be used for some nonallelic gene interactions.

Testing goodness of fit. It is appropriate at this point to consider that type of *a priori* probability which is concerned with

whether the data obtained agree with an expected ratio based on some genetic theory. As will be shown, the same observed results may actually fit more than one genetic model. In those instances either more data must be gathered or the results of some other type of cross must be observed before one can come to a decision about the genetic basis for the inheritance.

The most useful statistical test to compare experimentally obtained results with a given genetic model is the *chi-square* (χ^2) *test*. Let us consider once again the F_2 of a monohybrid cross involving simple dominance. We would expect a 3:1 ratio of dominant to recessive trait. Assume that we have observed 60 F_2 offspring and have found that 40 carry the dominant trait while 20 carry the recessive trait. Out of a total of 60 F_2 progeny, the theoretical model calls for 45 (¾ × 60) with the dominant trait and 15 (¼ × 60) with the recessive trait. The question arises as to whether 40:20 can occur strictly by chance when one expects a 45:15 ratio.

We could use the binomial distribution discussed earlier under *a priori* probability to solve this problem but it would be long and tedious. Chi-square offers a more convenient test. The chi-square formula is

$$\chi^2 = \Sigma \left[\frac{(X_{obs} - X_{exp})^2}{X_{exp}} \right]$$

where χ is the Greek letter chi from which

Table 3-3. Calculation of χ^2 for an F_2 monohybrid cross

	Dominant	Recessive	Total
Observed number	40	20	60
Expected number	45	15	60
Obs. − Exp.	−5	+5	0
(Obs. − Exp.)^2	25	25	
(Obs. − Exp.)^2/Exp.	0.555	1.666	$2.221 = \chi^2$

the chi-square test gets its name, X_{obs} is the observed number of cases with a particular outcome, X_{exp} is the number of cases expected to show this specific outcome, and Σ is the Greek capital letter sigma that in this formula means "summation of." It is important to note that the χ^2 formula is based on actual numbers and not on percentages. When data are reduced to percentages, the total automatically becomes 100, thereby eliminating the important factor of sample size in the evaluation. Table 3-3 shows the format and the results of the chi-square test for our problem.

In thus solving the χ^2 equation we have obtained a value of 2.221. However, before we can put this number to use, we must consider another aspect of the chi-square test called *degrees of freedom,* which takes into account the number of classes that are involved in a given cross. This is important since the larger the number of classes, the more opportunities for chance deviations to occur. Such chance deviations will cause the χ^2 value to be very large when the number of classes is increased. We must allow for the effect of the number of classes in assessing the significance of any χ^2 value. The number of degrees of freedom is one less than the number of classes. In our present example there are two classes (dominant and recessive). Therefore we have 1 degree of freedom. We may now interpret the χ^2 value in terms of probability. At this point we need to consult Table 3-4, which gives the probability of obtaining, by chance, various values for χ^2 under different degrees

of freedom. We again take 5% probability as our critical point. Reading Table 3-4 along the 1 degree of freedom line, we find that our χ^2 value of 2.221 lies between the χ^2 values for 0.05 and 0.20 probability. Since the 40:20 observed ratio in our F_2 has a greater than 5% probability of occurring under a 3:1 genetic model, we accept the hypothesis that our observed sample of organisms is an F_2 of a simple monohybrid cross. For us to reject our sample as an example of a 3:1 ratio, we would have had to obtain a χ^2 value greater than 3.841.

The question always arises as to why we should choose 5% as our cutoff point, rather than one which is more rigorous. The geneticist is extremely reluctant to discard a hypothesis of a genetic model for a particular type of inheritance even when the data have a relatively small chance of fitting the model. Being so cautious has a great deal of merit. Should an incorrect model be accepted for a given set of data and research in the field be continued on the basis of the false hypothesis, we can anticipate that —sooner or later—sufficient data will be gathered to prove the model wrong. However, if a true hypothesis is discarded, the research in that area will be concentrated on other genetic models because of the general hesitation to repeat experiments to "prove" hypotheses that have been "disproved." Any return to the discarded hypothesis will occur only after a great expenditure of energy has shown all other likely models to be untenable.

Table 3-4. Table of chi-square (p = probability)

Degrees of freedom	p = 0.99	0.95	0.80	0.50	0.20	0.05	0.01
1	0.000157	0.00393	0.0642	0.455	1.642	3.841	6.635
2	0.020	0.103	0.446	1.386	3.219	5.991	9.210
3	0.115	0.352	1.005	2.366	4.642	7.815	11.345
4	0.297	0.711	1.649	3.357	5.989	9.488	13.277
5	0.554	1.145	2.343	4.351	7.289	11.070	15.086
6	0.872	1.635	3.070	5.348	8.558	12.592	16.812
7	1.239	2.167	3.822	6.346	9.803	14.067	18.475
8	1.646	2.733	4.594	7.344	11.030	15.507	20.090
9	2.088	3.325	5.380	8.343	12.242	16.919	21.666
10	2.558	3.940	6.179	9.342	13.442	18.307	23.209
15	5.229	7.261	10.307	14.339	19.311	24.996	30.578
20	8.260	10.851	14.578	19.337	25.038	31.410	37.566
25	11.524	14.611	18.940	24.337	30.675	37.652	44.314
30	14.953	18.493	23.364	29.336	36.250	43.773	50.892

Modified from Table 3. Fisher, R. A., and Yates, F. 1963. Statistical tables for biological, agricultural and medical research. Oliver & Boyd, Ltd., Edinburgh. By permission of the authors and publishers.

A further examination of the chi-square table (Table 3-4) illustrates the role of the number of classes in affecting χ^2 values. You will notice that at the 5% level of probability, the critical χ^2 value increases dramatically as the number of degrees of freedom increases. This takes into account the increased sampling error that occurs with an increased number of classes.

The consideration of our 40:20 ratio as an example of a 3:1 ratio is not yet completed. The 40:20 ratio is obviously a perfect fit of a 2:1 ratio. Since the deviation of observed from expected is 0, the χ^2 value would be 0, and the probability that our observations would fit a 2:1 ratio is 100%. What decision can we make? First let us consider what we cannot do. We cannot argue for the acceptance of the 2:1 rather than the 3:1 ratio, on the basis of the higher probability of our data fitting the 2:1 ratio. Our entire consideration of the χ^2 test rests solely on the acceptance or rejection of a genetic model at the 5% level of significance. There is no aspect of "greater than" or "less than" in our acceptance or rejection. Probability values greater than 5% may be useful in guiding further research, but they cannot be used in deciding which of two hypotheses is correct. What then can we do? Let us see what will happen if we increase our number of observations. Let us assume that we now count 120 F_2 offspring and get a ratio of 80:40 (dominants to recessives). Table 3-5 shows the chi-square computations. Taking the computed χ^2 value of 4.444 and checking the chi-square table (Table 3-4), we find that for 1 degree of freedom, there is less than a 5% chance that our data fit the 3:1 genetic model. We therefore reject the hypothesis that our data represent the F_2 offspring of a simple monohybrid cross. Obviously our observed 80:40 continues to represent a perfect fit of a 2:1 ratio. We have, by gathering more data, been able to decide which genetic model fits our observed results.

However, merely saying that our data fit a 2:1 ratio does not explain the genetic mechanism controlling our results. We are

Table 3-5. Calculation of χ^2 for an F_2 monohybrid cross

	Dominant	Recessive	Total
Observed number	80	40	120
Expected number	90	30	120
Obs. − Exp.	−10	+10	0
(Obs. − Exp.)^2	100	100	
(Obs. − Exp.)^2/Exp.	1.111	3.333	4.444 $= \chi^2$

Table 3-6. Calculation of χ^2 for agreement of a two-factor segregation with a 9:3:3:1 ratio

Phenotype	Walnut	Rose	Pea	Single	Total
Observed number	240	75	65	28	408
Expected number	408($\frac{9}{16}$)	408($\frac{3}{16}$)	408($\frac{3}{16}$)	408($\frac{1}{16}$)	408
Expected number	229.5	76.5	76.5	25.5	408
Obs. − Exp.	+10.5	−1.5	−11.5	+2.5	0
(Obs. − Exp.)^2	110.25	2.25	132.25	6.25	
(Obs. − Exp.)^2/Exp.	0.480	0.029	1.729	0.245	2.483 $= \chi^2$

now faced with the problem of constructing a theory that explains a 2:1 ratio among the offspring of two heterozygous parents. The problem is easily solved, since there are many such examples known. If one postulates that the homozygous dominant genotype is lethal, the offspring of two heterozygous parents will then be in a ratio of 2:1 (dominants to recessives). This genetic model can be put to a number of tests. If the model is correct, then all offspring showing the dominant trait should themselves be heterozygotes. Backcrosses would easily demonstrate this. If the model is correct, one fourth of all fertilized eggs should fail to develop into viable offspring. This can be confirmed by observation in many animals and plants. We thus can check on our hypothesis by making crosses and observations other than those that first led us to postulate the genetic mechanism underlying

our unexpected findings. We shall discuss genes with homozygous lethal effects in Chapter 11.

The chi-square test can be used also in a cross of two, three, four, etc. factors. In the case of combs in fowls, we have a two-factor cross. Let us assume that the F_2 of Fig. 3-6 yielded the following results: 240 walnut, 75 rose, 65 pea, and 28 single. Again we have the question as to whether the offspring obtained fit the genetic model postulated for this type of inheritance. Table 3-6 shows the chi-square computations. Calculation of the χ^2 formula gives us a value of 2.483. Since we have four phenotypic classes, we have 3 degrees of freedom. An examination of Table 3-4 shows us that for three degrees of freedom, our chi-square value has a probability between 20% and 50% of fitting the hypothesized genetic model. We therefore accept our observed results as being consistent with a dihybrid cross.

Seed color in corn (9:7 ratio)

Corn, more properly called maize *(Zea mays)*, occurs in many varieties. Some of these varieties have colorless seeds, whereas some others have seeds of a purple color. The purple color, which represents the ancestral condition, is due to the presence of a purple substance known as *anthocyanin* in the aleurone layer of the seeds. Still other varieties of maize have seeds of various other colors. We shall restrict our consideration to a relatively simple facet of this complex problem, which was worked out by

R. A. Emerson of Cornell University and other geneticists.

A cross of two varieties of corn that had been characterized by colorless seeds was found to produce an F_1 that had purple-colored seeds. The F_2 produced colored-to-colorless seeds in a ratio of 9:7. The crosses are diagrammed in Fig. 3-7. In order to explain these results, it was postulated that for the production of anthocyanin the plant must carry dominant alleles at two different loci. Plants homozygous for the recessive alleles at either one of these loci would have no pigmentation in the seeds. Under this hypothesis each locus controls a step essential for the production of the pigment. Should either

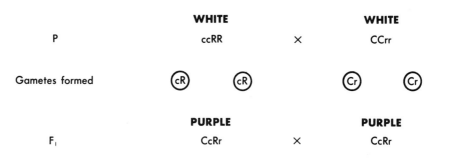

Fig. 3-7. The 9:7 ratio, showing the expected composition of the F_2 from a cross of two corn plants with colorless seeds that produced an F_1 with purple seeds.

biochemical reaction be missing, no anthocyanin will be produced. When two or more loci are so involved, we say that the actions of these independent genes are *complementary*. In our particular example the two colorless seed varieties of corn were homozygous for recessive genes at two different loci.

The inheritance of seed color in maize illustrates a number of interesting points. The situation in which two colorless seeds may be homozygous for two different genes once again shows that two identical phenotypes may be caused by different genotypes. Although we previously met this situation in cases of simple dominance, we see it now when recessive traits are involved. This highlights the problem of inferring genotype from phenotype and indicates the extreme caution that must be exercised in formulating genetic models. A second point deals with the observation that crosses between some true-breeding varieties give offspring that resemble a remote ancestor. This has been noted among domesticated animals and plants. The organisms produced have variously been called *throwbacks, atavisms,* or *reversions*. In the absence of any satisfactory explanation, the early breeders thought that some mysterious force caused the retention and subsequent reappearance of a remote ancestral trait. However, we now can postulate that, in the evolution of the differing varieties of these organisms, different alleles were lost. The cross of two varieties produces offspring whose genotype contains all the alleles of the ancestral stock, and the "reversion" to the ancestral type appears to have occurred.

Coat color in rodents (9:3:4 ratio)

A more complex case of gene interaction is seen in the coat color of rodents. Here many genes are involved, of which we shall consider but three. The first gene we have already discussed under simple dominance. This is the C locus gene that determines whether or not pigment will be deposited in the fur. The recessive allele, c, when homozygous, results in albinism. A second gene is

the B locus. The dominant allele causes a black pigment to be deposited in the individual hairs; the recessive allele, b, when homozygous, causes a brown pigment to be deposited. The third gene is the A locus. The dominant allele causes the deposition of a narrow yellow band of pigment near the tip of the individual hairs, and the recessive allele, a, when homozygous, results in the absence of this yellow band.

Following are examples of the many combinations that these three independent genes can form:

1. *CC BB AA* results in the coat color pattern called black agouti. This is the characteristic coat color of wild rodents. Only one dominant allele at each locus is needed to yield the black agouti pattern.
2. *CC BB aa* results in a solid black fur.
3. *CC bb AA* results in a pattern called brown agouti or cinnamon. The individual hairs are brown with a narrow yellow band near the tip of each hair.
4. *CC bb aa* results in a solid brown or chocolate fur.
5. *cc — —* results in an albino regardless of which alleles are present at the B and A loci. In this case the homozygous recessive, cc, covers up or hides the expression of the other genes. This phenomenon is called *epistasis*, and the genes involved are said to be *supplementary*. The gene that masks or prevents the expression of another is said to be *epistatic* to it, and the gene that is hidden is said to be *hypostatic*.

An interesting cross involving these three loci can be seen when one takes a black mouse of genotype *CC BB aa* and mates it to an albino mouse whose genotype is *cc BB AA*. The results in the F_1 and F_2 are shown in Fig. 3-8. Since all mice involved in the crosses are homozygous BB, these symbols have been omitted; but they must be considered in specifying the phenotypes. The F_1 of this cross shows the phenomenon of reversion to the ancestral coat color. The

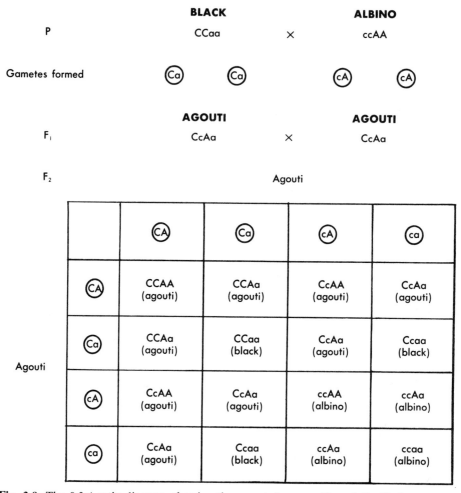

Fig. 3-8. The 9:3:4 ratio diagram, showing the expected composition of the F₂ from a cross of black and albino mice, which produced all agouti animals (wild type) in the F₁.

F₂ yields a ratio of 9 black agouti:3 black:4 albino. Due to epistasis, all four albinos have the same phenotype despite the fact that they have much different genotypes at the A locus. This last term of the 9:3:4 ratio really represents the last two terms of the 9:3:3:1 ratio and indicates that in this situation these two ordinarily different phenotypic classes of the dihybrid F₂ zygotes cannot be distinguished.

ELECTROPHORETIC MOBILITY

In Chapter 1 it was pointed out that genes have two direct functions, replication and transcription. The latter function, trans-

cription, leads to translation, which yields both structural and enzymatic proteins. The different proteins produced by an organism vary from one another in their amino acid composition and sequence. Each protein will, because of its particular amino acid composition, have a characteristic size and shape. In addition, each protein will have a plus or minus *net electrostatic charge* of a particular strength on its surface, depending on the pH of its surrounding medium. Differences in size, shape, and net charge can be used to experimentally separate and identify different types of proteins.

One experimental method used to study

proteins is *electrophoresis.* Its operation depends on the fact that charged molecules that are suspended in a conducting medium will migrate in an electric field. The conducting medium consists of a solution of appropriate chemical compounds, usually called a *buffer,* which is held in a porous supporting medium such as filter paper, cellulose acetate, or a gel made of agar, starch, or polyacrylamide. In mixtures of charged particles, each type of charged particle will migrate at a characteristic rate dependent on the strength of the electric field, the size, shape, and net charge of the particle, and the particular conducting medium. Identical molecules will migrate at the same rate and will, after a period of time, become concentrated at some definite point in the medium, forming a sharply defined band. With sufficient time for migration, a clear-cut separation of different types of protein molecules can be achieved. Electrophoretic mobility represents a relatively new phenotype whose inheritance and variability can be studied. As will be pointed out in our discussion of sickle cell anemia in Chapter 10, this experimental technique is sufficiently sensitive to distinguish between polypeptides that differ from one another in only a single amino acid.

The fact that each protein is composed of one or more polypeptide chains has been known for a long time. If a protein consists of a single polypeptide chain, it is called a *monomer.* When two polypeptide chains make up the protein, it is called a *dimer,* and quite obviously—by extension—*trimers, tetramers,* or any alternative form of *multimer* is possible. Although each allele produces a different polypeptide chain, any protein that is multimeric in composition has the possibility of being formed of either identical or nonidentical polypeptide chains.

Although the electrophoretic mobilities of various mouse, maize, and human proteins have been studied, research work on the fruit fly, *Drosophila melanogaster,* has given us the best information on the patterns of inheritance of this phenotype. In such studies the flies are killed and ground up

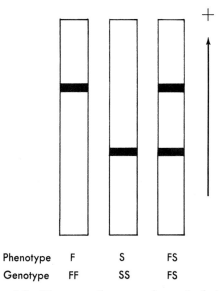

Phenotype	F	S	FS
Genotype	FF	SS	FS

Fig. 3-9. Diagrammatic comparison of electrophoretic phenotypes obtained in study of esterase-6 enzymes in *D. melanogaster.* Arrow indicates direction of migration of the proteins.

in a buffer solution, and the homogenate is subjected to electrophoresis. A study of one of the *esterase* enzymes (est-6) in *D. melanogaster* was reported by T. R. F. Wright in 1963. His results are diagrammed in Fig. 3-9. He found that certain stocks of flies differ in the electrophoretic mobility of their esterase-6 enzyme. Some stocks of flies have a type of est-6 enzyme that moves quickly in a given amount of time and forms a band called the "F" (fast) band. Flies that are found to have only this form of the enzyme are considered to be *FF* in genotype. Other stocks of flies have a type of est-6 enzyme that moves slowly, in the same amount of time, and forms a band called the "S" (slow) band. Flies that are found to have only this form of the enzyme are considered to be *SS* in genotype. A mating between flies from the above two stocks produces an F_1 that contains both types of est-6 enzyme, as seen in the appearance of both F and S bands. It is quite apparent that in the heterozygotes *(FS),* there is no combination of the different protein units or subunits produced by each allele. Under the experimental conditions just described, the protein appears to be a monomer. It could, of course, be a dimer, trimer, or any other multimer, provided it were made up of identical polypeptide chains. Under these conditions, information on the quaternary structure of a protein can come only from other types of studies. The equal and independent expression of both alleles indicates that we have here a condition of *codominance*.

A study of another enzyme, *alkaline phosphatase,* in *D. melanogaster* was reported by Beckman and Johnson in 1964. Their results are diagrammed in the first three columns of Fig. 3-10. The striking difference of these results from those of Wright is in the F_1 *(FS)* offspring. Here we see the appearance of a band of intermediate (FS) mobility. The FS band represents the occurrence of a hybrid protein. These results are consistent with a hypothesis that the enzyme is a dimer, which in homozygotes consists of identical polypeptide chains and which in heterozygotes can be formed of either identical or nonidentical subunits. The formation of a hybrid protein of intermediate mobility in addition to the fast and slow bands indicates that we are dealing with a condition of *codominant interacting alleles*.

The research on alkaline phosphatase and on other proteins is continuing. In 1966 F. M. Johnson reported the discovery of an allele at this locus that in homozygous condition results in the complete absence of a protein band. This is shown in the fourth column of Fig. 3-10. The discovery of the *O* allele indicates that we are dealing with a multiple allelic series (i.e., *F, S,* and *O*). Matings of *OO* flies with *FF* and *SS* flies gave the results shown, respectively, in the fifth and sixth columns of Fig. 3-10. The *FO* offspring show *simple (complete) dominance* of the F phenotype over the O. However, the *SO* offspring show not only an S-type band but also a new band that is different in location from all those previously observed. The presence of the new band in *SO* heterozygotes could result from the combination of an S subunit with an O polypeptide chain. It would follow that the O subunit by itself cannot dimerize nor can it form a combination with an F polypeptide chain. The SO phenotype would be an example of *partial (incomplete) dominance*.

In closing this discussion of electrophoretic mobility, we should indicate the importance of this relatively new research field. This phenotype is the closest we have come, in many cases, to studying the primary end products of gene action. In a number of instances we have, for the first time, been able to observe the results of allelic interactions in a diploid organism as they relate to protein formation. Another important aspect of this phenotype is that it permits the study of a huge area of genetic variation that has hitherto been relatively unknown. Finally, our ability to detect, in some instances, single–amino acid differences in polypeptide chains promises to bring us closer to the study of the effects of single nucleotide changes on the functioning of the proteins of the body.

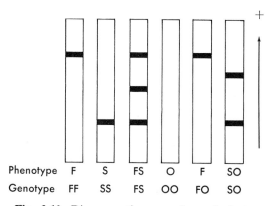

| Phenotype | F | S | FS | O | F | SO |
| Genotype | FF | SS | FS | OO | FO | SO |

Fig. 3-10. Diagrammatic comparison of electrophoretic phenotypes obtained in the study of alkaline phosphatase enzymes in *D. melanogaster.* Arrow indicates direction of migration of the proteins. (From Johnson, F. M. 1966. Science **152:** 361-362. Copyright 1966 by the American Association for the Advancement of Science.)

SUMMARY

In this chapter we have reviewed some of the basic patterns of chromosomal inheritance. These include (1) allelic interactions such as simple dominance (e.g., pigment production in mammals), partial dominance (e.g., brachydactyly in humans), and co-dominance (e.g., AB blood groups in humans); and (2) nonallelic gene interactions such as those in which both genes express themselves (e.g., combs in fowls), complementary interaction (e.g., seed color in corn) or supplementary interaction (e.g., coat color in rodents). Electrophoretic mobility of proteins, a relatively new field of research, was discussed. We also reviewed the concepts of *a priori* and *a posteriori* probability and discussed one of the statistical tests (chi-square) used to decide whether a body of data fits a particular genetic model. We shall now consider the mode of inheritance of those traits that depend on the action of many genes for their phenotype.

Questions and problems

1. Explain the following terms:
   a. Allele
   b. Phenotype
   c. Genotype
   d. Homozygous
   e. Heterozygous
   f. First filial generation
   g. Second filial generation
   h. Multiple alleles
   i. Codominance
2. The gene *B* (for brown eyes) is dominant in man to the gene *b* (for blue eyes). Diagram the following crosses:
   a. BB × bb
   b. Bb × Bb
   c. Bb × bb
   d. Bb × BB
   What is the chance that a particular offspring from each cross will be (a) blue-eyed? (b) brown-eyed?
3. A cross is made between two heterozygous walnut-combed fowls.
   a. Diagram the cross.
   b. What phenotypic classes are found among the offspring?
   c. What is the chance that a particular offspring will inherit the single-comb genotype?
   d. What genetic principle, with reference to gene action, is reflected in the progeny of a cross between Wyandotte and Brahma fowls?
4. Parents in which the mother has type A blood and the father has type B blood have a son whose blood is analyzed as being type O.
   a. What is the chance that their next child will be a girl with type AB blood?
   b. What type of probability is acting in this case?
5. In man, phenylketonuria (a metabolic disorder) is inherited as a recessive autosomal gene. A man and wife, each heterozygous for the gene in question, plan to have five children. What is the probability that of the five offspring three will be normal and two will have phenylketonuria?
6. A paternity suit is filed against a man whose alleged child has type A blood. The child's mother has type B blood and the "accused" has type AB blood. Does the "defense" have a chance? Explain!
7. Out of the first ten plants whose seeds were counted by Mendel, the one with the largest percentage of green seeds had thirteen green and twenty-three yellow seeds. Does this differ significantly from 25% green?
8. From a cross between round yellow plants and wrinkled green plants, Mendel obtained the following phenotypic seed classes:
   31—round, yellow   26—wrinkled, yellow
   26—round, green   27—wrinkled, green
   a. How "good" is Mendel's hypothesis that the genes were segregating in a 1:1:1:1 ratio?
9. A cross is made between albino, *cc*, and pigmented, *CC*, mice. In the F_2, twelve pigmented and eight albino mice were observed.
   a. Is this in agreement with a hypothesized 3:1 ratio? Explain!
   b. If eleven pigmented and nine albino had been observed in the F_2, what would you say?
10. How would you explain the molecular basis for the phenomena of partial dominance?
11. A female, heterozygous for the albinism gene and exhibiting the heterozygous brachydactyly syndrome, marries a male having the identical genotype with respect to these two factors.
    a. Theoretically, how many different phenotypic classes of offspring—and in what proportion—are possible from this marriage?
    b. What factor is responsible, in this case, for the deviation from the usual 9:3:3:1 ratio?
12. Refer to question 11 and answer the following:
    a. What is the probability that the first child will be a crippled albino?
    b. What is the probability that the third will be a normally pigmented individual with short toes and fingers?
    c. What is the probability that the fifth child will be normal?
    d. What is the probability that the second child will be an albino with normal toes and fingers?
13. Considering the types of gene interactions discussed in this chapter, diagram a specific mating in humans—involving two different

genes—that could result in the following ratio: 4:2:2:1:2:1:2:1:1.

14. In a mating between a single-combed fowl and a heterozygous walnut-combed fowl, what is the probability that of the first six offspring, four will be rose combed?

15. A cross is made between two varieties of corn plants, each having colorless seeds.
   a. What will the F_1 phenotype(s) be if one parent is ccRR and the other is CCrr?
   b. How is the presence of this (these) phenotype(s) explained, considering the seed color of the parental generation?
   c. If two F_1 corn plants are crossed and 100 progeny are produced from this mating, what phenotypes will, ideally, be found in the F_2 generation, and in what proportions?

16. In *Drosophila* the dichaete mutation results in an abnormal wing. Crosses between dichaete-winged flies and normal flies always result in progeny of which 50% are normal-winged and 50% are dichaete. Crosses between dichaete flies always result in progeny of which two thirds are dichaete and one third are normal winged. Explain these results.

17. In a hypothetical organism, gene *A* acts to suppress eye formation, and its allele, *a,* is inactive in this regard. A second gene, *B,* is responsible for eye pigment formation, and its allele, *b,* is inactive. A cross is made between *AAbb* and *aaBB*.
   a. What phenotypic classes will be present in the F_2, and in what proportions?
   b. How do the terms "epistatic" and "hypostatic" apply in this instance?

18. Two black agouti mice are mated. Among the progeny of this cross, three phenotypic classes are present, in the following proportions: 9 black agouti:3 black:4 albino.
   a. What were the genotypes and phenotypes of the progenies' grandparents?
   b. How do the terms "epistasis" and "supplementary genes" apply in this instance?

19. A cross between a black mouse and a black agouti mouse results in the following:

   9/32 black agouti
   9/32 black
   3/32 cinnamon
   3/32 brown
   8/32 albino

   What is the genotype of each parent?

20. Assess the importance of gene interaction as a genetic principle.

References

Bateson, W., and Punnett, R. C. 1906. Comb characters. Report to Evolution Committee of the Royal Society of London **II:**11-16. (Discussion and interpretation of the interaction of P and R genes.)

Castle, W. E. 1951. The beginnings of mendelism in America, Chapter 4. *In* Dunn, L. C. [ed.] Genetics in the 20th century. The Macmillan Co., New York. (General discussion of various types of genetic interactions with specific examples of modified 9:3:3:1 ratios.)

East, E. M. 1910. A Mendelian interpretation of variation that is apparently continuous. Amer. Natur. **44:**65-82. (Development of the multiple-gene hypothesis.)

Levene, H. 1958. Statistical inference in genetics, Chapter 29. *In* Sinnott, E. W., Dunn, L. C., and Dobzhansky, T. Principles of genetics. McGraw-Hill Book Co., New York. (Excellent discussion of the use of statistics in genetics.)

Scheinfeld, A. 1950. The new you and heredity. J. B. Lippincott Co., New York. (Popularly written book on human genetics.)

Senders, V. L. 1958. Measurement and statistics. Oxford University Press, Inc., New York. (Comprehensive treatment of statistical methods.)

Shaw, C. R. 1965. Electrophoretic variation in enzymes. Science **149:**936-943. (A general discussion of the importance of electrophoretic studies and a review of the organisms and enzymes studied.)

Sinnott, E. W., Dunn, L. C., and Dobzhansky, T. 1958. Principles of genetics. McGraw-Hill Book Co., New York. (Excellent discussion of classical genetics.)

Stern, C. 1960. Principles of human genetics. W. H. Freeman & Co., Publishers, San Francisco. (Pertinent chapters, including Chapter 5 on probability and Chapter 9 on genetic ratios.)

Stern, C., and Sherwood, E. R. (eds.). 1966. The origin of genetics. A Mendel source book. W. H. Freeman & Co., Publishers, San Francisco. (A collection of the pertinent publications relating to Mendel's experiments.)

Sturtevant, A. H. 1913. The Himalayan rabbit case, with some considerations on multiple allelomorphs. Amer. Natur. **47:**234-239. (A general discussion of multiple alleles by one of the pioneers in the field.)

Sturtevant, A. H., and Beadle, G. W. 1962. An introduction to genetics. Dover Publications, Inc., New York. (Reprinted from the original publication by W. B. Saunders Co., outlining the field of genetics as of 1939.)

Weiner, A. S., and Wexler, I. B. 1958. Heredity of the blood groups. Grune & Stratton, Inc., New York. (Genetics of blood group inheritance.)

4 Multiple-factor inheritance

Up to this point, we have considered types of inheritance in which we could distinguish the action of the individual genes involved. As the number of loci affecting a trait increases, it becomes difficult to determine how much of the trait is attributable to the action of a given gene. When the number of loci becomes too great, we are no longer able to specify the genotype of the individual. We can only characterize the population as to the average size or color of the trait in question and the variability of the trait in the population. Types of statistical methods different from those already discussed are necessary for handling the data obtained, and the inferences that one can make from phenotype to genotype become general rather than specific.

MULTIPLE-FACTOR TRAITS

The study of multiple-factor inheritance is an ever-increasing field of research because many important traits (stature, weight, intelligence; egg, milk, and meat production; yields of fruits and seeds; etc.) are determined by multiple genes.

Color in wheat kernels

The early workers in the field of wheat-kernel color included the Swedish geneticist H. Nilsson-Ehle (1908) and the American geneticist E. M. East (1910, 1916). One of the first analyses of this type of inheritance involved the results of crosses between red-kerneled and white-kerneled wheat. The F_1 seeds were intermediate in color between those of the two parents. At first this was considered to be a case of incomplete dominance. However, the F_2 zygotes showed five different classes of color intensities in the ratio of 1:4:6:4:1. The crosses involved are shown in Fig. 4-1. These results are interpreted as follows. Two pairs of genes control the color of the wheat kernel. Each gene has two alleles. One allele produces a given quantity of the red pigment, whereas its counterpart does not produce any pigment. All alleles are equally potent in the production or lack of production of pigment. An examination of the F_2 zygotes in Fig. 4-1 shows that only $\frac{1}{16}$ of the kernels have four pigment-producing genes and hence are red; $\frac{4}{16}$ have three genes for pigmentation and are dark; $\frac{6}{16}$ have two pigment-producing genes and are medium (same genotype and phenotype as the F_1); $\frac{4}{16}$ have only one gene for pigmentation and are light; the remaining $\frac{1}{16}$ have no pigment-producing genes and are white.

This example illustrates a very simple case of quantitative inheritance. In other crosses, between different red- and white-kerneled wheat plants, a more variable F_2 was observed that contained seven different classes in the ratio of 1:6:15:20:15:6:1. To explain these last results, the involvement of three pairs of genes was postulated. Here the red parent contained six genes for pigment production, and the white parent contained the alleles that prevented pigment production. It is evident that one of the three pairs of genes that were segregating in this latter cross must have been homozygous in both parents in the cross described in Fig. 4-1 and, hence, not shown in that diagram.

Our description of these dihybrid and trihybrid crosses indicates that as the number of segregating alleles increases, the probability of obtaining F_2 zygotes as extreme as either parent must necessarily decrease. The theoretical expectations for different numbers of pairs of segregating alleles have been worked out and are presented in Table 4-1. The denominator of that fraction of F_2 which is as extreme as either parent increases as the power function $(4)^n$. This is in keeping with our earlier pattern, shown in Table 3-2, which indicated that the number of ways of forming an F_2 zygote follows this same power function.

Skin color in man

Shortly after the genetic basis for multiple-factor inheritance had been postulated, attempts were made to analyze the human traits that appeared to follow this pattern. One of the earlier investigations dealt with human skin color and was conducted by

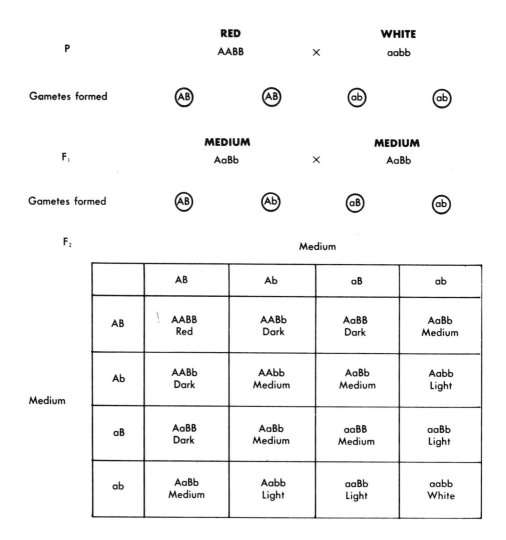

Summary of F_2: 1/16 red, 4/16 dark, 6/16 medium,

4/16 light, 1/16 white

Fig. 4-1. Diagram of a cross between a wheat variety with red kernels and another variety with white kernels.

C. B. Davenport (1913). He estimated the amount of pigment in a person's skin by visually matching the color of the skin to a color standard that he had devised. Working with Negroes and Caucasians and the hybrids between them, he postulated that two pairs of genes are involved in the production of the skin color pigment, melanin. This would yield results similar to those of Fig. 4-1, with the F_2 in a cross between the

races consisting of $\frac{1}{16}$ black, $\frac{4}{16}$ dark, $\frac{6}{16}$ intermediate (mulatto), $\frac{4}{16}$ light, and $\frac{1}{16}$ white-skinned individuals. The above hypothesis appears to postulate too few segregating alleles to account for the distribution of skin color types among Negro Americans.

The question of skin color inheritance in man has been reexamined more recently by Harrison and Owen (1964). They utilized a "reflectance spectrophotometer" to estimate the amount of pigment in a person's skin. This photoelectric instrument measures the amount of light of different wave lengths that is reflected from the skin. Using this instrument, they examined Negroes and Caucasians, the hybrids between them, and the backcross offspring to both parental types. Their findings are diagrammed in Fig. 4-2. It is easily seen that the F_1 hybrids are intermediate between the two parental types and that the respective backcross offspring are intermediate between the F_1 hybrids and the respective parental type. The data clearly indicate a multiple-gene type of inheritance.

A statistical analysis of their data led

Table 4-1. Probability of occurrence of F_2 individuals as extreme as either parent

Pairs of segregating alleles	Fraction of F_2 as extreme as either parent
1	1/4
2	1/16
3	1/64
4	1/256
5	1/1024

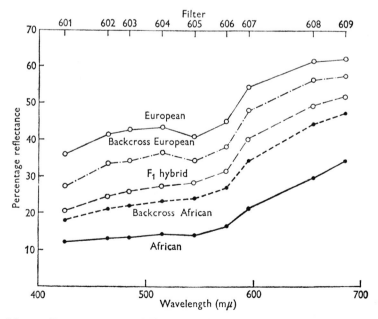

Fig. 4-2. Mean reflectance curves of European, African, and various hybrid groups. (From Harrison, G. A., and Owen, J. J. T. 1964. Ann. Hum. Genet. **28:**27-37.)

Harrison and Owen to conclude that the number of gene-pairs responsible for human skin color was "between 3 and 4." With a three-gene–pair model, the F_2 offspring derived from Negro and Caucasian grandparents would include seven different genotypes in the ratio of 1:6:15:20:15:6:1. With a four-gene–pair model, F_2 offspring would include nine different genotypes in the ratio of 1:8:28:56:70:56:28:8:1. Either of these models fits the known distribution of skin color types among Negro Americans better than the original two–gene pair hypothesis. The development of a more exact model for human skin color inheritance has been hindered by the considerable probability that not all pigment-producing genes are equally potent. A number of major genes and a number of minor genes may be involved. These minor genes may not only be important in the phenotypes of the hybrids studied by Davenport and others but are probably the explanation of the great variability in skin pigmentation found among Caucasians, Mongolians, and Negroes. These minor genes, also called *modifiers,* appear to play a large role in all quantitative traits. The modifiers are in reality normal segregating genes. Our confusion about the number and importance of such genes stems from our inability to easily detect and follow their individual effects on the phenotype.

Transgressive variation

The examples of multiple-factor inheritance thus far considered have dealt solely with genetic models in which the original parents were at the extremes for the trait in question. A different situation was revealed when R. C. Punnett (1908) crossed a Hamburgh chicken (large-sized breed) with a Sebright Bantam (small-sized breed). The F_1 was intermediate in size between the two parental types. However, the F_2 contained some birds that were larger and some that were smaller than the parental strains, although most of the F_2 were intermediate between the parental varieties. On the basis of the frequencies of the extreme types, four

genes were postulated to be involved in determining size in these chickens. The Hamburgh chickens were hypothesized to have the genotype $++$ $++$ $++$ dd, whereas the Sebright Bantams had the genotype aa bb cc $++$. The F_1 would then be heterozygous for all four genes, $a+$ $b+$ $c+$ $d+$, and of intermediate size. The F_2 would contain some individuals who were homozygous for all the $+$ alleles. These birds would be larger than the parental Hamburgh chicken. The F_2 would also contain some individuals who were homozygous for all the recessive alleles. These birds would be smaller than the parental Sebright Bantam. This situation, in which the extremes of the F_2 exceed those of the parents, is called *transgressive variation.* It is usually found in crossing inbred lines, when the inbred lines do not represent the extremes possible for the species.

Polygenic inheritance

In our discussion of skin color in human beings, reference was made to major and minor (modifier) genes. The action of major genes are relatively simple to study and detail. However, it is extremely difficult to follow and characterize the effects of minor genes. The problem is compounded by the great number of such genes that affect many traits of the individual and the relatively small effect of each minor gene. If a trait is affected by a large number of minor genes, we characterize it as being controlled by *polygenic inheritance.* The term "polygene" was suggested to describe those genes whose alleles produce small phenotypic differences of about the same order of magnitude as the differences caused by the usual environmental fluctuations. How much of a phenotype is determined by polygenes and how much is determined by environment is often impossible to estimate. Polygenic inheritance can be studied only through statistical analysis of the trait in a population. The methods employed will be outlined in the next section of this chapter. We may characterize polygenes as (1) segregating at a very large number of loci, (2) having individual phenotypic effects that are usually small and additive, (3) showing little phenotypic effects of allelic substitutions, (4) showing little phenotypic effects of gene ·substitutions at nonallelic loci, and (5) being genes whose phenotypic expression is greatly affected by even very small changes in the environment. Most characteristics of the organisms are affected by numerous genes and are therefore controlled by polygenic inheritance. Most genes affect more than a single trait of the body and affect the different traits unequally. The same gene may act as a major gene for one characteristic and as a polygene for a number of other parts of the organism's phenotype.

Quantitative inheritance

A summary of the types of results to be obtained from crosses involving single factors versus multiple factors is shown in Fig. 4-3. The sharp point of difference lies in the

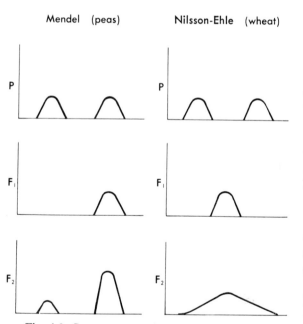

Fig. 4-3. Curves representing the results of single-factor versus multiple-factor inheritance. Left: **P**, **F₁**, and **F₂** from Mendel's crosses with garden peas. Right: **P**, **F₁**, and **F₂** from Nilsson-Ehle's crosses with wheat plants. The ordinate represents the number of plants, and the abscissa the range in expression of the trait.

occurrence of discrete classes in the F_2 of single-factor crosses as opposed to the one continuous class in the F_2 of multiple-factor crosses. An examination of the results of single-factor and multiple-factor crosses shows that as the number of genetic factors increases, the ability to distinguish separate classes decreases until the individual genotypes lose their identity and become fused into a single continuous class. The occurrence of discrete classes versus one continuous class implies that all variations found in a population may take one of two forms: discontinuous or continuous. When variation is discontinuous, the individuals are assignable to one or another of a comparatively few classes, as in the case of sex, blood type, coat color of mice, etc. Measurement in these cases consists of counting the frequency with which individuals are assigned to the separate classes and proposing a genetic model to explain the results of crosses. When variation is continuous, the number of classes to which individuals can be assigned, according to the grade of expression of a trait, is limited only by the sensitivity of the method of measurement. Without regular discontinuities, there is no natural means of grouping observations into different classes. Hence, it becomes necessary to use statistical quantities like means and variances, to replace frequencies, in describing continuous variation. In these cases none but the vaguest genetic models can be proposed to explain the results of crosses. The type of probability involved in quantitative inheritance is *empiric probability*.

We shall examine a sample consisting of one hundred men and consider the trait of height. It is quite generally understood that although human beings may be considered as short, tall, or average in height, no really sharp discontinuities exist to separate these groups. We further know that we may measure the height of a man to the nearest foot, inch, half-inch, quarter-inch, etc. However, we have no natural classes into which we may place our observations. For the sake of convenience, we shall take our measurements to the nearest

Table 4-2. Heights in inches of a random sample of 100 adult men, arranged in increasing order

60	65	67	69	71
62	66	67	69	71
63	66	67	69	71
63	66	67	69	71
63	66	68	69	71
63	66	68	69	71
64	66	68	69	71
64	66	68	69	71
64	66	68	69	72
64	66	68	69	72
64	66	68	69	72
64	66	68	69	72
65	67	68	70	72
65	67	68	70	72
65	67	68	70	73
65	67	68	70	74
65	67	68	70	74
65	67	68	70	74
65	67	68	71	75
65	67	69	71	77

Table 4-3. Chart illustrating the steps in handling data involving quantitative inheritance (heights of men in inches from Table 4-2)

X	f	fX	$(X - \overline{X})$ or "d"	$(X - \overline{X})^2$ or "d^2"	$f(X - \overline{X})^2$ or "fd^2"	$f(X - \overline{X})^2(d)$ or "fd^3"	$f(X - \overline{X})^3(d)$ or "fd^4"
60	1	60	-8	64	64	-512	4096
61	0	0	-7	49	0	0	0
62	1	62	-6	36	36	-216	1296
63	4	252	-5	25	100	-500	2500
64	6	384	-4	16	96	-384	1536
65	9	585	-3	9	81	-243	729
66	11	726	-2	4	44	- 88	176
67	12	804	-1	1	12	- 12	12
68	15	1020	0	0	0	0	0
69	13	897	+1	1	13	+ 13	13
70	6	420	+2	4	24	+ 48	96
71	10	710	+3	9	90	+270	810
72	6	432	+4	16	96	+384	1536
73	1	73	+5	25	25	+125	625
74	3	222	+6	36	108	+648	3888
75	1	75	+7	49	49	+343	2401
76	0	0	+8	64	0	0	0
77	1	77	+9	81	81	+729	6561
Totals	100	6799	—	—	919	605	26,275

inch. The data are shown in Table 4-2. These same data are shown in grouped form, with the frequencies of each class, in Table 4-3.

STATISTICAL METHODS

In studying quantitative data of this type, one may investigate different characteristics of the data. These characteristics are called *descriptive measures* and include the following measures: central tendency, variation, symmetry, and peakedness.

Measures of central tendency

The term *central tendency* implies that we are trying to locate the center or the middle of a set of data. Its vagueness is quite appropriate because there are actually a great many different ways in which we can define the center of a distribution. One such measure of central tendency is the *mode*. The mode is defined as the value that occurs with the highest frequency. It tells us which value is the most common. In our table of heights of men, the mode is 68 inches. A second measure of central tendency is the *median*. The median defines the center of a set of data as the point (number) that divides the measurements, arranged according to size, into two groups, each of which contains an equal number of the measurements. This means that there must be as many measurements below the median as there are above it. In our data of heights of men, the median is calculated to be 67.9 inches. The significance of the closeness of values of mode and median will be discussed in our consideration of measures of symmetry. The median and the mode are not used to any great extent in the statistical tests involving quantitative genetics but are useful in characterizing bodies of data. The third measure of central tendency, and the one almost universally used in quantitative genetics, is the *arithmetic mean,* or simply the *mean*. The mean is the sum of the measurements divided by the number of individuals in the sample. If X stands for the measurement in question, then the sample mean is always denoted by $\overline{X}$ (read X bar), and if N stands for the sample size, we have the formula:

$$\overline{X} = \frac{\Sigma X}{N}$$

As before, Σ (read summation of) directs us to add all the values of X. In our example the mean height of men is 67.99 inches, which we shall consider to be 68 inches. Refer to Table 4-3 for the steps involved in making calculations. Note that f stands for the frequency with which a certain measurement has occurred, $\Sigma f = N$, and $\Sigma fX = \Sigma X$ of the raw data of Table 4-2. We shall return to use our mean value in comparisons that we shall make later of two samples of heights of men.

The mean, median, and mode each supplies a single number that we may substitute for an entire set of measurements. They supply a specific kind of information about our data but unfortunately do not give us a complete picture. There are other features of our distribution that we must also describe. These other features are involved with the variation, spread, or dispersion of our data and are collectively called *measures of variation*.

Measures of variation

Measures of variation will tell us about the distribution of our data—to what extent they are spread out or bunched together. As in the case of the measures of central tendency, we can use several different statistical measures of variation. The most widely understood measure of variation is the *range*. The range of a set of numbers is defined simply as the difference between the largest and the smallest numbers that belong to the set. The range of height of men in our sample is 17 inches. Unfortunately, the range is a poor measure of the dispersion of a set of data because it is based exclusively on the two extreme values and tells us nothing about the distribution of the intervening measurements. A second measure of variation is called the *mean deviation*. It is the average of the deviations of all the observations from the mean of the data. This measure of variation in our example of height of men is given in the $(X - \overline{X})$ column of Table 4-3. Unfortunately, one finds that the sum of the mean deviations of the raw data always equals zero, thereby eliminating its usefulness. Efforts to use this statistic, by disregarding the plus and minus signs and using absolute values, do not really produce a useful statistic. The main reason for discussing the mean deviation is that it is the logical steppingstone to the most important measure of variation, the *standard deviation*. The standard deviation (s) is defined as the square root of the mean of the squared deviations. This indicates that in order to calculate s, we first find the mean, then the deviations of all the observations from the mean of the data and their squares, and finally the square root of the average of the squared deviations. Symbolically we can write the standard deviation as follows:

$$s = \sqrt{\frac{\Sigma(X - \overline{X})^2}{N - 1}}$$

The factor $(N - 1)$ is used rather than N in order to account for the fact that we have data for only a sample rather than for the entire population. When we group our data as we have done in Table 4-3, our formula becomes:

$$s = \sqrt{\frac{\Sigma f(X - \overline{X})^2}{N - 1}}$$

The steps involved in its calculation are shown in Table 4-3. The standard deviation for our data is s = 3.05. We shall consider later the use to be made of the standard deviation. For the present, let us point out that the standard deviation is read in terms of the unit of measurement used in the experiment. In our own case the standard deviation is s = 3.05 inches. It must be noted that s^2, the square of the standard deviation, though in the wrong units to measure spread, is nevertheless of such theoretical importance in the study of variability that it has its own name, the *vari-*

ance. In our example the variance $s^2 = 9.28$.

Measures of symmetry

Symmetry refers to the similarity of form or arrangement on either side of a dividing line. If our data are distributed symmetrically, the mean, median, and mode will generally coincide. In our analysis of the height of a sample of 100 men, we found that the mode equals 68 inches, the median equals 67.9 inches, and the mean equals 68 inches. If the distribution is not symmetrical, the median will usually fall somewhere between the mean and the mode. If the curve is positively skewed (tail of the distribution at the right) as in Fig. 4-4, *A*, the mean $(\overline{X})$ will generally exceed the median (M), which in turn can be expected to be larger than the mode. This order will be reversed in a negatively skewed curve as in Fig. 4-4, *B*.

A better test of symmetry than the above rule of thumb can be made by calculating a statistic called *alpha three*. It is found by using the following equation:

$$\alpha_3 = \frac{\frac{1}{N}\Sigma(X - \overline{X})^3}{s^3}$$

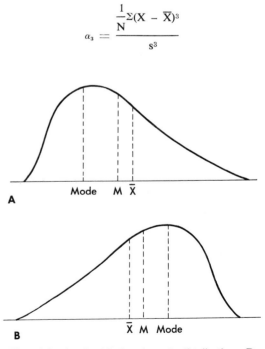

Fig. 4-4. A, Positively skewed distribution. B, Negatively skewed distribution.

The equation reads that α_3 is the average of the cubed deviations from the mean divided by the cube of the standard deviation. For a symmetrical distribution, $\alpha_3 = 0$. When data for a quantitative trait are not symmetrically distributed about the mean, one can look for factors that are preventing the random distribution of the segregating genes. Asymmetry may occur because certain combinations of genes are lethal and hence one end of the curve is nonexistent; or it may be caused by incomplete linkage of certain genes controlling the trait or by epistasis and nonadditivity of effects. Whether or not a body of data fits a normal curve is important to ascertain, since different statistical tests must be used depending on the distribution of the data. We shall consider the normal curve in greater detail later in this chapter. We now, however, turn our attention to the measures of peakedness.

Measures of peakedness

Peakedness, or *kurtosis,* refers to the manner in which the data are arranged on both sides of the highest point of the distribution. This statistic is called *alpha four* and is defined as follows:

$$\alpha_4 = \frac{\dfrac{1}{N}\Sigma(X - \overline{X})^4}{s^4}$$

In this equation the numerator is the mean of the fourth powers of the deviations from the mean and the denominator is the fourth power of the standard deviation. The α_4 of a normal curve is always equal to 3.

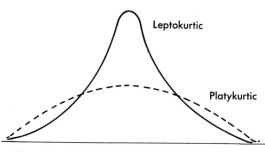

Fig. 4-5. Examples of curves demonstrating different types of peakedness.

When a curve is very peaked, as the solid curve in Fig. 4-5 is, the value of α_4 will be greater than 3, and the curve is said to be *leptokurtic.* If the curve is flatter than the bell-shaped normal curve, as is the dotted curve in Fig. 4-5, its α_4 will be less than 3, and it is said to be *platykurtic.* Data that are distributed in leptokurtic fashion usually represent traits that are controlled by relatively few segregating genes, whereas data distributed in platykurtic fashion usually represent traits that are controlled by many segregating genes.

At this point we shall consider further the normal curve and its relationship to our quantitative character: the height in inches of a sample of 100 men. The last two columns of Table 4-3 show the calculations of the cubed deviations from the mean and the fourth powers of the deviations from the mean. Using the formula for α_3, we obtain a value of 0.213, which is in excellent agreement with the zero value expected for a normal curve. The value for α_4 is calculated to be 3.04, which is also in excellent agreement with the value of 3 expected for a normal curve. These findings tell us that our sample is normally distributed and will follow the rules of such distributions. What may we then say about our data and the normal curve?

Normal curve

The normal curve is a bell-shaped curve that extends indefinitely in both directions. Fortunately, the two tails of the curve need not concern us because the percentages of cases in these tails are so small that they can be safely ignored. In our consideration of heights of men, we found that the mean of the sample (measure of central tendency) was 68 inches and the standard deviation (measure of variation) was 3.05 inches. It is easy to appreciate that the mean is that point which divides the normal curve, or any other symmetrical curve, into two equal parts. We do, however, have to discuss the relationship of the standard deviation to the normal curve. Considering the total area under the normal curve as equal to 100%,

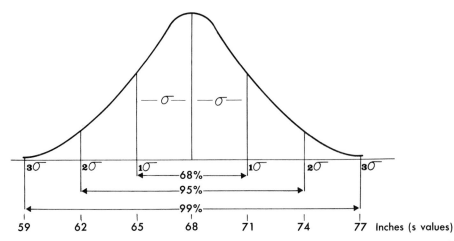

Fig. 4-6. Diagram of the normal curve showing expected distribution of heights of men, based on the sample in Table 4-2.

we find that the area of the curve that lies within one standard deviation to the right of the mean contains a little more than 34% of the total area. The same is true of the area one standard deviation to the left of the mean. We designate the standard deviation of the normal curve by the small Greek letter sigma (σ). We also find that 95% of the curve lies within $\pm 1.96\sigma$ of the mean and 99% of the curve lies within $\pm 3\sigma$ of the mean. It will be recalled that earlier in this chapter we designated the standard deviation by the letter "s." The basis for this distinction is as follows. If we have studied every member of a normally distributed population (i.e., the equivalent of the theoretical normal curve), we use the designation "σ" to represent the true standard deviation of the population. If we have studied only a sample of a normally distributed population, we use the designation "s," which implies that we have no more than an estimate of the true standard deviation of the population.

Returning to our sample of height in men, we now know that its normal distribution means that 34% of the men will be found to measure somewhere between 68 and 71 inches tall. Another 34% of the men will be found to measure somewhere between 65 and 68 inches tall. Appropriate values can

be calculated for 95% and 99% of the men in our sample. These distributions are illustrated in Fig. 4-6.

The shape of any symmetrical curve is determined by the amount of variation in the sample as indicated by the standard deviation. If the area is kept constant, a small s is associated with a high curve and a large s with a flat curve. If s is small and the curve is high, most of the observations are clustered around the mean. If s is large and the curve is flat, the observations are spread away from the mean. The standard deviation of our sample can be used to yield an estimate of the mean of the entire population. This estimate is called the *standard error* ($s_{\bar{x}}$) and is actually the standard deviation of the means of a number of samples drawn from a single population. It is not necessary to go through the process of taking several samples from the same population, since it can be shown that the standard error is inversely proportional to the square root of the sample size, N. The formula for the standard error is as follows:

$$s_{\bar{x}} = \frac{s}{\sqrt{N}}$$

In our sample, $s_{\bar{x}} = 0.31$ and our mean would be reported as 68 ± 0.31 inches. The standard error says that we are about two-

Table 4-4. Probability (p) for values of t (vertical columns) and various degrees of freedom (d.f.) (the degrees of freedom are one less than the number of classes)

p d.f.	0.5	0.4	0.3	0.2	0.1	0.05	0.01
1	1.000	1.376	1.963	3.078	6.314	12.706	63.657
2	.816	1.061	1.386	1.886	2.920	4.303	9.925
3	.765	.978	1.250	1.638	2.353	3.182	5.841
4	.741	.941	1.190	1.533	2.132	2.776	4.604
5	.727	.920	1.156	1.476	2.015	2.571	4.032
6	.718	.906	1.134	1.440	1.943	2.447	3.707
7	.711	.896	1.119	1.415	1.895	2.365	3.499
8	.706	.889	1.108	1.397	1.860	2.306	3.355
9	.703	.883	1.100	1.383	1.833	2.262	3.250
10	.700	.879	1.093	1.372	1.812	2.228	3.169
15	.691	.866	1.074	1.341	1.753	2.131	2.947
20	.687	.860	1.064	1.325	1.725	2.086	2.845
25	.684	.856	1.058	1.316	1.708	2.060	2.787
30	.683	.854	1.055	1.310	1.697	2.042	2.750
50	.680	.849	1.047	1.299	1.676	2.008	2.678
100	.677	.846	1.042	1.290	1.661	1.984	2.626
∞	.674	.842	1.036	1.282	1.645	1.960	2.576

Modified from Table 4. Fisher, R. A., and Yates, F. 1963. Statistical tables for biological, agricultural and medical research. Oliver & Boyd Ltd., Edinburgh. By permission of the authors and publishers.

thirds (68%) confident that the population mean lies somewhere between 67.7 inches and 68.3 inches and we are about 95% confident that the population mean lies between 67.4 inches and 68.6 inches. Such estimates of the range within which the population mean should be found are called *confidence intervals*. They must always specify the level of confidence (68%, 95%, 99%, etc.) with which we make the estimate. When the sample size, N, is large (20 or more), we can use the standard error directly for obtaining our confidence interval without incurring too great an error.

However, a more accurate method, especially when N is less than 20, includes the use of a table of probabilities known as Student's distribution or the "t" distribution. This is shown in Table 4-4. The statistic "t" is defined as the difference between the sample mean and population mean divided by the standard error of the sample.

This is shown in the following formula:

$$t = \frac{\overline{X} - \mu}{s_{\overline{x}}}$$

The Greek letter mu (μ) is used as the symbol for the population mean. The equation may now be solved for μ.

$$\mu = \overline{X} \pm ts_{\overline{x}}$$

The upper confidence limit is given by the "+" side of the sample mean and the lower confidence limit is given by the "–" side of the sample mean. Table 4-4 shows the distribution of t values. The column at the left lists the degrees of freedom, which in this case is one less than the number of observations (N – 1). If we wish to be 95% confident in our limits, we would look in Table 4-4 under d.f. = 100 (the closest number to our N – 1 = 99) and at the P column 0.05. There we would read that t = 1.984. Substituting in the above equation, we would get two values for μ:

$$\mu(\text{upper}) = 68 + 1.984 \ (0.31) = 68.61$$
$$\mu(\text{lower}) = 68 - 1.984 \ (0.31) = 67.39$$

These values for the confidence limits agree very closely with those crudely estimated solely from the standard error. This is the result of our N being much larger than 20. The standard error is actually an indication of the reliability with which the sample mean $\overline{X}$ estimates the population mean μ. The smaller the standard error, the more reliable is the estimate. From the formula for the standard error we see that as the sample size N increases, the value of the standard error tends to decrease, and vice versa. This relation illustrates the advantage of large samples for obtaining the best possible estimates of population values. The basic problem of biological statistics is to make inferences about characteristics of the population on the basis of observations on a sample. Values for the mean, standard deviation, etc. calculated from a sample are called *statistics,* and the corresponding quantities for the population are called *parameters.* In most cases we never know the parameters of a population and can only estimate them from the statistics of the sample.

A quantitative inheritance problem

We have considered in detail the height in inches of a sample of 100 men. Let us take another sample of 100 men. Without repeating the laborious analysis, we shall assume that the mean height of this second sample of men is 67 inches and the standard deviation is 2.06 inches. We shall also assume that the distribution of heights fits a normal curve. Using the formula for the standard error, we obtain a value of 0.21. When one obtains two samples of individuals for a trait involving quantitative inheritance, the question always arises as to whether the two samples are statistically different from one another. The implication of a statistical difference is that there is either an environmental or a genetic factor that resulted in the statistical dif-

ference. A glance at the closeness of the mean values would ordinarily lead one to expect that there will be no statistically significant difference between these samples. The test for statistical differences again uses Student's "t" table and is given by the accompanying formula:

$$t = \frac{\overline{X}_{(1)} - \overline{X}_{(2)}}{\sqrt{(s_{\overline{x}_1})^2 + (s_{\overline{x}_2})^2}}$$

The formula instructs us to subtract the mean of one population from the mean of the other and divide this number by the square root of the sum of the squares of the two standard errors. Substituting the values we have obtained for both samples, we get:

$$t = \frac{68 - 67}{\sqrt{(0.31)^2 + (0.21)^2}} = \frac{1}{\sqrt{0.1402}} = 2.7$$

Turning once more to Student's distribution as shown in Table 4-4, we would take our d.f. $= \infty$ (the closest to our $N_{(1)} + N_{(2)} - 2 = 198$). Our value of 2.7 is off the table. Hence, our two samples are statistically different from one another. This might appear surprising in view of the closeness of the means of the two samples. However, the standard errors, which reflect the variability of the data, are decidedly different from one another. It should be noted that even if we had taken our d.f. $= 100$, the samples would still be significantly different.

We then conclude that the two samples of men have been drawn from two distinctly different human populations. These populations may differ from one another as to the alleles they contain for height, or they may contain the same alleles but have different environments (nutrition, disease, etc.) that affect the mean and variance of the height of their men. Which of these two factors is involved or what combination of them is involved can only be elucidated from other studies. Our statistical analysis has shown us that the samples come from two different populations. The causative factor or factors cannot be gotten from statistical analysis but must come from different types of investigations.

SUMMARY

In this chapter we reviewed multiple-factor inheritance. As examples of this type of inheritance, we discussed kernel color in wheat and skin color in humans. We then noted that as the number of genes involved in a particular trait increased, individuals could no longer be considered as belonging to a particular discrete class but rather became part of a single continuous group. We could no longer specify the genotype of the individual but were forced to describe the population. This required that we consider such characteristics of quantitative data as (1) measures of central tendency (mode, median, mean), (2) measures of variation (standard deviation, standard error), (3) measures of symmetry, and (4) measures of peakedness. We then discussed the characteristics of the normal curve and the kind of information it gives us about normally distributed data. The chapter closed with a consideration of a quantitative inheritance problem, including the use of Student's "t" test. We shall now turn our attention to the sex chromosomes, the genes they carry, and the role they play in sex determination.

Questions and problems

1. A cross is made between a Hamburgh rooster (large-sized breed) ($a^+a^+b^+b^+c^+c^+dd$) and a hen obtained from a cross between a Hamburgh and a Sebright Bantam (small-sized breed whose genotype is $aabbccd^+d^+$). What are the expected phenotypes of the progeny of this cross if each of the "+" alleles results in 1 pound of weight and each of the other alleles results in ½ pound of weight of the progeny?

2. Define the following terms:
   a. Modifier genes
   b. Transgressive variation
   c. Polygenic inheritance
   d. Empiric probability

3. Distinguish between continuous and discontinuous variation.

4. Define the following statistical terms:
   a. Mode
   b. Median
   c. Mean
   What do these terms measure?

5. In an F_2 population of corn being investigated for kernel color, 1 zygote among a total of 5000 was found to be phenotypically as extreme as one of its grandparents. How many gene pairs for kernel color are segregating in this instance?

6. Assume that the size of rabbits is determined by genes with an equal and additive effect. A cross was made between large and small varities of rabbits, and a total of 2012 F_2 individuals were obtained. Of these, 7 were as small as the average small grandparent variety and about 9 were as large as the average large grandparent variety. How many gene pairs were operating?

7. The following figures are the heights, in inches, of a random sample of 100 adult women.

48	50	60	62	63	64	65	65	66	66
48	50	60	62	63	64	65	65	66	66
48	50	60	62	63	64	65	65	66	67
49	51	60	62	63	64	65	65	66	67
49	51	61	62	63	64	65	65	66	67
49	51	61	62	63	64	65	66	66	67
49	51	61	63	64	64	65	66	66	67
50	51	61	63	64	64	65	66	66	67
50	60	62	63	64	64	65	66	66	67
50	60	62	63	64	65	65	66	66	68

   a. What is the mean of this population?
   b. What is the median of this population?
   c. What is the mode of this population?

8. Define the following statistical terms:
   a. Range
   b. Mean deviation
   c. Standard deviation
   d. Variance
   e. Standard error

9. Refer to question 7.
   a. What is the range of the population sample?
   b. What is the mean deviation of the sample?
   c. What is the standard deviation of the sample?
   d. What is the variance of the sample?

10. A sample of 20 plants was measured, in inches, and tabulated as follows: 10, 11, 9, 10, 9, 11, 12, 10, 8, 10, 11, 9, 9, 10, 12, 12, 10, 9, 8, 11. Calculate the following:
    a. Mean
    b. Standard deviation
    c. Standard error

11. Using two different statistical techniques, evaluate the normalcy of the distribution of the data presented in question 7.

12. What factors might be responsible for a quantitative characteristic's not fitting a normal curve?

13. Define the following statistical terms:
    a. Normal curve
    b. Leptokurtic
    c. Platykurtic
    d. Kurtosis

14. Given two different tables of data, each listing the heights of a random sampling of 100 women, how could you determine whether the 200 women had been picked from the same or from two different populations?

15. Two pure types, differing in size, are crossed.

Is it possible for F_2 individuals to be more extreme than either grandparent? Explain!

16. Assume that the difference between strains of oats, one yielding about 4 grams per plant and one yielding about 10 grams per plant, is due to 3 equal and cumulative pairs of genes, *XX YY ZZ*. Cross the two strains of oats and give the expected F_1 and F_2 phenotypes.

17. Two pairs of genes *(AA BB)* are believed to influence the size of corn in certain varieties. These gene pairs have equal and additive effects. A short variety, averaging 2 feet, is crossed with a tall variety averaging 6 feet. If the genotype *aabb* is reflected in the short variety's phenotype:

    a. What is the effect of each gene that increases the height of the corn above 2 feet?

    b. Diagram the cross described above; classify the expected F_2 phenotypes.

18. On the basis of Davenport's studies, it has been suggested that skin color in humans is determined by two different pairs of genes: *A²* and *B²*, which produce dark pigmentation, and *A¹* and *B¹*, which form no gene product. Accordingly, Caucasians have no *A²* or *B²* alleles; light brown-skinned Negroes have 1 of these alleles; medium brown-skinned Negroes have 2 of these alleles; dark brown-skinned Negroes have 3 of these alleles; and black-skinned Negroes have 4 of these alleles. Can a cross between a Caucasian *(A¹A¹B¹B¹)* and a black-skinned Negro *(A²A²B²B²)* result in white-skinned offspring? black-skinned offspring? Why?

19. Refer to question 18. Can a mating between a Caucasian and a medium brown-skinned Negro produce black-skinned children?

20. Diagram the cross and classify the expected phenotypes found among the progeny of a mating between two medium brown-skinned Negroes.

References

Davenport, C. B. 1913. Heredity of skin color in Negro-white crosses. Carnegie Inst. Wash. Publ. 188. (Explanation for the inheritance of skin color based on two pairs of genes.)

Dunn, L. C., and Charles, D. R. 1937. Studies on spotting patterns. Genetics **22**:14-42. (Analysis of multiple-factor inheritance of pied spotting in the house mouse.)

East, E. M. 1910. A Mendelian interpretation of variation that is apparently continuous. Amer. Natur. **44**:65-82. (Development of the multiple-gene hypothesis.)

Emerson, R. A., and East, E. M. 1913. The inheritance of quantitative characters in maize. Nebraska Agr. Exp. Sta. Res. Bull. 2. (Description of quantitative inheritance mechanisms in corn.)

Falconer, D. S. 1960. Introduction to quantitative genetics. The Ronald Press Co., New York. (Clear presentation of modern statistical techniques.)

Fisher, R. A. 1947. Statistical methods for research workers. 10th ed. Oliver & Boyd, Edinburgh. (Methods and tables for the analysis of quantitative data.)

Harrison, G. A., and Owen, J. J. T. 1964. Studies on the inheritance of human skin colour. Ann. Hum. Genet. (London) **28**:27-37. (Analysis of skin color inheritance in man, using spectrophotometry to determine skin pigment.)

Hoel, B. G. 1960. Elementary statistics. John Wiley & Sons, Inc., New York. (General statistics book that covers many of the statistical tests used in quantitative genetics.)

Levene, H. 1958. Statistical inference in genetics, Chapter 29. *In* Sinnott, E. W., Dunn, L. C., and Dobzhansky, T. Principles of genetics. McGraw-Hill Book Co., New York. (Excellent discussion of the use of statistics in genetics.)

Mather, K. 1943. Polygenic inheritance and natural selection. Biol. Rev. **18**:32-64. (Review and discussion of the fundamental concepts of quantitative inheritance.)

Punnett, R. C. 1923. Heredity in poultry. The Macmillan Co., New York. (Discussion of the inheritance of quantitative characteristics in poultry, including the phenomenon of transgressive variation.)

Stern, C. 1960. Principles of human genetics. 2nd ed. W. H. Freeman & Co., Publishers, San Francisco. (Chapter 8, discussion of polygenic inheritance.)

5 Sex chromosomes and sex determination

In our discussion of the physical basis of inheritance, it was stated that chromosomes are usually found as homologous pairs. When we considered the fundamentals of genetics, we added the concept that the homologous chromosomes are equal in their total genic content and able to carry any of the alleles of a gene. We have now come to the point where we must consider in detail an important exception to the concept of paired and equal chromosomes. The exception is the sex chromosome. In some cases we shall find that the sex chromosomes need not be arranged in pairs whereas in other instances they will be paired but will not be equal in their total genic content. In both of these situations the expression of recessive genes will differ from that discussed in Chapter 3.

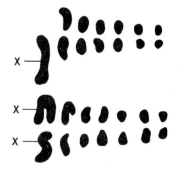

Fig. 5-1. Chromosomes of *Protenor*. Upper: Male chromosomes. Lower: Female chromosomes. They are lined up according to size—that is, in an ideogram. This insect illustrates the XO method of sex determination.

THE XO SITUATION (NOT PAIRED)

Historically, the first studies of sex chromosomes involved organisms in which the sex chromosomes are not found in pairs. These included the grasshoppers, some of the bugs, etc. The first well-demonstrated case was reported by Wilson and Stevens in 1905. They studied in detail the chromosomes of the genus *Protenor,* an uncommon group of insects related to the boxelder bug. In these insects, different numbers of chromosomes were observed in the cells of the two sexes, the diploid number in the female being 14 and that of the male, 13. By following oogenesis, these investigators found that all the eggs contained 7 chromosomes. A similar study of spermatogenesis showed that two types of sperm were formed, those with 7 chromosomes and those with 6 chromosomes. The conclusion was that eggs fertilized by sperm containing 7 chromosomes produced females whereas eggs fertilized by sperm containing 6 chromosomes produced males. The chromosomes of *Protenor* are shown in Fig. 5-1. That chromosome of which the females had 2 but the males had 1 had earlier been called the "X" body, and this designation was retained in calling it the X chromosome.

THE XY SITUATION (PAIRED BUT NOT EQUAL)

In 1905 Wilson and Stevens reported also on the chromosome arrangement in the milkweed bug *Lygaeus turcicus.* In this insect the same number of chromosomes was present in the cells of both sexes, the diploid number being 14. In the female, all the chromosomes were paired, and the homologues were equal in size. In the male, all the chromosomes were paired, but the chromosome identified as the homologue to the X was distinctly smaller and was called the Y chromosome. A study of oogenesis revealed that all the eggs contained 7 chromosomes with one of the chromosomes being the X. A similar investigation of spermatogenesis showed that all the sperm contained 7 chromosomes, some having the X chromosome and others having the Y. It was concluded that eggs fertilized by sperm carrying the X chromosome would produce females, whereas eggs fertilized by sperm carrying the Y chromosome would produce males. The chromosomes of the milkweed bug are shown in Fig. 5-2.

As investigations included an ever-increasing variety of organisms, the XY situation was found to be more prevalent than the XO. The XY condition is found in the fruit fly, *Drosophila*, and in man. It must not be assumed that in all cases the Y chromosome will be smaller than the X. In man it is; but in *Drosophila* the Y chromosome is somewhat larger than the X.

All the examples of sex chromosome differentiation thus far considered have had the females as XX *(homogametic)* and the males either XO or XY *(heterogametic)*.

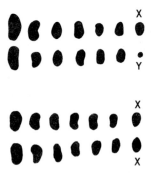

Fig. 5-2. Chromosomes of the milkweed bug *Lygaeus turcicus.* Upper: Chromosomes (2n) of a male cell. Lower: Chromosomes (2n) of a female cell. This illustrates the XY type of sex determination.

In birds, moths, some fish, and undoubtedly other species, the female is heterogametic (XO or XY), and the male is homogametic (XX). As we shall discuss next in this chapter, sex chromosome linkage in these groups will be the reverse of that in man, *Drosophila,* etc.

SEX CHROMOSOME LINKAGE

With the discovery of the *sex chromosomes,* it became necessary to use some term to designate the other chromosomes in the *karyotype* (complete set of chromosomes). The term *autosome* is used to refer to any chromosome other than the sex chromosomes. The possible interactions of alleles in the autosomes have been discussed in Chapter 3. These same interactions can occur for alleles located in the sex chromosomes of the homogametic sex (♀ ♀ in man, etc.; ♂ ♂ in birds, etc.). However, for the heterogametic sex, a different situation exists. In the XO case, any gene located on the X chromosome will be fully expressed in the phenotype of the organism, regardless of whether it acts as a dominant or as a recessive in the XX individual. In the XY situation, most of the genes located on the X chromosome will also be fully expressed due to the fact that the Y chromosome acts as if it were devoid of most genes.

Sex chromosome linkage in Drosophila

The first X-linked trait to be firmly established was studied by T. H. Morgan in 1910. It was a white-eyed mutant and appeared in a single male in a culture of normally red-eyed *Drosophila*. This white-eyed male was mated to a red-eyed female. All the F_1 flies were red-eyed. Males and females from the F_1 were mated to one another, and an F_2 was obtained. In the F_2, all the females were red-eyed, whereas half of the males were white-eyed and half were red-eyed. The interpretation of these results postulated that this gene for white eyes was located on the X chromosome and, although it was a recessive gene, it was capable of expression in the male. This implied that the Y chromosome was devoid of this gene and that the expression of the white-eyed gene in the male represented a case of "pseudodominance." The term *hemizygous* is used to describe individuals that have only one gene for a given characteristic. A diagram of the crosses involved in Morgan's X-linkage experiment is shown in Fig. 5-3.

The critical characteristic of X-linked inheritance is the absence of male-to-male (father-to-son) transmission. This is a necessary result of the fact that the X chromosome of the male is transmitted to none of his sons, yet to all his daughters. Among the offspring of a male affected by an X-linked recessive trait, all the sons are unaffected and all the daughters are carriers. Should the daughters also receive the X-linked recessive gene from their mother, the daughters will be homozygous for the allele and will exhibit the trait.

An important variation in X-linkage in *Drosophila* was reported by C. B. Bridges in 1916. He crossed a white-eyed female to a red-eyed male. Such a female with an X-linked recessive is expected to produce male offspring, all of which would inherit the mother's X chromosome and would show

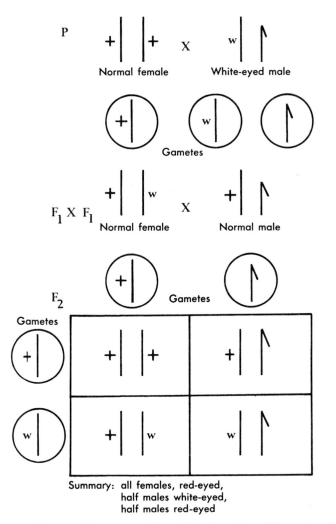

Fig. 5-3. Diagram of a cross between a red-eyed female and a white-eyed male, illustrating X-linkage.

the recessive character carried by it. However, in this cross, some males were produced that had red eyes. It was hypothesized that the two X chromosomes of the female had failed to segregate and enter separate eggs. As a result, some of the eggs contained two X chromosomes, but others had none, although all the eggs contained one of each pair of the autosomes. When an egg without an X chromosome was fertilized by an X-carrying sperm from a red-eyed male, a male was produced who inherited his X-linked genes from his father rather than from his mother. The nonseparation of homologous chromosomes in meiosis was called *nondisjunction*. The Bridges hypothesis was later proved to be correct by observations of the chromosomes of the flies involved. As we shall see later in the chapter, nondisjunction may also occur during meiosis in humans. A diagram of the Bridges cross involving nondisjunction is shown in Fig. 5-4.

Earlier we stated that the Y chromosome acts as if it were devoid of most genes. An interesting case of Y chromosome inheritance in *Drosophila* was demonstrated by C. Stern in 1926. There are many genes known that affect the bristles of the animal. One of these, located on the X chromosome, is called "bobbed." It is recessive and in the homozygous condition causes certain of the bristles to be shortened. A cross of females homozygous for bobbed to males with normal bristles produced sons with normal bristles. Nondisjunction was ruled out as a possible explanation. The hypothesis was therefore advanced that the allele for normal-length bristles, which is known to be dominant, must be present in the Y chromosome. Genes of this kind, which occur in the X chromosome and have a correspond-

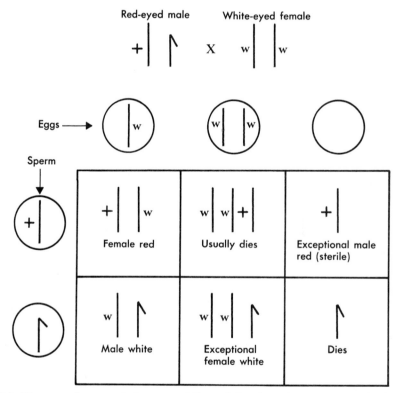

Fig. 5-4. Diagram of a cross between a white-eyed female and a red-eyed male, illustrating nondisjunction.

ing gene locus on the Y chromosome, are said to be *incompletely X-linked*. The number of such genes in *Drosophila* is very small.

Sex chromosome linkage in man

There are some eighty X-linked traits known in man. Most of the traits are pathological and produced by recessive genes. A particular type of red-green color blindness *(deutan)* is caused by the recessive allele *rg*. The classic bleeder's disease, *hemophilia type A,* is due to the X-linked recessive gene *h*. A form of *hemolytic anemia* appears in apparently healthy individuals on ingestion of the antimalarial drug primaquine, fava beans, or any of a large number of other hemolytic compounds. These individuals are either hemizygous or homozygous for one of the recessive alleles that cause a deficiency in the enzyme glucose-6-phosphate dehydrogenase (G-6-PD). A nonpathological trait that has proved very useful in studies is the *Xg blood group* system.

The distinctive crisscross pattern of inheritance, from father through daughter to grandson, of some human traits was known long before the twentieth century. It was apparently well understood by the Jews of the Middle Ages, as illustrated in the talmudic prohibition against the circumcision of any male child born to a woman whose father, older male child, or even brother was a bleeder. Especially significant was the stipulation that the existence of a bleeder-maternal-uncle was sufficient grounds for being fearful about the well-being of the child. This stipulation implies the realization that females may act as carriers of X-linked traits for any number of generations, handing the gene down from mother to daughter without exhibiting any of its effects, yet always capable of transmitting it to a male offspring.

Although the overall pattern of X-linked traits was known to older civilizations, the chromosomal mechanism of it was not understood until T. H. Morgan in 1910 published his findings on the white-eyed mutant

in *Drosophila*. Thereafter, the same explanation was clearly seen to apply equally well to human X-linked traits. An example of such inheritance is shown in Fig. 5-5, for hemophilia.

The most celebrated cases of hemophilia occurred in the royal families of England, Russia, and Spain, all of which were interrelated by marriage. The pedigree begins with Queen Victoria of England, who was a carrier of the gene. Since none of her an-

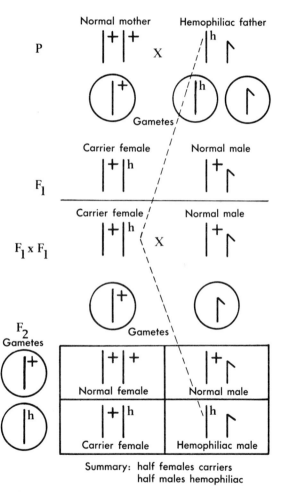

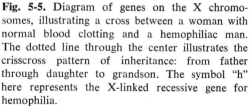

Fig. 5-5. Diagram of genes on the X chromosomes, illustrating a cross between a woman with normal blood clotting and a hemophiliac man. The dotted line through the center illustrates the crisscross pattern of inheritance: from father through daughter to grandson. The symbol "h" here represents the X-linked recessive gene for hemophilia.

cestors or relatives were known to be hemophiliacs, a mutation is assumed to have occurred in an X chromosome that was contained in one of the gametes of her parents or possibly in an embryonic gonadal cell in her body. One of her sons, Leopold, Duke of Albany, died of hemophilia at the age of 31. The birth of an affected son rules out Victoria's husband as the source of the gene for hemophilia, since male-to-male inheritance does not normally occur in X-linked traits. At least two of Victoria's daughters were carriers for hemophilia, since several of their male descendants were hemophiliacs.

The occurrence of hemophilia in the son of the last Czar of Russia and in the princes of Spain, all descendants of Victoria, had considerable political consequences.

For a female to have hemophilia, or any other X-linked recessive trait, she must receive a gene for the trait from each parent. This can occur in hemophilia, and a number of such cases have resulted from the marriage of hemophilic males and carrier females. As in the case of affected males, hemophilic females may bleed either continuously or for very long periods of time after only a slight scratch. For the hemo-

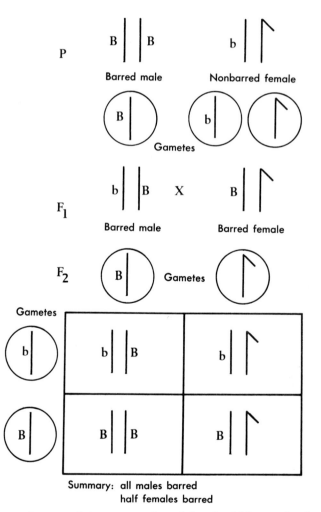

Fig. 5-6. Diagram of a cross between a nonbarred female chicken and a barred male. (In chickens the female is heterogametic.)

philic female, the menstrual flow is an added hazard, and these affected females often require transfusions to replace the great loss of blood accompanying the menses.

Attempts have been made to discover Y-linked traits in man. The most extensive studies conducted thus far have been on the inheritance of ear hair. There are men who have very hairy ear pinnae, and the trait appears to be transmitted from father to son. A number of independent studies of different family pedigrees have failed to resolve whether the trait is Y-linked or is autosomal. To date, we have no unequivo-

cally demonstrated case of a Y-linked gene in man.

Sex chromosome linkage in birds

In birds, the male is the homogametic sex (XX), whereas the female is the heterogametic sex (XY). The inheritance of barred plumage in poultry will illustrate X-linkage in birds. The barred pattern, as seen in such breeds as the Barred Plymouth Rock, is dominant over black or red unbarred plumage. The gene for barred plumage is represented by B and that for nonbarred by b. The cross between a nonbarred

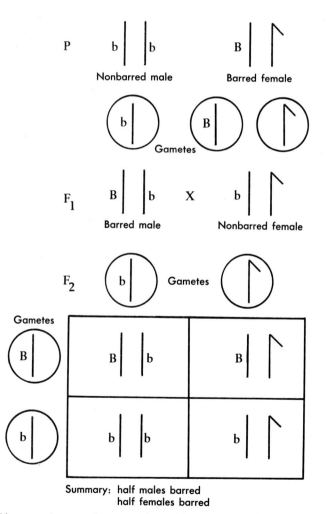

Summary: half males barred
half females barred

Fig. 5-7. Diagram of a cross between a barred female chicken and a nonbarred male.

hen and a homozygous barred cock produces only barred offspring of both sexes. When inbred, these produce only barred males in the F_2, whereas half the F_2 hens are barred and the other half nonbarred. This cross is diagrammed in Fig. 5-6. The reciprocal cross of barred hen and nonbarred cock results in F_1 barred males and nonbarred females. The F_2 yields equal numbers of barred and nonbarred birds in both sexes. This cross is illustrated in Fig. 5-7.

Barring in poultry thus follows a crisscross mode of inheritance, as does the character white eyes in *Drosophila.* The difference is that in the fowl the X chromosome goes from mother (XY) to her sons only, whereas the father (XX) transmits X chromosomes both to his sons and to his daughters. The gene follows the X chromosome in both cases.

Electrophoretic mobility

In Chapter 3 we discussed, at some length, electrophoretic mobility as a phenotype that provides exciting prospects for research and knowledge. From our study of sex chromosomes and their genes, we should not be surprised to learn that we can study the electrophoretic patterns of proteins whose allelic determinants are located on the X chromosome. Again, *D. melanogaster*

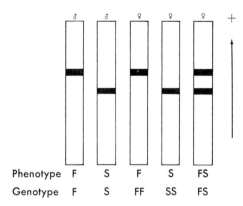

| Phenotype | F | S | F | S | FS |
| Genotype | F | S | FF | SS | FS |

Fig. 5-8. Diagrammatic comparison of electrophoretic phenotypes obtained when studying glucose-6-phosphate dehydrogenase enzymes in *D. melanogaster.* Arrow indicates direction of migration of the proteins.

will give us one of the clearest examples of this type of experimental approach. An enzyme that has been well studied is glucose-6-phosphate dehydrogenase (G-6-PD). The kinds of banding patterns that are found can be seen in Fig. 5-8.

As indicated earlier in this chapter, male *Drosophila* are hemizygous and have only one X chromosome. As expected in such cases, a male fly exhibits only one electrophoretic band for G-6-PD, either fast (F) or slow (S). In contrast, a female *Drosophila,* being normally diploid, shows one of three banding patterns. If homozygous, she exhibits either the F or S pattern. If heterozygous, she shows both the F and S bands. From these findings, it is clear that G-6-PD acts as a monomer-type protein and, as a result, heterozygous females show *codominance* of their alleles.

In 1964 Young and his co-workers reported on their study of the inheritance of the banding patterns of G-6-PD. Various combinations of single-pair matings were made. All S by S matings gave only S type offspring. All F by F matings gave only F type offspring. In cases of F by S matings, the male offspring always reflected the maternal phenotype, and the female offspring were invariably double-banded. With increased study, more complicated situations will undoubtedly be reported. In the study of human X-linked proteins, the electrophoretic patterns follow those found in *Drosophila.* When birds and other species in which the female is heterogametic and the male is homogametic are investigated, a reversal of the *Drosophila* protein banding patterns is found.

SEX-INFLUENCED GENES

There are a number of known autosomal genes whose expression will be either as a dominant if the individual is a male or as a recessive if the individual is a female. Such genes are said to be *sex-influenced.* In man, baldness appears to be such a sex-influenced autosomal trait. Heterozygous males become bald, but the gene must be in a homozygous state for baldness to occur in women. However, baldness can occur in a heterozygous woman who develops a masculinizing tumor of the ovary.

SEX-LIMITED GENES

The sex-limited genes are those autosomal genes which can be expressed in only one of the two sexes. Genes that control the characteristics of the male reproductive system have obviously no opportunity to function if the individual is a female, and vice versa. In birds whose plumage pattern varies with the sex, there are a large number of genes for feather colors; some of these function only in the male, and others only in the female.

SEX DETERMINATION

Our discussion of sex chromosomes and of sex-linked inheritance would naturally lead to the assumption that the determination of the sex of the individual is solely a function of the sex chromosomes and must therefore be unalterably fixed at fertilization. However, a consideration of the existence of hermaphroditic species will make it apparent that one does not need differences in chromosomes in order to get differences in reproductive systems. As we shall also see, some species of organisms, although not hermaphroditic, change from one sex to another with age or with alterations in the environment. Sex determination in all organisms should not, however, be assumed to be purely a matter of internal or external environment. For many species a quite rigid genetic system exists that determines sex. As we have seen in all our previous discussions, no genetic system is simple, and each has a large variety of forms. The same is true of sex determination. We shall therefore consider separately a few of the many known mechanisms, both genetic and environmental, of sex determination.

Genic balance system of sex determination

In *Drosophila,* there are two differences in the chromosomal constitutions of the two sexes. The female has one more X than the male, whereas the male has a Y chromosome that is absent in the female. An examination of Fig. 5-4 will quickly indicate that the Y chromosome does not, by itself, determine the sex of the fruit fly possessing it. This can be seen in the fact that the XXY is a fully fertile female but the XO is a sterile male. Thus the Y chromosome cannot make the individual a male, although it does control the fertility of the male. This leaves the X chromosomes and the autosomes to be considered in any hypothesis of sex determination in *Drosophila.* The fundamental work here, too, was done by C. B. Bridges who in 1925 reported his *genic balance* theory of sex determination in *Drosophila.* It was shown that female-determining genes were located in the X chromosome, while male determiners were in the autosomes. No specific loci were identified, and many chromosomal segments appear to be involved. The distribution of male-determining genes in the autosomes and female-determining genes in the X chromosomes meant that all individuals carry genes for both sexes. A tipping of the balance toward femaleness or maleness in a developing zygote would be determined by whether the particular sperm involved in the fertilization process had or had not carried an X chromosome. The development of the specific reproductive system would be a function of the sex-limited genes, as we discussed earlier. Bridges succeeded in experimentally producing various combinations of X chromosomes and sets of autosomes in *Drosophila.* The results of the different combinations produced are shown in Table 5-1.

From Table 5-1 we can establish a *sex index* by which various combinations of sets of autosomes and different numbers of X chromosomes can be evaluated as to their male- and female-determining capacities. Two sets of autosomes have sufficient male-determining genes to overbalance the female-determining genes contained in one X chromosome, and thus the XY + AA individual is a normal male. His sex index

Table 5-1. Sex index and sexual type in *Drosophila melanogaster*

Phenotypes		No. X chromosomes	No. sets of autosomes (A sets)	Sex index $= \dfrac{\text{No. X's}}{\text{No. A sets}}$
Superfemale		3	2	1.5
Normal female	tetraploid	4	4	1.0
	triploid	3	3	1.0
	diploid	2	2	1.0
	haploid	1	1	1.0
Intersex		2	3	0.67
Normal male		1	2	0.50
Supermale		1	3	0.33

would be 0.50. The XO + AA *Drosophila* also has a sex index of 0.50 and is a male. However, due to the lack of the Y chromosome, he is sterile. Two X chromosomes have sufficient female-producing genes to overcome the male-producing genes of the autosomes. Therefore the XX + AA (diploid) individual is a normal female. Her sex index is 1.0. The lack of function of the Y chromosome in sex determination is seen in the fact that the XXY + AA *Drosophila* also has a sex index of 1.0 and is a fully fertile female. Table 5-1 also shows that individuals that contain haploid (X + A) tissues in their bodies exhibit female traits in these tissues, regardless of the sex of the fly. The sex index also applies to triploid and tetraploid individuals, who are females if their sex indexes are 1.0 and males if their sex indexes are 0.5.

The actual proof of the genic balance theory of sex determination is seen, in Table 5-1, in the formation of intersexes and supersexes. Both types of individuals are sterile and thus represent genetic dead ends. However, their importance for the present discussion lies in their sex indexes. The intersexes are XX + AAA and have a sex index of 0.67. Their appearances cover a wide range, from those that look like normal males through all possible intermediate types to those that look like normal females. Experimentally, one can get a high propor-

tion of malelike intersexes by allowing the flies to develop at low temperatures and a correspondingly high percent of femalelike intersexes by allowing the flies to develop at high temperatures. Since raising normal *Drosophila* at different temperatures does not affect their maleness or femaleness, the fact is apparent that the intersex represents a state of genic imbalance in which environmental factors such as temperature can have large effects on the functioning of the masculinizing genes of the autosomes and the feminizing genes of the X chromosomes.

Supermales and superfemales resemble their normal counterparts. However, as mentioned earlier, they are sterile. In addition, they have rather poor viability, especially the superfemale, which usually dies before adulthood. These supersexes also indicate that sex indexes other than 0.50 and 1.0 represent states of genic imbalance in which the organism cannot function normally.

Bridges' explanation of sex determination in *Drosophila* through the mechanism of genic balance is undoubtedly correct for the fruit fly, all the XO male species, and many of the other XY male species. However, there are other mechanisms of sex determination even where, as in man, the male is XY. We shall now discuss sex determination in man. This mechanism appears also to determine sex in most other species of mammals, although variations are known.

The Y chromosome in sex determination

Sex determination in man was originally thought to follow the same pattern as that of *Drosophila.* However, this erroneous idea was due to the lack of adequate human chromosome preparations. The first discovery of significance for an understanding of the mechanism of human sex determination occurred in 1949 when Barr and Bertram found a deeply staining body in the nuclei of the nerve cells of female cats. This body, eventually to be labeled the *Barr body,* was absent from the nerve cells of male cats. A flurry of investigations showed that the Barr body was a sex indicator in many organisms, man included. Most types of cells in the human body exhibit this characteristic chromatin pattern: a Barr body in the female, who is also designated as "chromatin positive," and none in the male, who is also designated as "chromatin negative." The most commonly used preparations for this type of study are skin biopsy, oral epithelium smear, and blood smear. During a deliberate search for sex indicators in human blood cells, Davidson and Smith in 1954 discovered another difference between the sexes. In stained blood smears from females, they noted a small bulblike structure attached by a thin thread to the nucleus of some of the polymorphonuclear leukocytes. They called these accessory nuclear lobules *drumsticks.* Drumsticks are rarely, if ever, present in normal males. Barr bodies and drumsticks are shown in Figs. 5-9 and 5-10, respectively.

Problems soon arose in the use of Barr bodies and drumsticks as sex indicators in certain cases involving sterility. Some human females exhibit what is called *Turner's syndrome.* These females are sterile as a result of a complete lack of development of their ovaries and are also characterized by an

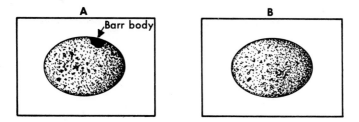

Fig. 5-9. **A,** Diagram of a nucleus of normal female squamous epithelium showing Barr body (arrow). **B,** Diagram of a nucleus of normal male squamous epithelium.

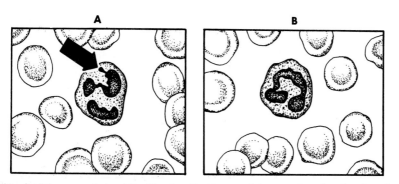

Fig. 5-10. **A,** Diagram of a white blood cell of normal human female, showing drumstick (arrow) on nucleus. **B,** Diagram of a white blood cell of normal human male.

underdevelopment of their breasts and the presence of a small uterus. Females exhibiting Turner's syndrome lack Barr bodies and drumsticks in their appropriate cells and are designated as chromatin negative. Some human males exhibit what is called *Klinefelter's syndrome*. They are sterile because of a defective development of their testes and are also characterized by a tendency toward femaleness, such as having enlarged breasts and underdeveloped body hair. Males exhibiting Klinefelter's syndrome show Barr bodies and drumsticks and are designated as chromatin positive. These seeming contradictions were resolved as a result of the work of Tjio and Levan, who in 1956 reported a technique for the preparation of human chromosomes that permitted their separation and study. It was found that females exhibiting Turner's syndrome were XO and males exhibiting Klinefelter's syndrome were XXY, XXXY, or XXXXY. All these abnormal types are sterile. The role of the Y chromosome in man is therefore very different from its function in *Drosophila*. In man, the Y chromosome is absolutely male-determining. Individuals with XX, XXX, or XXXX are, in the absence of a Y chromosome, always female. Individuals carrying one Y chromosome are males regardless of the number of X chromosomes they carry. As with all generalities, there appear to be exceptions to our statements on sex determination in man. Cases have been recorded of XX individuals who have a male phenotype, and there are XY individuals who have a female phenotype. In both these conditions, all individuals are sterile. At present, it is not possible to determine whether such anomalous sexual developments ought to be attributed to hormonal imbalance, chromosomal aberration, developmental disturbance of unknown etiology, or the action of individual genes. Any of these factors could effect such conditions, and some will be discussed later.

Our ability to study the role of the Y chromosome in sex determination was advanced considerably in 1970 when Cas-

persson and his co-workers reported a technique for distinctively staining the Y chromosome both in oral epithelial and white blood cells. Involved in this procedure is the staining of the cells with the compound *quinacrine mustard* or *quinacrine dihydrochloride*. This chemical binds to the Y chromosome to a greater extent than it does to the other chromosomes of the cell. As a result, the Y chromosome exhibits a distinctive bright fluorescence when the cells are viewed under a microscope equipped with the proper light source. A study of both normal and Klinefelter's syndrome males showed that both types possess a single Y chromosome. An investigation of the XX type males, mentioned above, failed to demonstrate the presence of even a portion of a Y chromosome anywhere in the cell, including a possible attachment of the Y chromosome to another chromosome. A study of the XY type females, discussed above, demonstrated the presence of a normal appearing Y chromosome. As stated earlier, the etiology of both these conditions (XX male and XY female) remains unsolved.

An interesting variation in the expected sterility of organisms with sex chromosome anomalies occurs in the mouse. The mouse also shows a strict dependence on the Y chromosome for the production of males. However, the XO female mouse is found to be perfectly fertile as opposed to the sterile XO human female. Again we see that genetic systems follow different patterns in different species.

Sex determination in Xiphophorus maculatus

The sex-determining system in the southern platyfish, *Xiphophorus maculatus*, has been studied over a long period of time by many geneticists. In 1965 K. D. Kallman published an extensive review of the literature and reported his own findings on what had been previously thought to be two systems of sex determination within a single species. In certain strains of the fish, the females are the homogametic sex (XX

= ♀, XY = ♂), but in other strains the male is homogametic (WY = ♀, YY = ♂). The Y chromosomes in both strains appear to be identical in male-producing potencies, but the X and W chromosomes have different female-producing potencies.

Fish with the various sex chromosome combinations are found in the same seine haul, and breeding experiments with females indicate that XY and YY males fertilize all females, regardless of their sex chromosome constitution. The offspring of all crosses are fully fertile. Fish with the WY, WX, and XX constitutions differentiate into females, and the fish that are XY or YY differentiate into males.

This system has been hypothesized to represent a single, integrated sex-determining system. The W chromosome, with its strong female-producing capacity, is thought to have arisen from an ancestral X chromosome. Its evolutionary advantage lies in the fact that it results in a sex ratio favoring females. This would have a selective advantage in any species like the platyfish in which each male courts and inseminates many females. However, one must not assume that all species possess a similar genetic system for adjusting the sex ratio to some optimum value. Among well-studied species, the above situation is very rare. Other mechanisms such as reduced viability of males *(Drosophila)* and environmental control of sexual differentiation (*Bonellia,* to be discussed) are more frequently found involved in maintaining a proper sex ratio for a given species.

Male haploidy in sex determination

The insect order Hymenoptera includes the ants, bees, saw flies, and wasps. In a number of species within this order, males develop parthenogenetically (without fertilization) and have the haploid chromosome number. The females develop from fertilized eggs and have the diploid chromosome number.

Male haploidy, as seen in the Hymenoptera, contrasts sharply with what is found in *Drosophila* where haploid tissues always exhibit female traits. (Remember that the *Drosophila* findings were explained on the basis of a genic balance system.) In the Hymenoptera, some unknown mechanism associated with the haploid-diploid chromosome arrangement is involved in sex determination. The genetic system is complicated, as can be judged from the work of P. W. Whiting, who in 1945 reported on his studies of sex determination in the parasitic wasp, *Habrobracon juglandis.* In this wasp the diploid number is 20, and the haploid number is 10. Some *Habrobracon* males studied by Whiting came from fertilized eggs and were diploid, but others came from unfertilized eggs and were haploid. All females were diploid. The diploid males tended to be inviable. It was postulated that populations of *Habrobracon* contain a multiple-allelic series of a gene that determines sex. These alleles may be designated as *S-1, S-2, S-3,* etc. The haploid males are of as many kinds as there are *S* alleles in the population. A diploid zygote that contains any two different alleles (*S-1/S-2, S-1/S-3, S-2/S-3,* etc.) develops into a normal female. A diploid zygote that is homozygous for a given sex allele (*S-1/S-1, S-2/S-2,* etc.) develops into a poorly viable biparental male. This method of producing females involves a form of forced heterozygosity. As we shall see in a later chapter, it may result in the development of more vigorous females. However, the formation of haploid males should carry with it a serious problem. We would expect any recessive alleles that affect the organism in deleterious fashion to be expressed in the males. Research by W. Drescher on *Drosophila,* reported in 1964, showed that some X-linked deleterious recessives are not expressed in hemizygous males but are expressed in homozygous females. A similar situation is assumed to prevail in the Hymenoptera. The mechanism by which the functioning of deleterious genes is suppressed in the male is unknown.

Sex differentiation in Melandrium

Melandrium is a seed plant belonging to the pink family. Two distinct types of sporophytes exist in these plants. One type produces megaspores from which develop the female gametophytes, whereas the other type of sporophyte produces microspores from which develop male gametophytes. Thus ovules and pollen are borne on different plants. The ovule-producing plants have been found to possess two X chromosomes, and the pollen-producing plants to possess one X chromosome and a much larger Y. The chromosome picture here is very similar to that found in *Drosophila* and man. However, not all plant species characterized by two types of sporophytes possess demonstrable chromosome differences between the sporophytes.

Single genes and sex determination

In *Zea mays* (corn), there is one type of sporophyte. The microspores are produced by the staminate flowers in the tassels, and the megaspores are produced by the pistillate flowers in the ear. The recessive gene for barren plant *(ba),* when homozygous, makes the stalk only staminate, by eliminating the ears. On the other hand, the recessive gene for tassel seed *(ts),* when homozygous, transforms the tassel into a functional pistillate structure. A plant of the genotype *ba/ba ts/ts* is only pistillate (megaspore-producing), and one with *ba/ba ts+/ts+* or *ba/ba ts/ts+* is only staminate (microspore-producing). We can visualize the evolution of a plant like maize, with a single sporophyte, to a plant like *Melandrium,* with two types of sporophytes, by the establishment of the following two genotypes:

ts+/ts ba/ba—normal tassels, no ears = male
ts/ts ba/ba—pistillate tassels, no ears = female

The homozygosity of all plants for the *ba* allele, together with either heterozygosity or homozygosity for the *ts* allele, would yield two types of sporophytes producing, respectively, male and female gametophytes. In this fashion a hermaphroditic species like maize can be transformed into one like *Melandrium* in which the sexes are separate. A species in which the sexes are separate can also be presumed to evolve into a hermaphroditic species by appropriate allelic

changes in those genes controlling the development of the respective reproductive systems.

In *Drosophila,* there is an autosomal recessive allele known as *transformer (tra).* Homozygotes for *tra* always form males regardless of the X genes present, whereas the heterozygotes *(tra/tra+)* or homozygotes for the wild type allele *(tra+/tra+)* have their sex determined by the genic balance system, as discussed earlier in this chapter. A cross of an XY *tra/tra* (male) and an XX *tra/tra+* (female) produces one-fourth each XY *tra/tra* (males), XY *tra/tra+* (males), XX *tra/tra* (males who are transformed females), and XX *tra/tra+* (females). The males that are transformed females are sterile and hence represent genetic dead ends. However, the existence of the *tra* allele does indicate how a single gene may override the normal sex-determining mechanism of the species and may, by its action, determine the sex of the individual.

Cytoplasmic factors and sex determination

All the preceding discussions of sex determination have been involved with the role of whole chromosomes or single chromosomal genes. In the colon bacillus *Escherichia coli,* one finds that the sex of the bacterium is controlled by an extrachromosomal cytoplasmic factor. The sex factor takes one of many different forms, of which we shall consider two: F+ (malelike) and F- (femalelike). In Chapter 2 we discussed conjugation in bacteria and pointed out that conjugation can occur only between F+ and F- cells, and chromosome transfer occurs only when the F+ cells are in the Hfr state. If a male cell, in the F+ state, is placed in a culture of F- cells, the F+ cell will conjugate with one of the F- cells. In such a conjugation no chromosome material is transferred, but an F+ factor is transferred from the male to the female cell, converting the latter into an F+ cell. The original male cell remains F+, indicating that its F+ factor has been replicated and one of the replicas has been passed along to the female cell. The two F+ cells (original and transformed) will in turn conjugate and transform other F- cells.

This process will continue until all F⁻ cells have been converted to the F⁺ type.

An F⁺ strain can be made, over a number of cell divisions, to produce some F⁻ individuals among its members by the exposure of the F⁺ strain to the dye acridine-orange. Acridine dyes inhibit the replication of the F⁺ factor more than they inhibit the replication of the bacterial chromosome—hence the appearance of the F⁻ cells after a few divisions. It has been possible to isolate the F⁺ factor as a cell-free particle, and it is found to be composed of DNA.

This example of a transferable cytoplasmic factor that determines the sex of the individual must have its counterparts in many organisms. Knowledge as to how widespread this phenomenon may be must await future research.

Environmental control of sex determination

We shall now consider some examples in which the environment determines the sex of the individual. Under such conditions, every zygote must contain all the genes necessary for the development of both reproductive systems. Correspondingly, some mechanism, which may be unknown to us, must exist for inhibiting the action of some genes but permitting others to function. As we shall see, in cases of sex reversal, the particular genes to be inhibited and those to be permitted to function can be changed by either internal or external environmental circumstances.

Our first example of environmental control of sex determination concerns the marine annelid *Ophryotrocha*. All individuals of this species start out as males and on maturing produce sperm. When they become older and develop more than twenty segments, they change into females and produce eggs. If the female is reduced in size to less than twenty segments, by starvation or by amputation, it becomes a male again and produces sperm. When it again develops more than twenty segments, the same individual becomes a female and produces eggs. In *Ophryotrocha*, overall size and number of segments control sex determination.

A different environmental mechanism of sex determination is found in the marine worm *Bonellia*. The female is about an inch long and has a fairly complex anatomical organization. Males are of the size of large protozoa and have rudimentary organs. The males live as parasites in the uteri of the females. All larvae reared in isolation become females. Larvae released in water containing mature females can have one of two fates. Some of them will be attracted to females and become attached to the female proboscis. These develop into males and eventually migrate to the female reproductive tract, where they take up their parasitic existence. The remaining larvae, which do not make contact with mature females, develop into females. The proboscis of the female has been demonstrated to produce a hormonelike substance that influences the larvae toward maleness. This fact implies, although it does not demonstrate, that the male-producing substance inhibits the actions of female-developing genes of the larvae while permitting its male-developing genes to function.

The snail *Crepidula* combines some of the sex-determining patterns of both *Ophryotrocha* and *Bonellia*. All young specimens are males. If raised in isolation, the initial male phase is followed by a period of transition in which the male reproductive tract degenerates and the animal develops into a female. If the young animal becomes attached to a female, it will remain male as long as it is associated with the female. If such a male is isolated, it will develop into a female. The presence of a large number of males influences certain of the males to become females. When an individual becomes female, it will remain in that state. There would appear to be some hormonelike substance secreted by both sexes that influences the sex of the surrounding individuals. We may hypothesize some type of hormonal balance that can be shifted in either direction by the presence of sufficient members of either sex.

Hormones and sex determination. In many cases it has been possible to demonstrate that sexual differentiation is controlled by hormones. This represents a specialized case of environmental control of sex determination. As we shall see, sex reversal may occur here too, demonstrating that the particular karyotype is capable of developing either type of reproductive system. Which type is produced is determined by a switch mechanism, in this case known to be a hormone balance, that throws the development either toward maleness or toward femaleness.

Our first example involves sex reversal. In birds, only one gonad of a normal female develops into a functional ovary. The other gonad remains rudimentary. If the functional ovary of a hen is destroyed, the rudimentary gonad develops into a testis. Such sex-reversed hens can even father chicks, which will be expected to show a sex ratio of two females to one male, since in birds the female is the heterogametic sex. The interpretation of this type of sex reversal is as follows. During embryonic development, the XY genotype stimulates the pituitary gland to produce female hormones that cause the gonad of the hen to develop into an ovary. After the development of the ovary has been completed, the pituitary ceases to produce female hormones, due to inhibition of the pituitary by hormones secreted from the ovary, thus acting as a developmental feedback system. The high level of female hormones secreted sequentially by the pituitary and the ovary is sufficient to suppress the action of male hormone–producing cells of the body, such as the steroid-producing cells of the adrenal glands. When the ovary is removed, the cells of the adrenals are, for reasons unknown, able to function before those of the pituitary. As a result, the rudimentary gonad develops into a testis that itself is hormone-secreting. The high level of male hormones produced by the adrenals and the testis is, in turn, sufficient to suppress the action of the female hormone–producing cells of the pituitary.

A natural occurrence in cattle is similar, on the functional level, to what we have just discussed in birds. When twin calves of different sexes occur in cattle, the female member is usually a sterile intersex called a *free-martin;* with female external genitalia but internal organs more or less like those of the male. The male twin is usually normal. F. R. Lillie in 1917 suggested that the formation of a freemartin was due to a fusion of the fetal membranes of the twin calves while they were in the uterus of the mother. The fusion of the fetal membranes permitted the blood of each twin to circulate in the blood vessels of the other. The male hormones produced by the male twin are presumed to suppress the differentiation of the female internal sex organs of the co-twin. The hormonal influence occurs in only one direction. This would indicate that, in cattle, the female hormone is produced later in development than the male hormone.

SUMMARY

In this chapter, we have reviewed sex chromosomes and sex determination. This included a discussion of the types and combinations of sex chromosomes found in different organisms, and some examples of X-linked inheritance in *Drosophila,* man, and birds. We also studied the phenomena of sex-influenced and sex-limited inheritance. We then considered various genetic mechanisms of sex determination, such as the genic balance system in *Drosophila,* the Y chromosome system in man, the WXY chromosome system in the platyfish, male haploidy in the Hymenoptera, the XY system in *Melandrium,* and the cytoplasmic F⁺ factor in *E. coli.* The chapter was concluded with a discussion of the environmental factors that determine the sex of the individual in some species and the role of hormones in the sexual development of birds and mammals. In the next chapter we shall consider another aspect of the genetic system of a species—its chromosome number—and the effects of altering this characteristic of a species.

Questions and problems

1. Define the following terms:
   a. Homogametic sex
   b. Heterogametic sex
   c. Nondisjunction
   d. Hemizygous
   e. Incompletely X-linked genes

f. Sex-influenced genes   h. Freemartin
g. Sex-limited genes

2. What is the difference in the manner in which X-linked alleles are transmitted in poultry and *Drosophila?*

3. What is the "genic balance" theory of sex determination?

4. Refer to question 3. What is the expected sex of each of the following combinations of autosomes and sex chromosomes:

  a. 4A + 3X   d. 4A + 4X
  b. 3A + 2X   e. 3A + 1X
  c. 2A + 2X   f. 2A + 1X

5. What is the difference between the XY and the XO type of sex determination? Give an example of an organism whose sex is determined by each type.

6. What cytological identification is there for sex?

7. What is the nature of the sex-determining mechanism in the fish *Xiphophorus maculatus?* What is the evolutionary advantage of such a mechanism? What other mechanisms that serve the same end exist in other organisms?

8. What is the role of the Y chromosome in each of the following organisms?

  a. *Drosophila*   c. Mouse
  b. Man   d. Platyfish

9. Assess the evolutionary advantages and disadvantages in the mechanism of sex determination as it exists in the order Hymenoptera.

10. How important is the role played by the environment in determining sex?

11. How can sex-influenced and sex-limited traits be explained?

12. A white-eyed female *Drosophila* is mated to a red-eyed male.

  a. What phenotypes do you expect among the progeny of a cross between an F_1 female and her father?

  b. What phenotypes do you expect among the progeny of a cross between an F_1 male and his mother?

13. The gene that determines vermilion eye color in *Drosophila* is X-linked and is recessive.

  a. What phenotypes do you expect from a cross between a vermilion female and a wild type male?

  b. How would you explain the appearance of a few vermilion daughters and red-eyed sons among the progeny of the above cross?

  c. What classes of offspring would you expect to appear when these exceptional daughters are crossed with normal males?

14. A nonbarred cock is mated to a barred hen.

  a. What phenotypes do you expect when a daughter from the above cross is mated to her father?

  b. What phenotypes do you expect when a son from the above cross is mated to his mother?

15. What are the genotypes and phenotypes of the progeny of the following crosses involving the bobbed bristles (bb) trait of *Drosophila:*

  a. $X^{bb}X^{bb}$ by $X^{bb}Y^+$ c. X^+X^+ by X^+Y^{bb}
  b. X^+X^{bb} by $X^{bb}Y^+$ d. X^+X^{bb} by X^+Y^{bb}

16. What will be the sex of the progeny of each of the following bacterial crosses? Why?

  a. $F^+ \times F^-$ b. Hfr × F^-
  c. How can F^- cells be obtained from an F^+ culture?

17. In corn the recessive gene *ba* (barren stalk), when homozygous, eliminates the ears of the plant. The recessive gene *ts* (tassel seed), when homozygous, converts the tassel into a functional "pistil." How can these two genes be manipulated so as to convert corn from a hermaphroditic to a bisexual species?

18. A cross is made between two fruit flies, one genotypically *XXtra$^+$tra* and the other genotypically *XYtra tra*. What will be the genotypes and phenotypes of the resulting progeny?

19. A woman whose father was a hemophiliac but who herself is normal marries a normal man. What are the possible phenotypes of their children?

20. A woman whose mother was a carrier for the deutan syndrome and was otherwise "wild type" marries a man who is color-blind and whose red blood cells are sensitive to the drug primaquine. What phenotypes can be expected among this couple's offspring?

References

Barr, M. L. 1959. Sex chromatin and phenotype in man. Science **130:**679-685. (A report on cytological evidence of sexual differentiation.)

Bridges, C. B. 1916. Nondisjunction as proof of the chromosome theory of heredity. Genetics **1:**1-52, 107-163. (The original report on the phenomenon of the failure of X chromosomes to separate.)

Caspersson, T., Zech, L., and Johansson, C. 1970. Differential binding of alkylating fluorochromes in human chromosomes. Exp. Cell Res. **60:** 315-319. (Original report of a technique for the differential staining of the Y chromosome.)

Falkow, S., and Citarella, R. V. 1965. Molecular homology of F-merogenote DNA. J. Mol. Biol. **12:**138-151. (Isolation of the sex factor in *E. coli* and its identification as DNA.)

Haldane, J. B. S. 1939. Royal blood. Living Age **356:**26-31. (A discussion of the sex-linked gene for hemophilia in humans.)

Hayes, W. 1968. The genetics of bacteria and their viruses, Chapter 24. 2nd ed. John Wiley & Sons, Inc., New York. (An excellent account of sex determination in bacteria.)

Kallman, K. D. 1965. Genetics and geography of

sex determination in the poeciliid fish, *Xiphophorus maculatus*. Zoologica **50:**151-190. (Discussion of the complicated patterns of sex determination in the southern platyfish.)

Kerr, W. E. 1962. Genetics of sex determination. Annu. Rev. Entomol. **7:**157-176. (Survey of the literature on the genetics of sex determination up to 1960.)

Lillie, F. R. 1917. The freemartin: a study of the action of sex hormones in the fetal life of cattle. J. Exp. Zool. **23:**371-452. (An account of the role of hormones in sex determination.)

Morgan, L. V. 1922. Non crisscross inheritance in *Drosophila melanogaster*. Biol. Bull. **42:**267-274. (A report on the genetic ramification of attached-X chromosomes.)

Morgan, T. H. 1910. Sex-limited inheritance in *Drosophila*. Science **32:**120-122. (A discussion of X-linked inheritance and its mechanisms.)

Morgan, T. H., and Bridges, C. B. 1916. Sex-linked inheritance in *Drosophila*. Carnegie Inst. Wash. Pub. 237. (A description of X-linkage in the fruit fly.)

Sturtevant, A. H. 1945. A gene in *Drosophila melanogaster* that transforms females into males. Genetics **30:**297-299. (A report on an example of the role of genes in sex determination.)

Warmke, H. E. 1956. Sex determination and sex balance in *Melandrium*. Amer. J. Bot. **33:**648-660. (A discussion of the mechanism of sex determination in this species of plants.)

Young, W. J., Porter, J. E., and Childs, B. 1964. Glucose-6-phosphate dehydrogenase in *Drosophila:* X-linked electrophoretic variants. Science **143:**140-141. (Experiment demonstrating X-linked inheritance of G-6-PD electrophoretic banding patterns in *Drosophila*.)

6 Chromosome numbers

Every species has a characteristic number of chromosomes. For sexually reproducing forms one can even designate two basic numbers: diploid (2n), which is found in the somatic cells, and haploid (n), which is found in the germ cells. Asexually reproducing species have only one characteristic number, which can be either a diploid or a haploid number depending on the evolutionary origin of the species. Some examples of basic chromosome numbers are the following: *Escherichia coli,* 1 chromosome; various *Drosophila* species, three, four, five, or six pairs of chromosomes; maize *(Zea mays),* ten pairs; human beings *(Homo sapiens),* twenty-three pairs. The range of reported diploid numbers varies from one pair in one species of the roundworm *Ascaris* to over a hundred pairs in some crabs and butterflies. As is true of most biological phenomena, there are exceptions to the rule of a single diploid or haploid number of chromosomes for a given species. As discussed in Chapter 5, in a number of insect species the somatic cells of the female contain one more chromosome than do the somatic cells of the male (e.g., the squash bug, *Anasa tristis,* female with 22 chromosomes as its diploid number and the male with 21). In other insects such as the bees and wasps the female is diploid, but the male is haploid.

Changes in chromosome number may occur spontaneously, or they may be induced by radiations, chemicals, etc. A classification of changes in chromosome numbers revolves about whether whole haploid sets or only individual chromosomes are involved. Two main categories may be distinguished, euploidy and aneuploidy (*ploid,* Greek for unit; *eu,* true or even; and *aneu,* uneven). The term *euploidy* designates genomes containing whole sets of chromosomes. These include haploid (n), diploid (2n), triploid (3n), tetraploid (4n), etc. Any situation involving a multiple of the *n* number of chromosomes is called *polyploidy.* The term *aneuploidy* refers to genomes containing an irregular number of chromosomes. These include monosomic (2n − 1); double-monosomic (2n − 1 − 1); nullisomic (2n − 2), where both lost chromosomes are homologues; trisomic (2n + 1); double trisomic (2n + 1 + 1); tetrasomic (2n + 2), where both extra chromosomes are homologues; etc.

ANEUPLOIDY

An early study of aneuploidy, involving all the chromosomes of a species, was made by Blakeslee and Belling in 1924. They worked with the Jimson weed, *Datura stramonium,* which has twelve pairs of chromosomes as its diploid number. Through experimental breeding, they obtained twelve different trisomics, each having one of the chromosomes of the normal set present in triplicate and each recognizable by its appearance and distinguishable from other trisomics. The seed capsule of the normal plant and of each of the twelve trisomics is shown in Fig. 6-1. These diagrams imply that genes affecting the seed capsule are located on every chromosome of the genome. One might have expected to find a series of twelve monosomics, each having lost 1 of the 12 chromosomes of the normal set, but such monosomics are inviable in the

Jimson weed. Having an excess of chromosomes is generally found to be less deleterious than having a deficiency of the same chromosomes.

Aneuploids have been found in many species, including man. In humans, aneuploidy usually results in very serious physical or mental defects and sometimes both. As mentioned earlier, the diploid number of chromosomes in man is 46. This number was established as the correct value, as a result of the work of Tjio and Levan (1956). Previously, the procedures available for the preparation of human chromosomes for microscopic study did not permit an accurate count of their number. However, with the advent of tissue culture techniques, it was possible to obtain large numbers of cells. Two other laboratory innovations have served to complete the task. One is the treatment of cell cultures with colchicine, resulting in the accumulation of dividing cells in metaphase (the stage at which cells are best studied). The second technique is the treatment of the cell culture with a hypotonic solution that produces a swelling of the nuclei and a spreading of the chromosomes, thus facilitating their study when the cells are squashed between a slide and coverslip. After being squashed, the cells are stained and examined by ordinary light microscopy. Photographs are taken of cells in which the chromosomes are suitably spread, and the chromosomes are cut out from the photographs, matched in pairs of homologous chromosomes, and arranged in order of descending length.

In the classification of chromosomes, four categories are recognized according to the position of the centromere and the resulting relative length of the arms of the chromosomes on each side of the centromere. The four categories are listed below:

1. *Telocentric.* Centromere is at the tip of the chromosome, so that the chromosome consists of a single arm (note that *no* chromosome of man is telocentric).
2. *Acrocentric,* or subterminal. Centromere is near the end of the chromosome, so that one arm is very short.

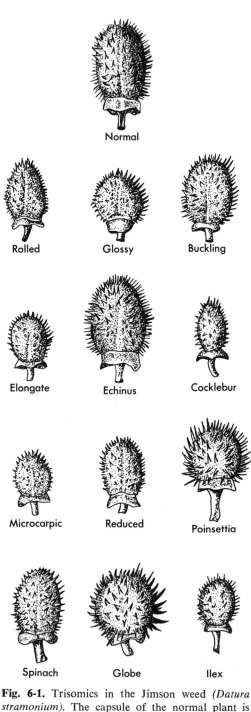

Fig. 6-1. Trisomics in the Jimson weed (*Datura stramonium*). The capsule of the normal plant is shown at the top, and the results of the triplication of each of the 12 chromosomes of this plant are shown, respectively, by the other twelve seed capsules. (From Blakeslee, A. F., and Belling, J. 1924. J. Hered. **15:**194-206.)

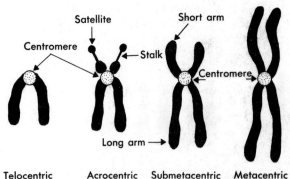

Fig. 6-2. Chromosome morphology. (From Mc-Kusick, V. A. 1964. Human genetics. Prentice-Hall, Inc., Englewood Cliffs, N. J.)

Telocentric   Acrocentric   Submetacentric   Metacentric

3. *Submetacentric,* or submedian. Centromere is nearer one end than the other, resulting in one short arm and one long arm.
4. *Metacentric,* or median. Centromere is in an approximately central position with both arms of equal length.

Examples of these four types of chromosome morphology are shown in Fig. 6-2. An outline drawing of a photomicrograph of the chromosomes of a single human white blood cells is shown in Fig. 6-3. The arrangement of chromosomes by homologous pairs, illustrating their relative sizes and appearance, is called an *ideogram* and is shown in Fig. 6-4 for the chromosomes of a male. Because the individual is a male, both the X and Y chromosomes are included. An ideogram of the chromosomes from a female would differ only in the absence of the Y chromosome and the presence of a second X chromosome. Human chromosomes can be further arranged in groups based on the ideogram. Such an analysis is shown in Table 6-1.

Autosomal aneuploidy

With the advancement in techniques for studying human chromosomes, a concerted effort was made to discover whether any of the known human defects were associated with aneuploidy. The first such association found was *Down's syndrome* (mongolism) in 1959. An individual with mongolism is characterized by mental retardation, short stature, stubby hands and feet, peculiarity

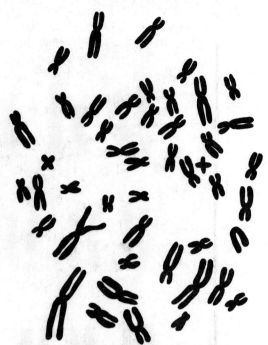

Fig. 6-3. Outline drawing of a photomicrograph of the chromosomes of a single human white blood cell in the metaphase stage of mitosis. (From Mc-Kusick, V. A. 1964. Human genetics. Prentice-Hall, Inc., Englewood Cliffs, N. J.)

of the palm prints, and congenital malformations, especially of the heart. One out of every 600 children born in the United States suffers from Down's syndrome. Since slightly more than 4 million children are born in the United States each year, this rate of incidence of Down's syndrome results in 6000 to 7000 such afflicted babies' being born annually. An analysis of the *karyotype* (appearance of the chromosomes) of individuals showing the Down's syndrome revealed that these people were trisomic for chromosome 21. Examining Fig. 6-4, we see that it is impossible to be certain whether the trisomy involves chromosome 21 or 22, since they are morphologically identical. However, it is reasonable to assume that only one of these pairs is involved and that in all cases it is the same pair. By convention and for the sake of convenience, the twenty-first pair is considered to be the affected one.

Other types of autosomal aneuploids are also known. One of them, Patau's syndrome, is considered to be a trisomy of chromosome

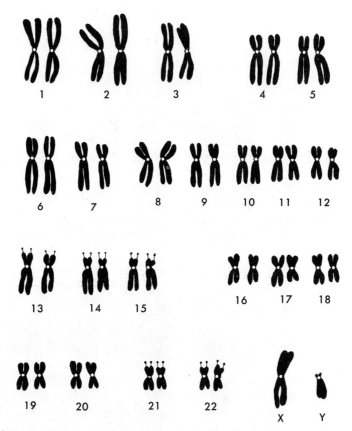

Fig. 6-4. Ideogram of the mitotic metaphase chromosomes of a human male. Note satellites on chromosomes 13 to 15, 21, and 22. (From McKusick, V. A. 1964. Human genetics. Prentice-Hall, Inc., Englewood Cliffs, N. J.)

Table 6-1. Human chromosome analysis by groups

Group	Size and centromere position	Ideogram number	Number in diploid cell
A or I	Large; metacentric/submetacentric	1-3	6
B or II	Large; submetacentric	4, 5	4
C or III	Medium; submetacentric	6-12 and X	15 (male) or 16 (female)
D or IV	Medium; acrocentric	13-15	6
E or V	Small; metacentric/submetacentric	16-18	6
F or VI	Small; metacentric	19, 20	4
G or VII	Smallest; acrocentric	21, 22, and Y	5 (male) or 4 (female)

13, although we cannot be certain that it is not 14 or 15. In most of these cases the infant dies within a few weeks after birth. The clinical picture includes deformities of the eyes, lips, palate, and ears. Polydactyly also may be present. Patau's syndrome occurs in one out of 7600 live births. The other autosomal aneuploid that we shall mention is Edward's syndrome, which involves a trisomy for either chromosome 17 or 18. In such individuals the head is long, ears are low set, and the chin is small. The fingers are held permanently flexed but with the distal joint extended. These individuals also have a very poor viability. Edward's syndrome occurs about once in every 4500 live births.

Sex chromosome aneuploidy

Aneuploidy for the sex chromosomes in man oftens leads to physical and mental defects. One well-known condition results from the addition of one or more X chromosomes to the XY karyotype that, by itself, produces a normal male. The addition of extra X chromosomes results in an abnormal development called *Klinefelter's syndrome*. It is characterized by a defective development of the testes, resulting in sterility of the affected males. In addition, males with this syndrome have tendencies toward femaleness, such as enlarged breasts and underdeveloped body hair. Most of these males are mentally retarded. This pattern develops regardless of whether one (XXY), two (XXXY), or three (XXXXY) extra X chromosomes are added to the normal XY karyotype. Even a male with an XXYY karyotype exhibits the typical Klinefelter's syndrome, demonstrating that the detrimental effects of the extra X chromosome cannot be mitigated by the addition of an extra Y chromosome. Klinefelter's syndrome occurs about once in every 500 male births.

In addition to those males who are aneuploids for one or more extra X chromosomes, one would expect to find males who are aneuploids for the Y chromosome. A number of such individuals have been discovered. They may be found either among the more violent cases in mental institutions or among people who appear normal both physically and mentally. The normal phenotype of this latter group makes the discovery of such individuals unlikely, and all the identified XYY individuals of normal appearance were initially investigated for other reasons. One such was a 44-year-old male who had been twice married and had a total of ten offspring, five of whom were defective. A study of his chromosomes was undertaken in the hope of finding the cause for the 50% incidence of defective offspring, which included two miscarriages, a baby who died at 3 days of age, a daughter lacking ovaries and uterus, and a daughter showing Down's syndrome. It was found that the man was of the XYY karyotype, and it was hypothesized that the extra Y chromosome may have been the result of the fertilization of a normal ovum by a YY sperm. This would imply that the man's XYY karotype was due to a nondisjunction of the Y in his father's gonad. The high incidence of defective children in his two marriages was in turn hypothesized to be caused by aneuploidy for different chromosomes, indicating that the extra Y chromosome may interfere with the normal segregation of other chromosomes during meiosis. This thesis was supported by the finding that in the cell cultures of the man's skin the frequency of nondisjunction was three times higher than normal.

Our consideration of sex chromosome aneuploidy has concentrated, up till now, on the male. We shall now consider the effects of aneuploidy on the female. The first such condition studied is called *Turner's syndrome*. Females with this syndrome are characterized by a complete lack of development of the ovaries, an underdevelopment of the breasts, and the presence of a small uterus. Other symptoms include a short stature, low-set ears, webbed neck, and a shieldlike chest. An examination of the chromosomes of such females reveals a total number of 45 with only one X. Turner's syndrome occurs about once in every 5000 female births. Other cases of sex chromosome aneuploidy in females include individuals with XXX or XXXX karyotypes. These females usually suffer from mental retardation as well as physical defects.

Mosaics and chimeras

As can be anticipated, the seemingly simple cause-and-effect relationship just described has a number of exceptions. One source of modification results from the presence in an individual of two or more different chromosomal types of cells. This can take one of two forms, mosaicism and chimerism.

In the first situation, there is the loss or gain of chromosomes in one or more cells of the body as a result of abnormal mitotic events. Should this occur in a developing embryo, the individual would be partially composed of cells that are aneuploids and partially of those that have a normal karyotype. This type of individual is called a *mosaic.* For reasons that will become apparent later in this discussion, it is important to stress that a mosaic is derived from a single zygote. The ratio of aneuploid to normal cells varies with the individual, and the phenotypic expression of the aneuploidy depends on which organs of the body contain the altered karyotype. Under such conditions, the individual may show only some of the typical characteristics of a particular syndrome. The great variability in phenotypic expression found in mosaics can be seen in XO/XY individuals. They are presumed to originate as XY zygotes and to have lost the Y chromosome from one of the early embryonic cells. They have been observed to range from typical cases of Turner's syndrome to phenotypically normal, though infertile, males. This great phenotypic variation is no doubt a consequence of the chance distribution of the two cell lines during embryogenesis, especially as related to the gonads. Possibly the most interesting illustration of the above was found in a pair of "identical" (monozygotic) twins. One twin is a Turner's syndrome female, and the other is a normal-appearing male. The evidence for monozygotic origin of the twins includes complete identity of blood groups and other genetic markers.

Among the mosaic types reported for the sex chromosomes one finds XO/XY, XO/XX, XO/XYY, XO/XXX, XX/XY, XO/XX/XXX, XO/XX/XY, XX/XXY/XXYYY, and so on. Presumably the extent of such mosaicism is almost limitless, when one considers that the percentage of the body cells containing a particular karyotype can vary from individual to individual. The problem of cause and effect in aneuploidy is further clouded by cases showing symptoms of Klinefelter's, Turner's, or Down's syndrome without any cytological proof of aneuploidy. An inability to demonstrate chromosomal mosaicism is rather frustrating, since one is always faced with the truism that no multicellular organism can ever be proved *not* to be a mosaic.

The second way in which individuals with two or more different chromosomal types of cells can be formed involves individuals whose cells have been derived from two or more distinct zygotic lines. These individuals are called *chimeras,* and they can be formed in one of two ways. In one case, there is an exchange of cells or small masses of cells between developing nonidentical (dizygotic) twins through an anastomosing of their embryonic blood vessels. The cells exchanged would have to be primordial cells that lodged in the correct organ and multiplied along with the cells of the individual. In man this type of transplantation has thus far been found to occur only with blood-forming tissues. In one such case involving twins of different sexes 86% of the male's red blood cells were group A and 14% group O. His sister had 99% group O cells and 1% group A. It has been assumed that the brother's blood is genetically group A and that his sister's is group O. When such individuals have reproduced, their offspring reflect the genotype of their parents and not that of their uncle or aunt.

The second method of forming chimeras includes the entrance of two sperm into an egg and the occurrence of two separate acts of fertilization, one involving the egg nucleus and the other most probably the second polar body. An example of this type of individual, first reported in 1962, involved an individual who was phenotypically a girl but whose karyotype was XX/XY. One of her gonads was a normal ovary, but the other was an ovotestis. An analysis of her blood groups and other genetic markers demonstrated that she was composed of two distinct cell populations. The number of genetic differences between the cell populations was so great that they could not be explained as resulting from abnormal mitotic events. These differences were best explained as resulting from a double fertilization. Other examples of this type of chimera formation have since been discovered.

Aneuploidy and dermatoglyphics

One of the benign side effects of the different types of human aneuploidy is the characteristic change that each causes in the dermatoglyphics of the affected individual. *Dermatoglyphics* is a collective name for all the skin patterns of the fingers, toes, palms, and soles. The skin of these areas is arranged in ridges that are separated by narrow grooves. On the fingertips the ridge pattern may be in the form of a whorl, loop, or arch as shown in Fig. 6-5. In a normal population, one would expect to find 25% of the fingertip patterns as whorls, 70% as loops, and only 5% as arches. The ridge pattern on the palms falls into six chief areas as shown in Figs. 6-6 and 6-7. An important diagnostic feature in dermatoglyphics is the *triradius*. A triradius can be defined as the meeting point of three spokes that demarcate three regions, each region containing a system of almost parallel ridges. On a fingertip, one triradius always accompanies a loop pattern, and two triradii always accompany a whorl. On the palm there are normally four triradii, one at the base of each finger, called a, b, c, and d, and another, known as t, near the base of the fourth metacarpal bone or at some point on its axis. At this time we might do well to point out that the dermal ridge lines are of an entirely different character from the flexion creases that have such an attraction for fortune-tellers.

By an analysis of families, the ridge patterns have been shown to be inherited and determined by many genes. All configurations are laid down permanently at a very early period of development, sometime during the third month of fetal life. One well-established dermatoglyphic measurement is called the *ridge count*. The measurement is made by counting the ridges crossed by a straight line drawn from the triradius to the core of a pattern. For fingertips, an arch scores 0, the loop in Fig. 6-5 would score 3, and the whorl in Fig. 6-5 would have two possible scores: 6 or 7. In practice only the larger of the two numbers obtained in a whorl is recorded. It has been found that the total score for all ten fingers averages about 145 in males and 127 in females. Another dermatoglyphic trait that has attracted attention is the measurement of the angle obtained by hypothetically connecting the "a," "t," and "d" triradii of the palm. This is known as the *atd angle*. In normal individuals it has a value of 48 degrees.

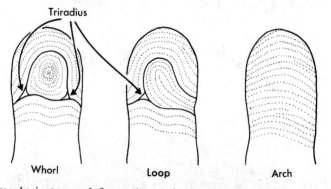

Triradius

Whorl   Loop   Arch

Fig. 6-5. Three basic types of fingerprints: whorl, loop, arch. (From Cummins, H., and Midlo, C. 1961. Finger prints, palms and soles. Dover Publications, Inc., New York.)

In studying the relationship of human aneuploidy to dermatoglyphics, we can once more consider the various types of aneuploids discussed earlier.

Down's syndrome (mongolism). On the fingertips one finds a strong tendency for every finger to possess a loop rather than a whorl or an arch. The total finger ridge count is less variable than among normal people. On the palm, the atd angle has a value of about 81 degrees.

Patau's syndrome. Unlike the individuals with Down's syndrome, the trisomics-13 show both arches and whorls as well as loops on their fingertips. Their most characteristic dermatoglyphic trait is the atd angle that has a value of 108 degrees.

Edward's syndrome. The fingertips almost always have their dermal patterns in the form of arches. In contrast to those of the other autosomal aneuploids, the palms of trisomics-17 show no obvious abnormality.

Klinefelter's syndrome. Males with an XXY karotype differ significantly in their fingertip and palm patterns from normal. Their fingerprints show a general tendency toward patterns with low counts, and arches are more frequently observed. The same is true for XXXY and XXXXY males.

As discussed earlier in this chapter, sex chromosome aneuploidy involving males may also include individuals with extra Y chromosomes. Measurements taken of their

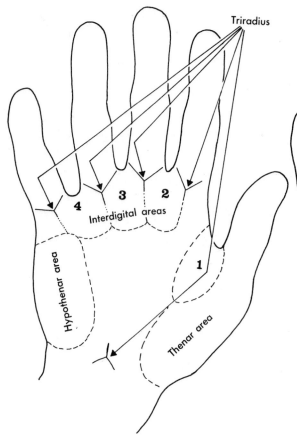

Fig. 6-6. Map of the six chief dermatoglyphic areas of the palm. (From Cummins, H., and Midlo, C. 1961. Finger prints, palms and soles. Dover Publications, Inc., New York.)

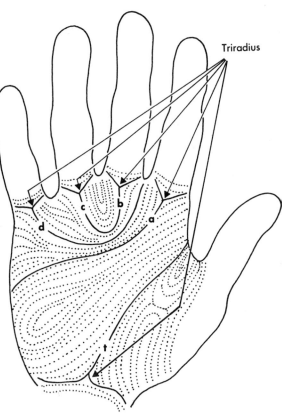

Fig. 6-7. Identification of the main lines (**a, b, c, d, t**) in a palm. (From Cummins, H., and Midlo, C. 1961. Finger prints, palms and soles. Dover Publications, Inc., New York.)

total ridge count show it to have an average value of 90. This is significantly less than that of normal males. The palmar patterns vary, depending on whether the individuals are XYY, XXYY, or XXXYY.

Turner's syndrome. Females with an XO karyotype have, on the average, greater than normal total ridge counts. In addition, large whorls are common. On the palms, the atd angle is about 66 degrees. A comparison of the atd angles for the normal individual and various aneuploids is shown in Fig. 6-8.

• • •

One general tendency appears to exist with regard to the effect of sex chromosome aneuploidy on dermatoglyphics. There

seems to be a correlation between an increased number of sex chromosomes and a decreased finger ridge count.

As an illustration in males, we can examine the mean ridge counts of the following four chromosomal types: (1) XY = 145, (2) XXY = 114, (3) XYY = 103, and (4) XXYY = 88. It is apparent that as sex chromosomes are added to the XY karyotype, a decrease occurs in the ridge count. It also would appear that adding a Y chromosome to a karyotype has a greater effect in reducing ridge count than adding an X chromosome. When one considers females, one finds that the XO (Turner's syndrome) individuals have a significantly higher mean ridge count (165) than normal

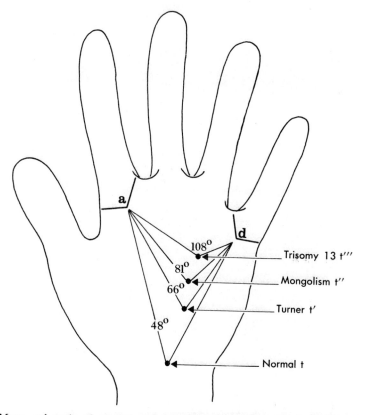

Fig. 6-8. Mean *atd* angle of normal and aneuploid individuals. (From Penrose, L. S. 1963. Nature **197**:933-938.)

females (127). Here, too, as X chromosomes are added, there is a tendency toward decreased ridge counts. At the extreme end of the spectrum, one finds that the mean ridge count for two known XXXXX females is only 17. The mechanism by which an addition of sex chromosomes in both males and females results in a reduction of ridge counts remains to be discovered.

It would be natural to infer from the above discussion that the genes for a particular dermatoglyphic pattern are located on the aberrant chromosome concerned. However, one must be cautious about such a conclusion. As an example of this type of error, let us consider the effect of some of the aneuploids on the height of the affected individuals. It has been observed that Turner patients are short whereas those with Klinefelter's syndrome are tall. The conclusion might then follow that the genes for stature must occupy loci on the X chromosome. However, previous studies on the inheritance of stature have shown it to be autosomal and not X-linked. The same reasoning precludes us from attributing the dermatoglyphic pattern of the affected person to the chromosome involved.

Origin of aneuploidy

In considering aneuploidy, one should question whether the gamete with too few or too many chromosomes can come from either parent. The answer appears to be that it can. This has been best studied for X-chromosome marker traits involving XO and XXY cases. Color blindness has, for example, been observed in XO individuals both of whose parents have normal vision. Since color blindness is an X-linked recessive trait, it is concluded that the mother is a heterozygous carrier for color blindness. She must have contributed to the offspring the X chromosome bearing the mutant allele for color blindness. This type of Turner's syndrome is designated as X^MO with the paternal sex chromosome missing. Cases of X^PO Turner syndrome have been identified by means of the blood type Xg^a, which is an X-linked dominant. In these cases the ma-

ternal X chromosome is missing. The origin of the XXY Klinefelter syndrome has been investigated by a similar approach, with similar findings as to the origin of the aneuploidy.

Investigations of the factors effecting aneuploidy have revealed a relationship between the incidence of births exhibiting aneuploidy and the age of the mother. This has been especially true of Down's syndrome in which there is a marked increase in the incidence of birth of afflicted children with the increasing age of the mother. It is thought that the increased age of the egg cell in older mothers may result in increased nondisjunction during meiosis. There also appears to be a different distribution of maternal age associated with XXY and XXX children than that found with other children. The three types of maternal age distributions are shown in Fig. 6-9.

One line of research on the etiology of Down's syndrome appears, in at least some cases, to implicate thyroid dysfunction in the mothers of affected individuals. The fact that Down's syndrome patients have a high incidence of thyroid disease has been known for some time. Various studies have been conducted on the families of these individuals. In a paper published in 1971, Fialkow and his co-workers reported that some form of thyroid disease was discovered in 30 (17%) of 177 mothers of patients with Down's syndrome. This frequency was significantly greater than the 11 (6%) of 177 control mothers who exhibited thyroid dysfunction. No significant difference was found in the incidence of thyroid disease between the fathers of the Down's syndrome and control groups. The presence of thyroid disease in women appears to be associated with an increased risk of Down's syndrome in their offspring. The mechanism by which this occurs remains to be discovered.

Before leaving this discussion dealing with human aneuploidy, we should point out that most of the known cases involve additions rather than losses of chromosomes. The only known monosomic is Turner's syndrome involving the X chromosome. No

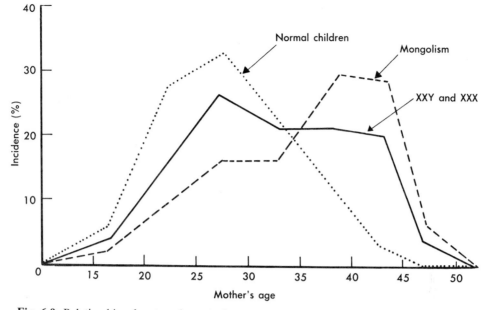

Fig. 6-9. Relationship of maternal age to frequency of normal and aneuploid offspring. (From Penrose, L. S. 1964. Ann. Hum. Genet. **28:**199-200.)

monosomic state involving an autosome has yet been discovered, although trisomics are well known. Apparently, to have too few autosomes is more disruptive than to have too many.

EUPLOIDY AND POLYPLOIDY

In the beginning of this chapter we defined euploids as individuals having chromosome complements composed of whole sets of chromosomes. These could be haploid, diploid, triploid, etc., with the multiples of the basic number (n) being called polyploids. This definition of polyploidy takes into account the fact that a diploid is actually a polyploid state for a normally haploid organism. Haploidy, sometimes called monoploidy, is relatively rare for animals although it is found in the male Hymenoptera. However, in all plants the gametophyte stage of the life cycle is haploid and therefore has (n) chromosomes as its basic number.

Polyploidy can be found in either some or all cells of an organism. In man, cancer cells can have more than 100 chromosomes. An examination of the liver of the rat re-

veals about 55% diploid cells, 40% tetraploid cells, and 5% octoploid cells. In the rectal glands of *Drosophila,* octoploid cells predominate. Investigations of plants frequently uncover polyploid cells and sectors of polyploid tissue in plant root tips and occasionally an entire polyploid shoot.

Production of polyploids

It is relatively easy to induce polyploidy in part of, or in an entire, organism. The mechanisms include injuring the tissue, heat or cold shocks, and specific chemicals. The tomato plant *(Lycopersicum esculentum)* has 24 chromosomes in the nuclei of its somatic cells. If a shoot of the plant is cut back, new shoots frequently arise from the callus of the wound. Usually about 7% of these shoots are composed mainly of tissue that has 48 chromosomes per nucleus and is therefore tetraploid. One can also cut back a tetraploid shoot and obtain further doubling of the number of chromosomes. This type of work was first done in the tomato by H. Winkler in 1907. Other workers experimented with the green alga *Spirogyra.* It was found that after the division of a nucleus, the daughter nuclei could be prevented from separating by treatment with refrigeration, ether, or chloroform. However, these treatments did not prevent the formation of a new cell wall. Thus, after treatment, cells with no nucleus occurred next to cells with two nuclei or with one nucleus of twice the normal size. From cells of this last type, polyploid strains were grown.

The most frequently employed technique for inducing polyploidy in plants involves the use of specific chemicals, either to suppress the formation of the spindle or to stimulate an extra replication of the chromosomes. Colchicine and acenaphthene are examples of spindle suppressors. In the presence of colchicine, the cells undergo a normal mitosis up to the end of prophase.

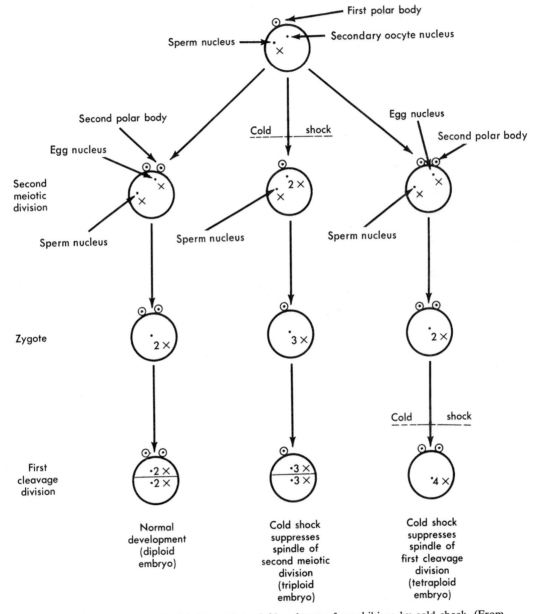

Fig. 6-10. Production of triploid and tetraploid embryos of amphibians by cold shock. (From Dawson, G. W. P. 1962. An introduction to the cytogenetics of polyploids. Blackwell Scientific Publications, Ltd., Oxford, England.)

The nuclear membrane then breaks down, but no spindle is formed. The centromeres divide, and a nuclear membrane is formed around the whole group of chromosomes. Although most plant species form polyploids in response to colchicine treatment, most animal species do not. The production of polyploid strains of *Amoeba* by the use of colchicine is a notable exception. In animals polyploidy is most frequently induced by giving temperature shocks to the egg or embryo. The Amphibia have proved very responsive to this type of treatment.

In amphibians the second meiotic division of the egg cell is not completed until after the sperm has entered the egg. A temperature shock shortly after the entry of the sperm disrupts the spindle of this division, and the two groups of chromosomes form a single diploid nucleus. When the sperm nucleus fuses with this diploid nucleus, a triploid zygote is formed that develops into a triploid embryo. However, if fertilization is allowed to proceed normally, a temperature shock can be administered before the first cell division. This will result in the suppression of the spindle of the first cleavage division, and the dividing diploid nucleus will be reconstituted as a tetraploid nucleus. Subsequent development will produce a tetraploid embryo. The production of triploid and tetraploid amphibian embryos by cold shock is shown in Fig. 6-10.

Temperature shock can be either a cold shock or a heat shock. The cold shock is usually given by plunging the eggs from room temperature into water at 1° C. and leaving them there for 30 minutes. The optimum temperature for heat shock is 37° C., which is usually given for only a few minutes. Production of polyploidy is generally more difficult in animals than in plants. This is due in part to the abnormal development and death of most animal polyploids.

Types of polyploidy

It is traditional to classify polyploids into autopolyploids and allopolyploids. *Autopolyploids* arise from a single individual or from an interbreeding population of individuals. All the haploid sets of chromosomes of an autopolyploid are homologous to one another. *Allopolyploids* arise from hybrids between more or less distantly related species. If the two haploid sets of chromosomes in each of the parents of the hybrid are represented by AA and BB, the chromosome groups of the hybrid are represented by AB. Such hybrids are often sterile due to the lack of homology between chromosomes of set A and those of set B. This lack of homology results in a haphazard distribution of the chromosomes at meiosis, causing most of the gametes to contain unbalanced groups of chromosomes. Most zygotes formed by such gametes will be inviable, if indeed the gametes are able to function at all. However, if the chromosome number of the hybrid is doubled, an allopolypoid (i.e., allotetraploid) will be produced, with a chromosome constitution that may be represented as AABB. During the meiosis of the gonadal cells of this type of individual, A chromosomes will pair with A, and B chromosomes will pair with B. The individual will behave cytologically like a diploid, and the resulting gametes will contain balanced sets of chromosomes A and B. Such gametes are normally fully functional, and the resultant zygotes are viable. The term *amphidiploid* is often used as a synonym for allotetraploid.

Problems in establishing polyploidy

In nature one finds both autopolyploids and allopolyploids. Polyploids are relatively rare among animals although they are quite common among plants. Fully one half of all known plant genera contain polyploids, and about two thirds of all the grasses are polyploids. The establishment of a polyploid population in a sexually reproducing, cross-fertilizing species is actually quite difficult. This is because most polyploids will occur as a single tetraploid individual in a population. The tetraploid will have to mate, or cross, with one of the surrounding diploids, thus forming triploid offspring. Triploids are usually viable, and the mitosis of triploid cells, as well as other types of polyploid cells, is essentially the same as in diploids.

The reason is that mitosis does not involve chromosome-pairing. However, at the meiosis of triploids, one finds that there are for autopolyploids three haploid sets of homologous chromosomes, while for allopolyploids there are two sets of homologous chromosomes and one extra group of nonhomologous chromosomes. In both cases, a normal segregation of two of the haploid sets of homologous chromosomes will occur, but the extra chromosomes will be distributed to the resultant cells in a random fashion. Most of the gametes therefore will not have a balanced complement of chromosomes. In plants very few such gametes will be viable. In animals such gametes may be functional, but the resultant zygote will be an aneuploid and often nonviable.

It is apparent that the establishment of a polyploid population must carry with it some radical modification of the normal life cycle. This, in fact, is the case. Most of the polyploid populations, plant as well as animal, exhibit some form of parthenogenesis. These populations have become essentially asexual in their life cycle. Without meiosis, extra sets of chromosomes present no hazard for reproduction.

Polyploidy in animals

Polyploidy is less frequent in animals than in plants, although an increasing number of animal species have been found either to contain polyploid populations or to be polyploids of some characteristic generic chromosome number. One of the first well-studied species was the brine shrimp, *Artemia salina*. The species was found to be composed of the following populations that differ from one another as to chromosome number and method of reproduction: (1) normal diploid bisexual populations, (2) parthenogenetic diploid populations, (3) parthenogenetic triploid populations, and (4) parthenogenetic tetraploid populations. The overwhelming number of offspring produced in the parthenogenetic populations are females. The existence of a parthenogenetic diploid population led to the hypothesis that the evolution of polyploidy in the brine shrimp consisted of two stages. The first stage is thought to involve the development of a parthenogenetic diploid population from one or more diploid individuals. The second stage is thought to involve the origin of parthenogenetic polyploids from parthenogenetic diploids. It is difficult to evaluate this hypothesis because not all species containing polyploid populations also contain a parthenogenetic diploid group.

More recently an increasing number of investigators have turned their attention to the vertebrates in search of naturally occurring polyploid groups. One example of such a condition has been found in some populations of salamanders related to the species *Ambystoma jeffersonianum*. In these populations there are two types of females: triploid females (42 chromosomes) with large erythrocytes and diploid females (28 chromosomes) with small erythrocytes. All males of these populations are diploid and consistently small celled. The offspring of triploid females are always triploid and female, while the offspring of diploid females are always diploid and may be either male or female. The populations containing triploid females appear to be well established and show every indication of persisting in their present form. We thus have an unusual situation of triploidy that is determined by the female parent. A study of the mechanism by which this occurs was reported by Macgregor and Uzzell in 1964. They found that early in oogenesis, the chromosomes of triploid females undergo an extra division (*endomitosis*). After the two meiotic divisions, ova with 42 chromosomes are produced. Sperm are necessary to stimulate the eggs to develop, but they do not contribute chromosomes to the triploid egg nucleus. It is not known whether these triploids are autopolyploids or allopolyploids.

A large number of laboratory experiments have been conducted on chromosome numbers in Amphibia by Kawamura and associates, starting in 1939 and continuing to the present. Autotetraploids were produced by heat-shock treatment of the fertilized eggs of the pond frog *Rana nigromaculata*. The

Table 6-2. Chromosome anomalies in twenty unselected studies of spontaneous abortions

	Trisomy	45, X	Triploid	Tetra-ploid	Other	Total abnor-mal	Total
Number	101	43	41	13	24	222	1026
Percent abnormal	45.5	19	18.5	6	11	100	
Percent total	9.8	4.2	4.2	1.2	2.3	21.7	

From Carr, D. H. 1971. Cytogenetics and malformation in abortions. Fed. Proc. **30**:102-103.

external characteristics of these polyploids were not different from those of the control diploids. The autotetraploids were all males and only partially fertile. When their sperm was used to fertilize the eggs of normal diploid females, triploid offspring were produced. There were both males and females among the triploids, with a preponderance of males. A similar series of experiments on the brown frog, *Rana japonica,* also resulted in partially fertile autotetraploid males. However, in contrast to the findings in the pond frog, the autotriploids formed in the brown frog were all males.

Allotetraploids (amphidiploids) were produced by heat-shock treatment of the eggs of the hybrids between two brown frog species, *R. japonica* (26 chromosomes) and *Rana ornativentris* (24 chromosomes). The allotetraploids (50 chromosomes) were very similar to the control diploid hybrids in external characters. Both males and females were found among the treated offspring, with a preponderance of males. The occurrence of females among the allotetraploids was in contrast to the situation found in diploid hybrids between these species, which were all males. The amphidiploid males were partially fertile and produced a small number of offspring by mating with *R. japonica* females. All these offspring grew into allotriploid (38 chromosomes) tadpoles, which developed into sterile males.

Our consideration of naturally occurring and laboratory-induced polyploidy in Am-phibia indicates a variety of effects of polyploidy in this group of animals. Each species must be considered separately, and there appear to be no general patterns of reactions to polyploidy. The evolutionary divergence of the different species has carried with it different effects of alterations in chromosome numbers.

In man, complete triploidy or tetraploidy have, with rare exceptions, thus far been found only in aborted embryos. It would appear that virtually any type of polyploidy that involves all the cells of the human body results in too many malformations to permit a complete development and birth of the individual. Triploid abortuses have 69 chromosomes in their cells: 66 autosomes and either an XXY or an XXX sex chromosome complex. Tetraploid abortuses have karyotypes of 92 chromosomes: 88 autosomes and either an XXYY or an XXXX sex chromosome complex. An interesting survey was reported by D. H. Carr in 1971 on the number of chromosomal anomalies that had been found in studies of spontaneous abortions. The data are shown in Table 6-2.

It is apparent from Table 6-2 that about 22% of the total of 1026 abortuses exhibited some form of chromosomal anomaly. The most frequent type of anomaly is some form of trisomy, which, for reasons unknown, most commonly involves chromosome 16. It is also clear that Turner's syndrome females constitute a large percentage of the chromosomally unbalanced abortuses.

It has been estimated from these and other data that XO individuals have a mortality of about 98% between fertilization and birth. As might be expected, based on the difference in the number of ways in which they can be formed, complete triploids are far more frequent than complete tetraploids. There have been occasional reports of liveborn triploid infants. However, these individuals are severely malformed and their chances for survival are extremely poor. In Table 6-2 the remaining 11% of "other" chromosomal anomalies includes various types of autosomal monosomics, mosaics, and chromosomal aberrations (to be discussed in Chapter 7).

Although, in the vast majority of cases, human polyploidy appears to be lethal, a number of cases of diploid/triploid mosaics have been found that are viable. The individuals may be either male or female. The triploid cells, in both cases, contain 66 autosomes. In addition, the triploid cells from males have XXY sex chromosomes, and those from the females have XXX. The individuals involved all suffer from some degree of mental retardation and syndactyly (joining of two or more digits). The developmental and biochemical causes of these defects are unknown.

Analyses have been made of the distribution of the triploid cells of these mosaics. The results of several independent cytogenetic studies have shown approximately 85% of cultured cells derived from skin and about 50% of cells derived from connective tissue were triploid. However, all cells derived from bone marrow and all leukocytes from the blood were diploid. The origin of the triploid cells in these human mosaics has not been established in all cases. However, when ABO blood tests and other genetic markers are available, they have indicated fairly well that the extra haploid set has been derived from the individual's mother. It may have come from a haploid polar body nucleus of the egg that was not extruded. This nucleus could then have become incorporated in one of the early-division cells, and the resultant triploid cell could have originated the triploid line. The absence of triploid cells in blood and blood-forming cells cannot, at this time, be satisfactorily explained.

Polyploidy in plants

As mentioned earlier in this chapter, polyploidy in plants is widespread in nature and relatively simple to produce in the laboratory.

Many species of wheats are considered to be allopolyploids, and we can examine the type of evidence that suggests this by comparing the following three species:

Triticum monococcum	$n = 7$	$2n = 14$
Triticum turgidum	$n = 14$	$2n = 28$
Triticum aestivum	$n = 21$	$2n = 42$

When *T. monococcum* is crossed with *T. turgidum,* the hybrid has 21 chromosomes, 7 from the former and 14 from the latter. At the meiosis of this species hybrid, 7 chromosomes from each species pair with one another, but the remaining 7 chromosomes from *T. turgidum* remain as univalents and seldom show any tendency to pair. We may conclude that 7 of the 14 chromosomes of *T. turgidum* are homologous with the 7 chromosomes of *T. monococcum.* This would imply that *T. turgidum* is an allotetraploid between *T. monococcum* and some other (presently unknown) species that also had a haploid number of 7 but whose chromosomes could not pair with those of *T. monococcum.*

T. turgidum can also be crossed to *T. aestivum.* At meiosis, 14 chromosomes from each species pair with one another, but 7 chromosomes from *T. aestivum* remain unpaired. Therefore it is concluded that *T. aestivum* had as one of its ancestors *T. turgidum,* which itself was an amphidiploid. In all these studies the ability to pair at meiosis is used as a test of chromosome homology.

The three wheat species discussed actually fall into separate groups. One group of species, known as *einkorn* wheats, all have 14 as their diploid number. They are small-grained, have a low yield, and are of comparatively little value as food. A second group of species, called *emmer* (durum) wheats, have 28 as their diploid number.

Table 6-3. Percentages of species that are probably polyploids in the angiosperm floras of eight regions of Europe

Region	Latitude	Percentage of polyploid species among	
		Monocotyledons	Dicotyledons
Schleswig-Holstein	54-55	63.1	43.5
Denmark	54-58	70.1	45.8
Sweden	55-69	74.6	45.9
Norway	58-71	74.3	45.9
Finland	60-70	77.4	45.4
Faeroes	62	76.1	51.7
Iceland	63-66	84.3	53.4
Spitzbergen	77-81	95.2	68.4

From Löve, A., and Löve, D. 1943. The significance of differences in the distribution of diploids and polyploids. Hereditas **29**:145-163.

These wheats are used mainly for macaroni, spaghetti, and stock feeds. The third group, called *vulgare* (spelt) wheats, have 42 as their diploid number and are used for breads. The probable origin of one of the bread wheats was demonstrated by McFadden and Sears in 1946. These investigators, using colchicine, doubled the chromosomes of an emmer wheat, *Triticum dicoccum* (2n = 28), and of a wild goat grass, *Aegilops squarrosa* (2n = 14). They then crossed the two autotetraploids, thereby producing an allotetraploid. The species hybrid had 42 chromosomes, 28 from the wheat parent and 14 from the goat grass. It was similar in appearance to the primitive forms of bread wheat. A genetic test was made by crossing the amphidiploid to a bread wheat, *Triticum spelta*. The hybrids were fertile and exhibited, at meiosis, normal chromosome-pairing and separation. It was therefore concluded that modern bread wheats arose as a result of the doubling of the chromosomes of the hybrid formed from a cross between an ancient emmer-type wheat and some form of wild goat grass. A similar series of events is thought to have resulted in the formation of modern cultivated cotton and maize.

In the laboratory a number of allopoly-ploids have been produced. Some of these actually constitute new species. One of the best examples of this is the production by G. D. Karpechenko (1927) of an allotetraploid named *Raphanobrassica*. It was the result of the doubling of the chromosomes of the hybrid between the radish, *Raphanus satinus* (2n = 18), and the cabbage, *Brassica oleracea* (2n = 18). These two species cross with difficulty; at meiosis the radish and cabbage chromosomes of the hybrid mostly fail to pair, resulting in spores that contain unbalanced sets of chromosomes. These spores usually degenerate, thereby making the diploid species hybrid, in effect, sterile. However, in some cases diploid spores were produced that were viable and subsequently produced functional diploid gametes. When fusion occurred between such diploid gametes, an allotetraploid was formed, which contained 36 chromosomes. The allotetraploid was both viable and fertile when crossed to other amphidiploids of the same genetic constitution. Backcrosses to both parental species resulted in infertile allotriploids. *Raphanobrassica* is therefore a new man-made species that is reproductively isolated from all related forms.

The experiments that produced *Raphanobrassica* demonstrated a mechanism, the production of new polyploid species, by which desirable features from two different species could be combined. Unfortunately, from a practical point of view, this particular experiment was a failure. The allotetraploid had the leaves of a radish and the roots of a cabbage. Nevertheless, the experiment raises the possibility that more practical results could follow from similar procedures involving other plants.

As we did with other genetic phenomena, we should ask whether there is any evolutionary value to polyploidy. The most apparent advantage of polyploidy is the increase in the amount of genetic material that is available for mutation and selection and, hence, for evolutionary divergence from the parental stock. This increase of available genetic material through polyploidy is especially useful because it comes in balanced sets. Increases in genetic material obtained through aneuploidy is detrimental to most organisms, while that obtained through polyploidy is not. Polyploids appear to be effective colonizers of new ecological niches at the periphery of the species distribution. This can be seen in the analysis of the percentage of polyploid species of angiosperms as one goes from central to northern Europe. The data are shown in Table 6-3.

The colonization of this section of Europe is thought to have followed the retreat of the glaciers after the last Ice Age. With the retreat of the glaciers, new ground became open. Any polyploids formed in this new habitat would have an advantage over the parental species in adapting to it, due to their increased genetic material.

SUMMARY

In this chapter we have considered various aspects of chromosome numbers. These included the effects of adding or deleting individual chromosomes (aneuploidy), as well as the addition of whole chromosome sets (polyploidy). In the latter case we reviewed two situations: one in which there is a doubling of the chromosome sets of an individual (autopolyploidy) and a second in which

there is a doubling of the chromosome sets of a species hybrid (allopolyploidy). Examples of the different ploidy situations have been cited, both from nature and from the laboratory, and the general problems associated with the establishment of polyploidy in a population have been discussed. Finally, we examined the distribution of polyploids among the angiosperms and hypothesized about the factors that might explain their distribution. We shall, in the next chapter, turn our attention to the spatial arrangement of genes in chromosomes and the various ways in which this arrangement can be changed.

Questions and problems

1. Define the following terms:
   a. Colchicine   g. Karyotype
   b. Telocentric   h. Mosaic
   c. Acrocentric   i. Klinefelter's
   d. Metacentric   syndrome
   e. Submetacentric   j. Turner's
   f. Ideogram   syndrome
2. Define the following terms:
   a. Dermato-   c. atd angle
      glyphics   d. Triradius
   b. Ridge count
3. How has dermatoglyphics been utilized in the study of human aneuploidy?
4. What is the evolutionary significance of *polyploidy*?
5. In *Zea mays*, how many chromosomes would be found in the root tip nuclei of the following:
   a. An autotetra-   e. A trisomic for
      ploid   10
   b. An octoploid   f. A nullisomic
   c. A triploid   for 9
   d. A monosomic
      for 1
6. The correlation between Down's syndrome and the development of leukemia in sufferers of this genetic condition is well documented. How would you explain this correlation?
7. How would you go about determining the somatic chromosome constitution of a given individual?
8. Discuss the following statement: All somatic cells derived from a diploid zygote are chromosomally identical.
9. Given the triploid and the tetraploid condition:
   a. Which would you expect to behave more regularly in meiosis? Why?
   b. Which would you expect to perpetuate itself more readily in nature? Why?

10. A plant species X, n = 7, was crossed with a related species Y, n = 9. Only a few pollen grains were formed by the hybrid. These gametes were used to fertilize the ovules of species Y. The resulting plants were few in number and were found to have 25 chromosomes. On self-fertilization a few plants were produced that had 32 chromosomes. What steps were responsible for this situation?

11. Why is irregularity in chromosome number often associated with low fertility in plants?

12. Distinguish between allopolyploidy and autopolyploidy.

13. Discuss the following statement: The establishment of a polyploid population must carry with it some radical modification of the organisms' normal life cycle.

14. Define the following terms:
    a. Monosomic
    b. Double monosomic
    c. Nullisomic
    d. Trisomic
    e. Double trisomic
    f. Tetrasomic

15. a. What type of genetic information can be obtained through the study of aneuploids?
    b. What are the limitations of this type of investigation?

16. Discuss the occurrence of autosomal aneuploidy in humans.

17. The radish and the cabbage each show 9 bivalents at meiosis. A cross between these two different plant genera produces a semi-sterile hybrid exhibiting 18 univalents at meiosis. This hybrid produces a few seeds from which a second-generation hybrid is obtained. This second-generation hybrid is seen to have 18 bivalents at meiosis and is vigorous and fertile. Explain this situation.

18. Describe two ways in which the trisomy causing Down's syndrome may originate.

19. Explain how each of the following ploidies occurring in unfertilized mammalian eggs might have originated:
    a. 1n   b. 2n   c. 3n   d. 4n

20. How can cytogenetic studies at meiosis be used as a test for chromosome homology, in plant species?

References

Blakeslee, A. F. 1934. New Jimson weeds from old chromosomes. J. Hered. **25**:80-108. (Effects of different chromosome numbers on plant phenotype.)

Blakeslee, A. F. 1941. Effect of induced polyploidy in plants. Amer. Natur. **75**:117-135. (Colchicine treatment for inducing polyploidy in plant cells.)

Book, J. A., and Santesson, B. 1960. Malformation syndrome in man associated with triploidy (69 chromosomes). Lancet **1**:858-859. (Abnormalities associated with polyploidy in humans.)

Bungenberg de Jong, C. M. 1958. Polyploidy in animals. Bibliogr. Genet. **17**:111-228. (Review of the literature to 1956.)

Clausen, J., Keck, J. D., and Hiesey, W. H. 1945. Experimental studies on the nature of species. II. Plant evolution through amphiploidy and autoploidy, with examples from Madiinae. Carnegie Inst. Wash. Pub. 564. (Review of the origin of many allopolyploids.)

Cleland, R. E. 1962. The cytogenetics of *Oenothera*. Advances Genet. **11**:147-237. (Excellent review of this complex cytogenetic system.)

Darlington, C. D. 1956. Chromosome botany. Allen & Unwin, Ltd., London. (Thorough consideration of plant chromosomes.)

Fialkow, P. J., Thuline, H. C., Hecht, F., and Bryant, J. 1971. Familial predisposition to thyroid disease in Down's syndrome: Controlled immunoclinical studies. Amer. J. Hum. Genet. **23**:67-86. (Correlation of thyroid disease and Down's syndrome.)

Ford, C. E. 1969. Mosaics and chimaeras. Brit. Med. Bull. **25**:104-109. (Review of the literature to 1968.)

Makino, S. 1956. Chromosome numbers in animals. Hokuryukwan, Tokyo. (Atlas of chromosome numbers in animals.)

Muntzing, A. 1936. The evolutionary significance of autopolyploidy. Hereditas **21**:263-378. (Discussion of the evolutionary aspects of polyploidy in several grasses.)

Patau, K., Therman, E., Smith, D. W., and De Mars, R. I. 1961. Trisomy for chromosome # 18 in man. Chromosoma **12**:280-285. (Description of a syndrome caused by trisomy for chromosome 18.)

Polani, P. E. 1969. Autosomal imbalance and its syndromes, excluding Down's. Brit. Med. Bull. **25**:81-93. (Review of the literature to 1968.)

Race, R. R., and Sanger, R. 1969. Xg and sex-chromosome abnormalities. Brit. Med. Bull. **25**:99-103. (Review of the literature on human sex-chromosome aneuploidy to 1968.)

Sears, E. R. 1948. The cytology and genetics of the wheats and their relatives. Advances Genet. **2**:235-270. (Evolution of different species of wheat as a result of alloploidy.)

Tjio, J. H., and Levan, A. 1956. The chromosome number of man. Hereditas **42**:1-6. (Report giving definitive evidence for the existence of 46 chromosomes in man.)

7 Genes and chromosomes

LINKAGE, CROSSING-OVER, AND GENETIC MAPS

As there are more genes than chromosomes, each chromosome must contain many genes. The genes that are located in the same chromosome are linked to one another and would be expected to be inherited as a single group. Thus each chromosome can be called a *linkage group*. There are as many linkage groups as there are pairs of chromosomes (4 for *Drosophila melanogaster,* 23

for man). Actual experiments have shown that linked genes are not always inherited as single units but merely tended to remain together. The hypothesis that linked genes tend to remain in their original combinations because of their residence in the same chromosome was advanced by T. H. Morgan in 1911. He further postulated that the degree or strength of linkage depends on the distance between the linked genes in the chromosome. This hypothesis resulted in

		Phenotypes		Ratio
	Wing	Body	Eye	
1/2 cn	dp	bl	cn	1/8
1/2 +	dp	bl	+	1/8
1/2 cn	dp	+	cn	1/8
1/2 +	dp	+	+	1/8
1/2 cn	+	bl	cn	1/8
1/2 +	+	bl	+	1/8
1/2 cn	+	+	cn	1/8
1/2 +	+	+	+	1/8

Fig. 7-1. Stick diagram showing expected results of testcross of three mutants if there is no linkage.

130

the theory of the linear arrangement of genes in the chromosomes and led to the construction of genetic, or linkage, maps of chromosomes.

An example of linkage can be seen in a consideration of the following three recessive mutants in *D. melanogaster:* (1) *dumpy* (wings reduced to two-thirds normal length), (2) *black* (black-colored body), and (3) *cinnabar* (orange-colored eye). If a fly homozygous for all three mutants is mated to a fly homozygous for the wild type alleles of the above genes, all the offspring will have the normal phenotype but will be triple heterozygotes in genotype. A testcross of the F₁ females to mutant males will yield different results depending on whether the above genes are linked to one another or not. If there is no linkage, one would expect to obtain eight phenotypic classes with equal frequency as shown in Fig. 7-1. If there is linkage and it is complete, one would expect to obtain two phenotypic classes with equal frequency as shown in Fig. 7-2. However, if there is linkage but it is incomplete, one would expect to obtain eight phenotypic classes with unequal frequencies. Typical results of an actual cross are shown in Fig. 7-3.

The data in Fig. 7-3 show eight phenotypic classes with very different frequencies. These findings indicate we are dealing with three genes, located on the same chromosome, that are not completely linked to one another but do tend to stay together. Those backcross offspring that show the original linkage of the three genes are called *parentals,* since they do in fact have one of the parental generation phenotypes. Those testcross offspring that show new linkage arrangements of the three genes are called *recombinants,* since they have new combinations of the parental phenotypes. The order of arrangement of the genes on the chromosome and the relative distances between them can be mapped by an analysis of the recombination data obtained from the above crosses. The degree of linkage between any two genes on the same chromosome can be represented by the proportion of progeny in which these genes (having become separated from one another) are no longer in their parental combinations but are now recombinants for the parental alleles of the two genes. This separation is effected by the occurrence of a *crossover (chiasma)* between the two genes during the first meiotic division, resulting in a reciprocal exchange between two of the four chromatids. The closer two genes are together on the chromosome, the less is the chance of a crossover occurring between them and the lower will be the recombination frequency.

In analyzing our data, it is essential to

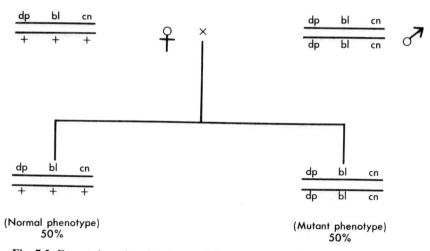

Fig. 7-2. Expected results of testcross of three mutants if there is complete linkage.

consider only two genes at a time. Fig. 7-3 lists the frequency of recombinants for the three possible associations of *dumpy, black,* and *cinnabar.* There were 367 offspring representing recombinants between the genes *dumpy* and *black* out of 1000 progeny. The frequency of recombination is then calculated as 36.7%, and the distance between *dumpy* and *black* is taken to be 36.7 crossover units. In similar fashion the recombination frequency between *black* and *cinnabar* is calculated as 10.7%, and the distance between these two genes is taken to be 10.7 crossover units. From these recombination data we can construct a so-called genetic map. We know that *dumpy* and *black* are 36.7 map units (one unit equals 1% of recombination) apart. We do not know whether *black* lies to the left or right of *dumpy,* but this is unimportant for the immediate consideration. The true spatial relationship of these two genes can be ascertained by other means. For the moment let

us assume that *black* lies to the right of *dumpy.* We also know that *cinnabar* lies 10.7 map units away from *black.* In this case we can be certain that *cinnabar* lies to the right of *black* since the distance from *dumpy* to *cinnabar* is calculated to be 45.0 map units. The genetic map for these three genes is as follows:

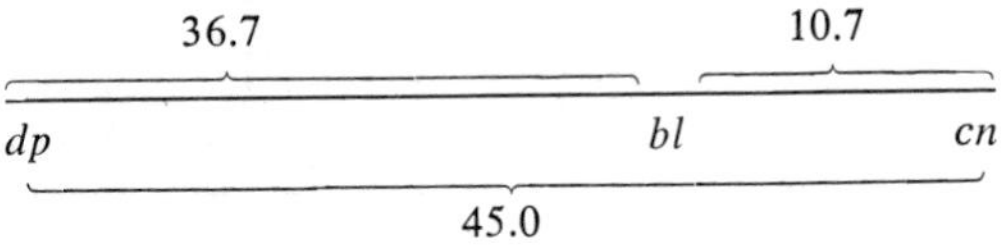

36.7	10.7

dp   *bl*   *cn*

45.0

A number of facets of the linkage phenomenon have to be considered further. The distance from *dumpy* to *cinnabar,* as calculated from the frequency of recombinants between these genes, is shorter than the sum of the distances between *dumpy* and *black,* and *black* and *cinnabar.* This discrepancy is no accident but is due to the occurrence of *double crossovers (double chiasmata)* between *dumpy* and *cinnabar.*

	dp bl cn			
	+ + +	♀ ×	dp bl cn / dp bl cn	♂

Phenotypes	Offspring	Recombinants		
		dp—bl	bl—cn	dp—cn
+ + +	261	—	—	—
dp bl cn	277	—	—	—
+ bl cn	173	173	—	173
dp + +	182	182	—	182
+ + cn	44	—	44	44
dp bl +	51	—	51	51
+ bl +	5	5	5	—
dp + cn	7	7	7	—
Totals:	1000	367	107	450

Fig. 7-3. Results of testcross of three mutants where there is incomplete linkage.

In Fig. 7-3, look at the F_1 female involved in the testcross. Consider the consequences of a break between the *dumpy* and *black* loci with an accompanying reciprocal exchange between chromosomes and, coincidentally, a break between the *black* and *cinnabar* loci with another simultaneous reciprocal exchange between the chromosomes. Such a double crossover will leave the *dumpy* and *cinnabar* genes in the parental association. The net effect of double crossovers between any two genes is to make the genes appear closer together on the chromosome than they really are. In constructing genetic maps, one always takes the sum of short distances as the best estimate for those genes which are widely separated on the chromosome. In our example we can demonstrate the role of double crossovers in making the genes *dumpy* and *cinnabar* appear closer together by the following calculation. Take the frequency of the flies that resulted from double crossovers, 1.2%, and double the number because each such individual fly represents two crossover events. Then add the 2.4% to the 45% obtained as the apparent recombination frequency between the two outside genes. The sum of 47.4% represents the distance in map units between *dumpy* and *cinnabar*. This distance is exactly equal to the map distance obtained by adding the observed frequency of recombinants between *dumpy* and *black* to that obtained between *black* and *cinnabar*. In a consideration of the phenomenon of multiple crossing-over, the fact should be borne in mind that only odd-numbered crossovers result in the recombination of those genes located at the extremities of the crossover region. An even number of crossovers results in the parental arrangement.

The relationship of single to double crossovers can be considered as a problem in probability. If we consider each crossover as occurring independently of the other, the probability of a double crossover should be the product of the frequencies of the single crossovers. In our case the chance of a double crossover is 3.9% (i.e., 0.367 × 0.107). However, our examination of the data has shown that there are only 1.2% double recombinants. This indicates that there must be some form of *positive interference* that prevents some of the expected double crossovers from occurring. A quantitative estimate of the correspondence of double recombinants obtained to those expected is called the *coefficient of coincidence*. It is expressed by the following fraction:

$$\text{Coefficient of coincidence} = \frac{\text{\% of double recombinants observed}}{\text{\% of double recombinants expected}}$$

In our example, the value of the coefficient of coincidence is 30.7% (i.e., 0.012 ÷ 0.039). This means that only 30.7% of all expected double crossovers actually occurred. Experiments have shown that interference usually increases as the distance between loci becomes smaller until a point is reached where no double crossovers are found. Under these conditions, the coefficient of coincidence = 0. Conversely, above a certain distance the observed number of double recombinants equals the expected. Under these conditions, interference disappears and the coefficient of coincidence = 1.

There is an exception to the above type of positive interference, which is called *negative interference* and is characterized by a coincidence value that rises above 1. Negative interference has been found in the fungus *Aspergillus,* in bacteriophage, in *Neurospora,* and in yeast, bacteria, and other organisms. In these organisms, there are minute chromosomal regions in which the coefficient of coincidence may rise to 20 or more, indicating the occurrence of double crossovers at a much greater than random frequency. Negative interference leads to a distortion in the genetic analysis of fine chromosome structure when small chromosomal distances must be expressed as a proportion of much longer distances, over which this type of interference does not operate. It is unlikely that this localized type of negative interference is determined by the same forces as the positive interference discussed earlier.

The phenomenon of gene recombination applies also for genes located on the X chromosome, provided it has another X chromosome or a pairing segment of the Y chromosome with which to engage in reciprocal exchanges. However, crossing-over does not occur in all sexes and species. One finds species in which little or no recombination of genes occurs in one of the sexes. This is true of the males in all species of *Drosophila* so far studied and of the females in the silkworm moth. The mechanism by which crossing-over is suppressed in these cases is not known. The effect of suppressing crossing-over is to completely link all the genes located on particular chromosomes. In study-

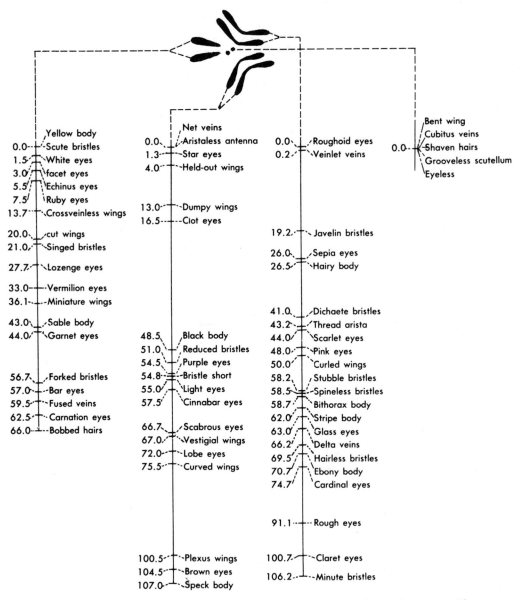

Fig. 7-4. A genetic or linkage map of the 4 chromosomes of *Drosophila melanogaster*. Figures refer to distances from the upper end of the chromosome. (From Sinnott, E. W., Dunn, L. C., and Dobzhansky, T. 1958. Principles of genetics. McGraw-Hill Book Co., New York.)

ing recombination in *Drosophila* one must rely solely on data derived from heterozygous females.

Chromosome mapping is based on recombination data. Examining the distribution of mutants of *D. melanogaster* (Fig. 7-4), we find that the mutants appear to be bunched together at certain points of the chromosome while other sections of the chromosome seem to be completely devoid of mutants. As will be shown later in this chapter, there are techniques that permit us to locate the exact position of the gene on the chromosome. When this is done, the distribution of genes on the chromosome is found to be an even one. This implies that recombination data do not give us the absolute distances between genes but do give us the relative order of genes on a chromosome.

Cytological demonstration of crossing-over

Earlier in the chapter, the statement was made that recombination of parental genes was due to a mutual break and reciprocal exchange between two chromatids. The proof that a recombination of genes was due to a reciprocal exchange of chromosome material was demonstrated in 1931 independently by Stern working in *Drosphila* and by Creighton and McClintock working in maize. Fig. 7-5 shows the experiment conducted by Stern. A strain of flies was ob-

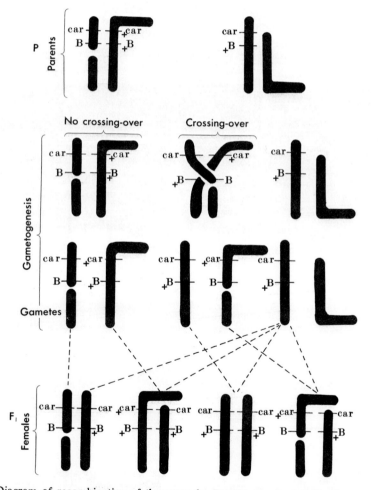

Fig. 7-5. Diagram of recombination of the genes for Bar (**B**, narrow eyes) and carnation (**car,** an eye color) in females in which the two X chromosomes are distinguishable in appearance under the microscope.

tained in which the females had one of their X chromosomes elongated because a portion of a Y chromosome was attached to it, while the other X chromosome was broken in two, with one portion of it attached to one of the fourth chromosomes. Under these conditions all parts of both X chromosomes were present in the cells of the body although divided into three portions. These females were also heterozygous for two sex-linked mutations located in the X chromosome: carnation, *car,* an eye color recessive mutant, and Bar, *B,* an eye-reducing dominant mutant. It was known from the way in which the stock had been made up that the *car* and *B* genes were in the broken X chromosome whereas their two normal alleles were in the X chromosome bearing the portion of the Y chromosome. Such females were bred to males having *car* and the + allele of Bar in their X chromosomes. The female offspring of the cross were found to form four classes: (1) carnation and Bar (noncrossover), (2) normal eye color and shape (noncrossover), (3) carnation and normal-shaped eye (crossover), and (4) normal color and Bar-shaped eye (crossover). A cytological study was made of the X chromosomes of these four classes of females. One X chromosome of each female comes from its father and should be normal in appearance. The other X chromosome comes from the mother and in the case of crossovers should have a different appearance from either of the mother's X chromosomes. Fig. 7-5 shows that the females with carnation-colored, normal-shaped eyes have two normal-looking X chromosomes but the females with normal-colored, Bar-shaped eyes have the portion of the Y chromosome attached to the fragmented X chromosome. These are the results that would be expected if a crossover and reciprocal exchange had occurred between both X chromosomes of the mother at some point between the *car* and *B* genes. It was thus shown that genetic recombination was accompanied by a reciprocal exchange of material between homologous chromosomes.

Stage at which crossing-over occurs. We now turn our attention to the evidence that genetic recombination is due to crossing-over between two of the four chromatids formed at the first meiotic division. Although evidence bearing on this problem has been obtained from *Drosophila,* it has shortcomings in that one cannot examine all the end products of meiosis individually. In *Drosophila* and most other organisms, the four sperm produced by each spermatogonium are mixed with the sperm from other spermatogonia, while in the female only one of the four products of meiosis is functional, since the other three products form polar bodies. For these organisms one obtains a sample of all possible gametic fusions and assumes that the sampling contains all the end products of meiosis and their various fusions in representative frequencies.

It would be much simpler and more reliable if one could collect all the end products of meiosis and examine them individually. This is possible in the red bread mold *Neurospora.* This fungus belongs to the class known as the Ascomycetes, to which also belong *Penicillium, Aspergillus, Saccharomyces,* and others. The life cycle of *Neurospora* can include asexual as well as sexual reproduction. *Neurospora* can grow by ordinary mitoses and form a threadlike mass of hyphae, which is called a *mycelium.* The nuclei of the mycelium are haploid and contain 7 chromosomes. Asexual reproduction occurs in the formation of spores called *conidia* that are budded off special stalklike hyphae. The conidia are usually dispersed by the wind, and if they fall on a suitable substrate, they grow and divide to form a new mycelium. Sexual reproduction involves a fusion of nuclei from mycelia of opposite mating types (*A* and *a*). The fusion of the nuclei results in a diploid zygote containing 14 chromosomes. The cell containing the zygote elongates to form a sac (*ascus*). The zygote immediately undergoes meiosis, so that at the completion of meiosis, the four haploid products are arranged in tandem as shown in Fig. 7-6, which demonstrates the life cycle of this mold. All divisions occur in the plane of the long axis of the ascus, and the nuclei remain in the order in which they were produced. As a result of this regularity of nuclear divisions, the two nuclei at one end of the ascus are derived from one of the first-division nuclei, and the two nuclei at the other end of the ascus are derived from the other of the first-division nuclei. Subsequently, each haploid nucleus divides once mitotically, again in tandem, so that each meiotic product is represented twice within the ascus. Each nucleus develops a hard case

around itself and is called an *ascospore*. The ascus may then be opened and the ascospores be removed individually and cultured in separate tubes. Appropriate tests can then determine the genotype of each ascospore.

We may now return to our question of the stage at which crossing-over occurs. Our analysis will center on the ascus with its four end products of meiosis. This method of studying the end products of meiosis is called *tetrad analysis*. Fig. 7-7 shows diagrammatically the meiosis of a *Neurospora*

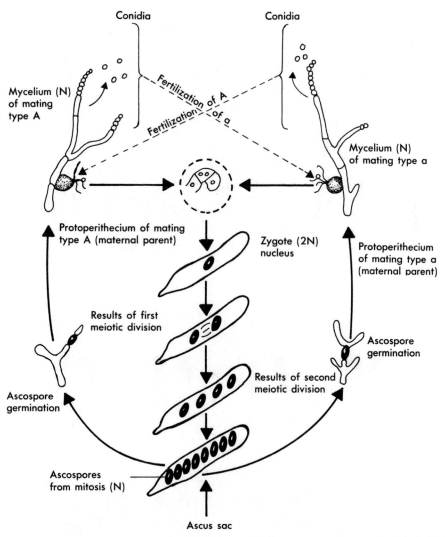

Fig. 7-6. The life cycle of bread mold, *Neurospora*. (From Wagner, R. P., and Mitchell, H. K. 1964. Genetics and metabolism. John Wiley & Sons, Inc., New York.)

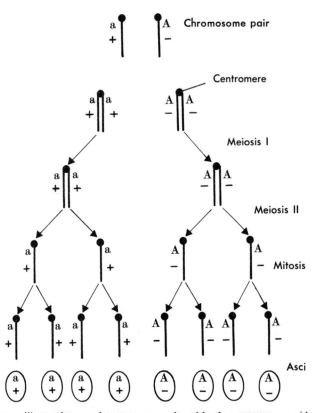

Fig. 7-7. Diagram illustrating a chromosome pair with the genotype a+/A−, at successive stages during the two meiotic divisions and one mitotic division in *Neurospora*. Since no crossing-over has occurred, there is a 4:4 distribution in an ascus.

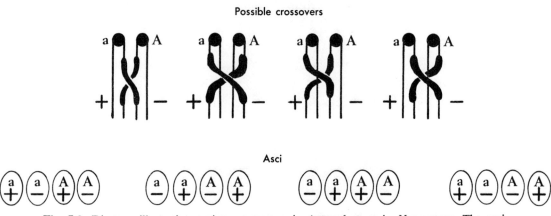

Fig. 7-8. Diagram illustrating various crossovers in 4-strand stage in *Neurospora*. The asci represent the completion of the meiotic sequence. The mitotic division is omitted.

zygote without crossing-over. Note that in this example the gene for *a* mating type is linked with the allele "+" and the gene for *A* mating type is linked with "–." The alleles + and – determine respectively whether the mycelium will be colored or colorless. When, as in this example, separate pairs of alleles are located at different points of the homologous chromosomes, we are able to detect crossing-over and to determine whether the process occurs at the two-strand or the four-strand stage. If crossing-over occurs at the two-strand stage, the asci in which crossing-over has occurred will contain only recombinant types of ascospores. However, if crossing-over occurs between two of the chromatids at the four-strand stage, the asci in which crossing-over has occurred will contain both parental and recombinant ascospores. Examination of the results of actual crosses, as shown in Fig. 7-8, reveals that crossing-over occurs in the four-strand stage but that only two of the four chromatids are involved in a single crossover.

A study of Fig. 7-8 shows that if a single crossover occurs between the mycelium-color gene and the sex-type gene, two of the resultant ascospores will be parental and two will be recombinant. If such a crossover occurs in every ascus, the total percent of recombination will be 50%. It is generally true that recombination frequency (map distance) is one half of cross-over frequency. However, an examination of Fig. 7-4 will show map distances between genes that exceed 50 map units. These larger numbers, which indicate long distances, are obtained by adding together the sums of the recombination values for intermediate genes.

Double crossovers. The establishment that crossing-over occurs between two of the four chromatids of the first meiotic division raises certain problems in connection with double crossovers. The genetic result of a single crossover is the same irrespective of which of the two paternal and two maternal strands are involved, since the two strands of each pair arise from the division of a single chromosome and are identical. On the other hand, the genetic outcome of a double crossover will vary, depending on whether the same or different strands participate in both events. The four possible types of double crossovers are shown in Fig. 7-9.

An examination of Fig. 7-9 shows that crossing-over between the same two strands on both occasions (two-strand double) gives two doubly recombinant and two parental strands. This is the situation we discussed in chromosome mapping, and as stated in that discussion, two-strand double crossovers appear genetically as parental types when genes outside the crossover region are considered. The two-strand double crossover tends to shorten the apparent map distance between such genes. Fig. 7-9 further shows that if only one of the two strands involved in the first crossover is also involved in the second (three-strand double), two singly recombinant strands, together with one double recombinant and one parental strand, are produced. The three-strand double crossovers will appear genetically as single crossovers when genes outside the crossover region are considered. The three-strand double crossovers will act to shorten the apparent map distance between such genes by only half the amount done so by two-strand double crossovers. A three-strand double crossover has only half the effect of a two-strand double because only one of the two exchanges is missed in the three-strand double whereas both the exchanges are missed in the two-strand double. The last possibility illustrated in Fig. 7-9 shows that if the two strands exchanging at the second crossover are different from those exchanging at the first (four-strand double), we get four singly recombinant strands and no parental strands. The four-strand double crossovers will appear genetically as single crossovers when genes outside the crossover region are considered. Since all four meiotic products will appear as single crossovers, the four-strand double crossovers will yield the true recombination map distance between such genes (i.e., the sum of the recombination values for the intermediate genes).

If the involvement of any strand in the

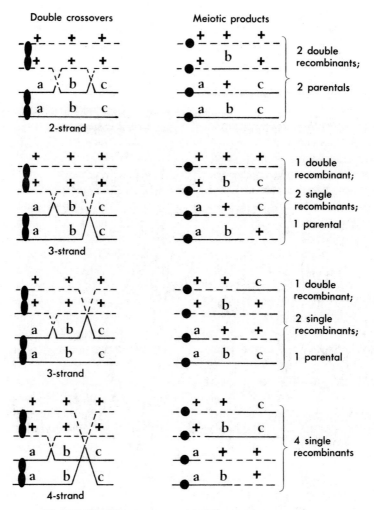

Fig. 7-9. Double crossovers and their genetic consequences.

two crossovers is entirely at random, two-strand, three-strand, and four-strand doubles will arise in the ratio of 1:2:1. This will result, on the average, in double crossovers yielding double recombinants, single recombinants, and parental strands in the ratio of 1:2:1. If the association of strands in neighboring crossovers is not random, the above ratios will be changed. Should the involvement of a strand in one crossover reduce the likelihood of its participation in a second, there will be a reduction in the frequency of two- and three-strand doubles and an increase in the frequency of four-strand double crossovers. Under these conditions the proportion of gametes carrying singly recombinant chromosomes will increase at the expense of those carrying parental type chromosomes. Such cases are known, and the phenomenon has been called *chromatid interference.*

Linkage studies in bacteria

In Chapter 2 we reviewed the three methods by which bacteria exchange genetic material: transformation, conjugation, and transduction. Although all three types of genetic exchange have been used in linkage studies, we shall restrict ourselves to conjugation. Three experimental approaches involving conjugation have been used to develop linkage maps in *E. coli.* These are "mapping by the gradient of transmission," "mapping by time units," and "mapping by counting recombinants." We shall consider each of these methods for ascertaining the linkage relationship of genes in bacteria.

Mapping by the gradient of transmission. In mapping by the gradient of transmission, Hfr male cells and F⁻ female cells, whose characteristics are shown in Table 7-1, were used. These two types of cells were mixed together in a liquid culture and allowed to remain there for 25 minutes. At the end of this time, the mixed cells were plated on a minimal medium containing streptomycin. An examination of Table 7-1 will indicate that the Hfr cells, being streptomycin sensitive, could not grow on this medium; the F⁻ cells were unable to grow on it because minimal medium lacks the amino acids threonine and leucine. The only cells that could grow on the medium were those recombinants that were *thr*⁺ and *leu*⁺ (genes obtained from the Hfr parent) and *str-r* (gene obtained from the F⁻ parent). These recombinant cells were then plated on various minimal media, each of which contained one of the following additives: so-

dium azide, bacteriophage T1, lactose as sole sugar, or galactose as sole sugar. The results demonstrated that among the colonies that were *thr*⁺*leu*⁺*str-r*, 90% were *azi-s*, 70% were *T1-s*, 40% were *lac*⁺, and 25% were *gal*⁺. These percentages represent the strength of linkage among the different genes and were used to develop a linkage map as shown in Fig. 7-10.

Mapping by time units. Linkage maps, as we have already discussed, can only indicate the relative position of genes on a chromosome. In order to determine the absolute location of specific genes, a different experimental approach must be used. Hfr and F⁻ cells of the above types were mixed together in a liquid culture. At intervals of time after mixing, samples were taken from the culture and agitated in a Waring blender. This resulted in a separation of the conjugants with the concomitant breakage of the donor chromosome. After the cells were separated, they were plated and tested to determine which genes from the donor had been integrated into the recipient's chromosome. The results, shown in Table 7-2, indicate that the genes of the donor chromosome

Percent of linkage

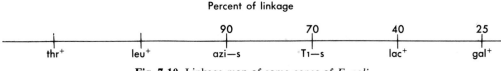

Fig. 7-10. Linkage map of some genes of *E. coli.*

Table 7-1. Characteristics of Hfr and F⁻ cells used in a linkage study

Characteristics	Hfr cells	F⁻ cells
Threonine synthesis	thr⁺	thr–
Leucine synthesis	leu⁺	leu–
Sodium azide sensitivity	azi-s	azi-r
Bacteriophage T1 sensitivity	T1-s	T1-r
Lactose fermentation	lac⁺	lac–
Galactose fermentation	gal⁺	gal–
Streptomycin sensitivity	str-s	str-r

Table 7-2. Time intervals required for the transfer of genes from Hfr to F⁻ cells

Minutes	Hfr genes transferred
0	0
8	*thr⁺*
8½	*thr⁺ leu⁺*
9	*thr⁺ leu⁺ azi-s*
11	*thr⁺ leu⁺ azi-s T1-s*
18	*thr⁺ leu⁺ azi-s T1-s lac⁺*
25	*thr⁺ leu⁺ azi-s T1-s lac⁺ gal⁺*

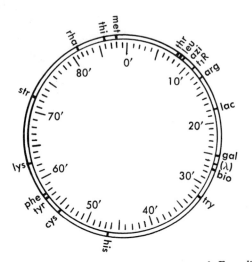

Fig. 7-11. The circular linkage map of *E. coli* K12, drawn to scale in time units (minutes). The symbol *thr* represents threonine; *leu,* leucine; *azi,* azide resistance; *T1ᴿ,* resistance to phage T1; *arg,* arginine; *lac,* lactose fermentation; *gal,* galactose fermentation; (λ), location of prophage lambda; *bio,* biotin; *try,* tryptophan; *his,* histidine; *cys,* cystine; *tyr,* tyrosine; *phe,* phenylalanine; *lys,* lysine; *str,* streptomycin resistance; *rha,* rhamnose fermentation; *thi,* thiamine; *met,* methionine.

enter the recipient cell in linear order. If the migration of the donor chromosome occurs at a constant rate, the information in Table 7-2 can be used to prepare a cytological map of the bacterial chromosome, which will show the exact location of the genes. Investigators have found that at 37° C. it takes about 90 minutes to transfer the whole *E. coli* chromosome. Since, as discussed in Chapter 2, the *E. coli* chromosome is in a ring form, it is always pictured in "time mapping" as a circle divided into time units, as shown in Fig. 7-11. More extensive investigations than those discussed here have mapped the entire *E. coli* chromosome and have also shown that the Hfr trait is always transferred last.

Mapping by counting recombinants. Chromosome mapping by this method must take into account the fact that in bacterial conjugation a temporary diploid state is achieved for only part of the chromosome. In order to be assured that the chromosomal section transferred includes the genes whose linkage relationship we wish to study, we must choose as a donor chromosome one that contains a gene known to be located distal to the gene or genes under investigation. Only those recombinants that contain this distal "marker" gene will be counted. As an example of this procedure let us consider the following experiment. We can mate Hfr

cells that are thr⁺leu⁺ to F⁻ cells that are thr⁻leu⁻. We know from our time-mapping data that the *leu* gene enters the F⁻ cell after the *thr* gene. Therefore any recombinant that includes leu⁺ must also have provided an opportunity for thr⁺ to become involved in a gene transfer. Recombinants that are thr⁺leu⁺ will occur if no crossing-over has taken place between these genes. If, however, a crossover does occur between these genes, the recombinant will be thr⁻leu⁺. To obtain an estimate of the crossover frequency between these two genes, one calculates the value of the proportion of recombinants thr⁻ leu⁺/thr⁺ leu⁺ plus thr⁻ leu⁺. For *thr-leu* this proportion has a value of 0.10, or a recombination frequency of 10%. In comparing recombination units to time-mapping units, one finds that there are about 20 recombination units contained within 1 time unit. The determination of linkage relationships by scoring recombination frequencies yields best results when one is dealing with genes that are less than 3 time units apart.

Genes that are farther apart than 3 time units are more than 60 recombination units apart and appear unlinked in recombination analysis.

. . .

Each of the above-mentioned experimental procedures for mapping bacterial genes yields useful and interchangeable information. If we consider the *azi-s* and *T1-s* genes (Fig. 7-10 and Table 7-2), we find by using the time-mapping method that they are 2 time units apart. The transmission-gradient approach indicates a linkage strength between them of 20%, or a ratio of 10:1 when compared to time units. A recombination-scoring frequency experiment yields a value of 40 map units, which is the expected ratio of 20:1 when compared to time units. Which experimental procedure is used in a particular study will depend on the kind of research problems the particular laboratory is investigating. However, once the information has been obtained, it can be converted to any other units of measuring linkage relationships in bacteria.

NATURE OF RECOMBINATION

The question of the mechanism by which recombinants occur has not been completely settled. In 1931 the cytologist J. Belling proposed a hypothesis that is called the *copy-choice* theory. In its modern modification, the theory postulates that homologous chromosomes are paired, at least at some points along their length, at the time of their replication. Under this hypothesis, the polymerase molecule that is replicating the paternal chromosome can switch, at a point of pairing, to the maternal chromosome and replicate the maternal one thereafter. When the polymerase molecule that is replicating the maternal chromosome comes to the switch point, it has no choice but to begin replicating the paternal chromosome from then on. As a result of this switch-over, two reciprocally recombinant chromosomes are formed. This model relates recombination to chromosome replication and does not tie recombination to any morphologically ob-

servable stage of meiosis (i.e., chiasma formation at pachytene).

The copy-choice theory has two main weaknesses. Under this hypothesis, only the two new strands can become involved in a crossover, but the two original strands remain intact. As was discussed earlier, though, three-strand and four-strand double crossovers are known to occur. A second shortcoming of the hypothesis is the requirement that recombination must be linked to chromosome replication. Experiments, which we shall not detail, have demonstrated that recombination can occur in the absence of chromosome replication.

A later and more probable hypothesis of the mechanism by which recombinants occur was advanced by the cytologist C. D. Darlington in 1935. This theory is called the *breakage and reunion* theory. Darlington postulated that the twisting of the chromosomes during the four-strand stage of the first meiotic division produces stresses on the chromatids. If the stresses are great enough, they will break one chromatid or more. Should more than one chromatid break, the possibility exists that the broken end of one chromatid will unite with the broken end of a different chromatid. If the breaks and reunion involve sister chromatids, then no genetic consequence is anticipated. However, should the breaks and reunion involve paternal and maternal chromatids, a recombinant would result. The breakage-reunion hypothesis represents the best explanation, to date, to account for the formation of recombinants. Diagrams showing the stages involved in the two proposed hypotheses of the mechanism by which recombinants occur are shown in Fig. 7-12.

Although it was originally thought that mechanical stresses produced the breaks that must precede recombination, it now appears more probable that some enzymatic process is involved. The production of a linear bacterial chromosome during conjugation indicates that the ring structure of DNA may be broken without any known mechanical stress. Also, no known twisting of chromosomes around one another occurs during the recombination processes involved in transformation, conjugation, and transduction. When considering the reunion of broken DNA strands, regardless of how the breaks may have been produced, one must hypothesize an enzymatic process in the formation of the covalent bonds necessary to repair the broken sugar-phosphate backbone of the DNA molecule. The idea that enzymes are involved in the breakage and joining of preformed polynucleotide strands was first put forth by Howard-Flanders and Boyce in 1964. A corollary of this theory was the prediction that mutants would occur in which the enzymes involved in the recombination process were defective. A successful search for such mutants was reported by Clark and Margulies in 1965. They were

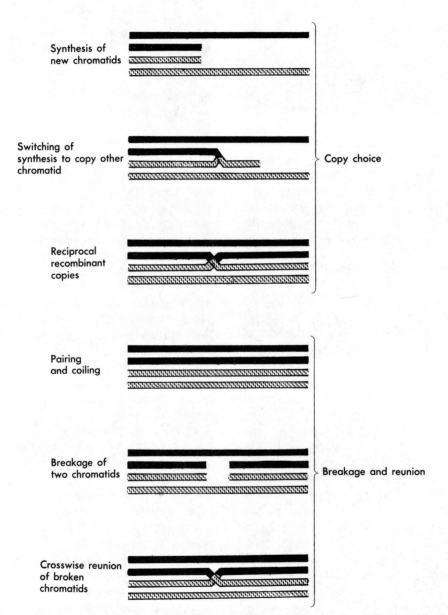

Fig. 7-12. Diagrammatic representation of two possible mechanisms of genetic recombination.

able to isolate two recombinationless (Rec⁻) mutants in an F⁻ strain of *Escherichia coli* K12. These mutants were able to conjugate with Hfr male cells. The conjugations included the transfer of the Hfr donor chromosomes to the Rec⁻ cells. However, the mutants were unable to form recombinants with the Hfr donor chromosomes. Subsequently, many other recombinationless mutants were discovered. What has been left unanswered is the question of what causes enzymatic breaks to occur in the DNA molecule.

Chromosome pairing

Our discussion of the mechanism by which recombination occurs must consider the pairing phenomenon (synapsis) in more detail. Recombination is an outcome of two processes: pairing and genetic exchange. Relative to the pairing process, two possible situations can be considered. In one, we can assume that the chromosomes, at synapsis, are equally close to one another all along their lengths *(continuous pairing)* and there is some low probability of genetic exchange, dependent on the amount of twisting and the types of breaks that result. The second possible situation assumes that the chromosomes, at synapsis, are not equally close to

one another all along their lengths *(discontinuous pairing)* and that there is a high probability of genetic exchanges which exists only in the paired regions. The number of paired regions is not specified nor must the chromosomes be paired at the same regions in every synapsis. These two possible pairing arrangements are diagrammed in Fig. 7-13.

The traditional view of what happens during meiosis is based largely on cytological evidence and has assumed that recombinants arise from crossing-over between chromosomes that are paired, point for point, along their entire lengths. Experimental work begun by R. H. Pritchard in 1955 on the phenomenon of negative interference in *Aspergillus,* which has also been shown to occur in other organisms, has shifted the balance of evidence in favor of discontinuous pairing as the main factor determining the formation of recombinants.

In *Aspergillus* no interference of any type is found between genes more than 0.5 map unit apart. However, over distances shorter than this a high degree of negative interference is observed, so that the frequency of double crossovers exceeds that expected by as much as 100 times. This negative interference is thus restricted to a very small re-

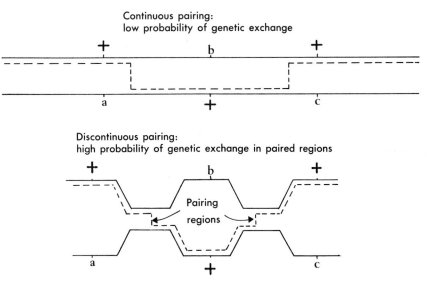

Fig. 7-13. Diagrams illustrating, respectively, continuous and discontinuous pairing.

gion of the chromosome. These results are interpreted to mean that (1) effective pairing between the chromosomes is restricted to very small regions of the chromosomes and (2) within these small regions the probability of recombination is high. Such an interpretation leads to the conclusion that there must be discontinuous pairing between homologous chromosomes at synapsis.

A second type of investigation that supports the concept of discontinuous pairing of synapsed chromosomes at meiosis comes from experiments on the transfer of genetic material from one bacterial cell to another. Of the three methods of such transfer (conjugation, transduction, and transformation), only the first two will be pertinent for our present discussion. In conjugation about $\frac{1}{5}$ to $\frac{1}{2}$ of the entire bacterial chromosome is transferred from donor to recipient. In transduction only about $\frac{1}{100}$ of the bacterial chromosome is transferred from donor to recipient. In *E. coli* strain K12 it is possible to transfer the same closely linked genes by either conjugation or transduction and to compare the results of recombination in the two systems. The three closely linked genes used are *thr* (threonine synthesis), *ara* (arabinose fermentation), and *leu* (leucine synthesis). The distance *thr-leu* is close to the maximum length of a transducible fragment. The diagrammatic representation of these three genes under a continuous-pairing versus discontinuous-pairing hypothesis is shown in Fig. 7-14.

If the pairing of homologous chromosomes is continuous, one would expect the frequency of recombination to be the same under conjugation or transduction. However, the closely linked donor genes appear to be more often separated by recombination in transduction than in conjugation. These findings are explainable on the basis of discontinuous pairing if it assumed that the two crosses are identical except for the disparate lengths of the two donor chromosome fragments. The shorter chromosome fragment, which was transferred by transduction, is restricted to pairing points close to the mutant genes. This results in a high frequency of recombination. The longer chromosome fragment, which was transferred by

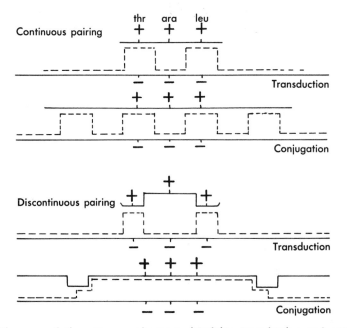

Fig. 7-14. Diagrams of the outcome of crosses involving transduction and conjugation, if pairing of the donor fragment of chromosome is (top) continuous or (bottom) discontinuous.

conjugation, can have pairing points outside the region bounded by the mutant genes. The crossovers that occur at the more distant points would not result in recombinations of the mutants being studied.

Molecular model of crossing-over

We have not yet discussed recombination in terms of the DNA molecule. For crossing-over to result in an exact exchange of equal and homologous chromosomal sections, it would appear necessary that the breaks must occur between the same corresponding two nucleotides in both strands of each of the DNA molecules. Any set of unequal breaks would involve additions or deletions of nucleotides in the resultant recombinants and would produce altered and probably nonfunctional proteins. However, it appears very unlikely that any set of breaks, whether of a mechanical or enzymatic origin, would occur at precisely the same internucleotide points in both strands of two different DNA molecules. A second problem that any theory on the molecular basis of crossing-over must consider is that no known attractive force exists that would tend to bring together the broken ends of homologous double-helical DNA molecules. The only attractive force known to operate on a DNA molecule in-

volves hydrogen bond formation between complementary regions of single-strand DNA chains.

A hypothesis of how crossing-over occurs, which takes into account the facts just mentioned, was proposed in 1963 by H. L. K. Whitehouse. According to this model, the steps involved in the process are shown in Fig. 7-15 and would be as follows:

1. Complementary nucleotide chains of opposite polarity and from homologous DNA molecules break at nearly, but not exactly, the same points (Fig. 7-15, *A*). Special enzymes, called *endonucleases,* are known to have this type of action. What is not known are the conditions that cause an endonuclease enzyme to nick a particular chain of a DNA double helix at a particular point.

2. The parts of the two broken chains, for reasons that are not at all clear, uncoil from their complementary chains. These single DNA chains are attracted to one another and unite by complementary base-pairing, thus forming a crossover molecule (Fig. 7-15, *B*).

3. DNA polymerase molecules begin to synthesize replacements for the chains that have crossed over, using in each case the unbroken strand of the respective DNA molecule as a template. This process is indicated by the broken lines in Fig. 7-15, *C,* and is essentially a DNA repair–type process.

4. For reasons totally unknown, the newly synthesized replacement chains are thought to also uncoil from their complementary chains. These newly synthesized single DNA chains are attracted to one another and unite by complementary base-pairing to form the other crossover molecule. This process is indicated by the broken lines in Fig. 7-15, *D*.

(An alternative to steps 3 and 4 would be a repetition of steps 1 and 2, involving the remaining two DNA strands of the paired chromatids.)

5. In either series of events, the last step, according to this model of crossing-over, would be the enzymatic excision of the

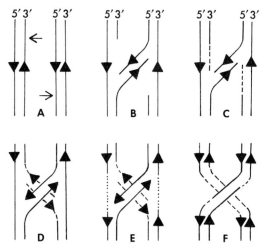

Fig. 7-15. A model illustrating crossing-over between two DNA molecules. (From Whitehouse, H. L. K. 1963. Nature **199:**1034-1040.)

nucleotides that remain unpaired in the parental strands. This is shown by the dotted line in Fig. 7-15, *E*, resulting in the typical chiasma in Fig. 7-15, *F*. The sugar-phosphate backbone of each crossover strand must also unite with the corresponding residual strand of the DNA double helix. Special enzymes, called *ligases*, are known to have this type of action.

• • •

Whitehouse's model for the molecular basis of crossing-over is thought to apply not only to the formation of recombinants during the meiosis of higher organisms but is also considered to be the mechanism involved in genetic recombination resulting from conjugation, transformation, and transduction. Geneticists generally recognize that the steps just outlined will not guarantee a foolproof crossover each time. Some of the "errors" that appear to occur will be discussed in Chapter 9.

Factors affecting recombination frequency

A number of factors can affect the frequency of recombination. These include the environmental conditions under which the experiments are conducted and the genotypes of the organisms studied. Among the environmental factors, it has been found that recombination frequency can be increased by a rise of temperature, an exposure to irradiation, or by injection of actinomycin D. The experiments with actinomycin D are very interesting, since this antibiotic inhibits the formation of mRNA. This would imply that regions that are functionally active are incapable of crossing-over. Such a situation would account for the unequal distribution of known mutants along the chromosome as measured by recombination frequency. However, such a hypothesis presents difficulties when the centromere region of the chromosome is considered. The centromere region shows very little crossing-over, yet is generally thought to be inactive in transcription.

Genetic factors affecting recombination may include a single gene, a whole chromosome, or the entire genotype. In *D. melano-*

gaster a single recessive gene *(c3G)* in the third chromosome, when homozygous, can suppress crossing-over in all the chromosomes of the females. With respect to the effect of a whole chromosome on recombination, it has been found that chromosomal rearrangements in the autosomes can increase the frequency of crossing-over in the X chromosomes of *Drosophila*. Such chromosomal rearrangements include deficiencies, translocations, and inversions. Their nature will be discussed later. For the present it is important to note that structural changes in one pair of chromosomes can alter the frequency of crossing-over in a different pair of chromosomes. This implies that the pairing (synapsis) and breaking of homologous chromosomes at meiosis are affected by the other chromosomes of the cell. We have already mentioned the effect of an entire genotype on recombination, in pointing out that in all species of *Drosophila* crossing-over is absent or exceedingly rare in males, while in the silkworm moth it appears to be absent in females.

One of the experimental methods used in studying the genetic control of a trait is to see if one can modify the trait in succeeding generations of a population through the selection and breeding of extreme pheno-

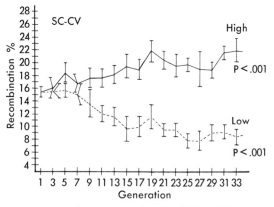

Fig. 7-16. Effect of selection for high and low recombination frequency in the *scute-crossveinless* region of the X chromosome in *D. melanogaster*. Data shown are the mean recombination values and the 95% confidence limits. (From Chinnici, J. P. 1971. Genetics **69:**71-83.)

types in every generation. An interesting example of this type of study, involving recombination frequency, was reported in 1971 by J. P. Chinnici. He examined the linkage relationship of two X-linked genes in *D. melanogaster,* namely "scute bristles" and "crossveinless wings" (Fig. 7-4). In his foundation stock of flies, he found that the genes had a recombination frequency of 15.4%. He then carried out a selection and breeding experiment for 33 generations in one group of flies for increased recombination frequency, and in a different group for decreased recombination frequency. His results are shown in Fig. 7-16.

It is evident that the experiment produced two lines of flies differing significantly in recombination frequency between the two genes. The relatively gradual response to selection in each line indicated a polygenic control of this trait. A series of "interline crosses" (i.e., high × low) yielded an F_1 that showed roughly intermediate recombination values. The intermediate nature of the F_1 values indicated a codominance relationship between the factors from the high and low lines. In further experiments, the effect of each chromosome of the selected lines on recombination frequency was studied. This is done by crossing individuals from each of the selected lines, separately, to flies from a laboratory "tester stock" that contain various homozygous mutations on each of its chromosomes. A pattern of matings is then followed that results in the isolation of each chromosome from the selected lines in a separate test stock. With such isolation, the effects of each chromosome on recombination frequency can be measured. All the chromosomes were found to contain genes that affect recombination frequency. We thus see that the linkage relationship between two genes can be affected by the rest of the individual's genotype. How this is accomplished is not known.

Somatic crossing-over

The discussion of linkage and recombination has thus far considered only gonadal cells. Crossing-over can also occur in somatic cells. The phenomenon is known to occur in a number of organisms, including *Drosophila* and maize. In the case of *Drosophila,* somatic crossing-over has been demonstrated as follows. A female is chosen who carries in one of her X chromosomes the

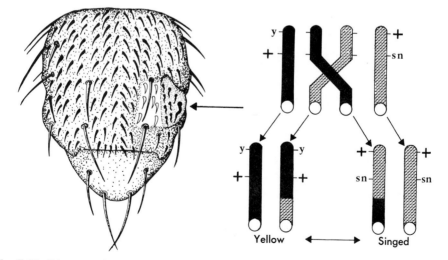

Fig. 7-17. Diagram of somatic crossing-over in *Drosophila melanogaster.* Left: Thorax of a fly heterozygous for the mutations yellow body *(y)* and singed bristles *(sn)* with a pair of twin spots in which the effects of these recessives are visible—yellow hairs and a yellow bristle in the left spot, singed hairs and bristles in the right spot. Right: Assumed mechanism of somatic crossing-over leading to the formation of the twin spots.

gene for singed bristles, *sn,* and in her other X chromosome the gene for yellow body, *y.* All other genes of her genotype are +. Such a female will have normal bristles and a gray body and have the genotype for these two genes as follows: + *sn/y* +. Should crossing-over occur between the two X chromosomes in meiosis-like fashion, as shown in Fig. 7-17, one would find, after a mitosis, one cell that is homozygous for *y* and another that is homozygous for *sn.* Their respective genotypes would be *y* + */y* + and + *sn/*+ *sn.* The tissue that resulted from further mitoses of these cells would appear as an abnormal "twin spot" on the body of the fly. One spot would be yellow in color but contain normal bristles, while its companion spot would be of normal color but contain singed bristles.

Studies on somatic crossing-over have revealed that its frequency can be controlled by both environmental and genetic factors. The environmental factors studied include temperature and irradiation, while the genetic factors studied include specific genes, a whole chromosome, or the entire genotype. These factors will be easily recognized as the same ones that affect meiotic crossing-over. Their effects on both types of recombination are similar, implying that the two processes are alike in nature but different in frequency due to the difference in amount of pairing of homologous chromosomes during the two processes. The consequences of somatic crossing-over may affect the individual both structurally and physiologically. Since most individuals are heterozygous for many genes, somatic crossing-over may result in the malfunctioning of parts of organs because the cells of the affected parts carry recessive deleterious genes in homozygous condition. This could seriously affect, and might even be fatal to, the organism concerned.

CHROMOSOMAL REARRANGEMENTS

Thus far we have considered the occurrence and consequences of chromosome breaks that resulted in the exchange of segments between homologous chromosomes. We shall now turn our attention to other types of chromosome breaks and their consequences.

Single chromosome breaks

If a single chromosome breaks and its ends reheal, there is no genetic indication that the event has occurred and there appear to be no detrimental consequences to the cell. However, if the chromosome forms chromatids after the break has occurred but before rehealing, the possibility exists for homologous portions of the two chromatids to fuse. Such a series of events is shown in Fig. 7-18. In stage D one of the healed fragments lacks a centromere *(acentric)* and as a result cannot become attached to the spindle formed at the next mitosis. The acentric fragment never becomes incorporated into the nucleus of either resultant cell at cell division and is eventually ejected from the cell. The other healed fragment contains two centromeres *(dicentric).* At cell division, both centromeres are attached to the spindle and are pulled in opposite directions. The chromosome between them forms a bridge. As anaphase of mitosis proceeds, the tension on the chromosome is usually sufficient to break the bridge. The broken chromosomes, each in its own cell, can then repeat the process of duplication, fusion, and breakage at the next cell cycle. In repeated breaking, each break is likely to be in a different position, and deletions and duplications of chromosome material will result. This will be in addition to the loss of chromosome material caused earlier by the formation of the acentric fragment.

Cells in which this type of healing of single chromosome breaks has occurred are eventually lost to the organism, as is the tissue derived from them. Should this occur in a fertilized egg, the organism will fail to develop normally and will usually die.

Two breaks in the same chromosome

Deficiency. When two different breaks occur in the same chromosome, two end fragments and a middle fragment are formed.

In healing, it is possible for the two end fragments to heal together, leaving the middle fragment as an acentric piece, as pictured in Fig. 7-19. The acentric piece, as mentioned earlier, will fail to be incorporated into any cells that result from subsequent mitoses. The healed segments will form a chromosome that is *deficient* for some of the genes normally found in it. The degree and type of damage to the descendant cells caused by the deficiency will depend on its size and importance and on the extent to which those genes present in the undamaged homologous chromosome are able in single dose to perform functions ordinarily carried out by the two genes. If the genetic imbalance thus produced is too great, the cell will die. Should the cell be a fertilized egg, the net effect will be the production of a dominant lethal. The dominant lethal effects produced by chemicals and radiations in the offspring of parents exposed to these mutagenic agents are most frequently due to deletions. However, if the deficiency is small enough, the cell can persist.

A deficiency is characterized by two effects, one genetic and the other cytological, both of which may be used to detect its presence. The genetic effect of a deficiency is called *pseudodominance*. It refers to the fact that if any recessive genes are located in the part of the homologous chromosome that corresponds to the deficiency, they will be expressed in the phenotype of the organism just as if they were dominant. If any of these corresponding genes tend to have a lethal effect in double dose, they will now be lethal in single dose. The cytological effect of a deficiency can be seen in cells in which a normal and a deficient chromosome are synapsed (e.g., gonadal cells during meiosis;

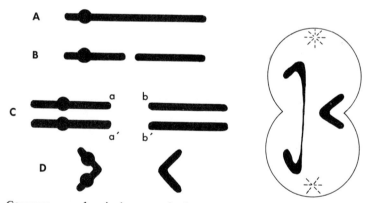

Fig. 7-18. Consequences of a single nonrestituting chromosome break (breakage-fusion-bridge cycle).

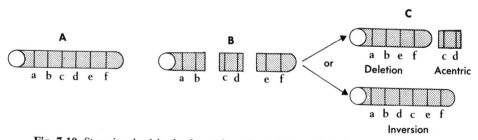

Fig. 7-19. Steps involved in the formation of a deficiency (deletion) or an inversion.

salivary gland cells of *Drosophila*). Since the chromosomes tend to pair gene by gene during synapsis, a *buckling* will occur in the synapsed chromosomes at the point of the deficiency. An example of this in *Drosophila* salivary gland cells is shown in Fig. 7-20.

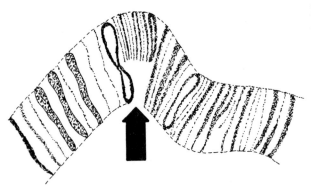

Fig. 7-20. Portions of salivary chromosomes of *Drosophila melanogaster* showing a deficiency of ten or more bands. (From Sinnott, E. W., Dunn, L. C., and Dobzhansky, T. 1958. Principles of genetics. McGraw-Hill Book Co., New York.)

The combination of pseudodominance and buckling of synapsed chromosomes can be used to determine the exact location of the gene showing pseudodominance. The genes that, on the basis of pseudodominance, are known to be missing in the deficiency must be located in the part of the chromosome that forms the buckle in the deficiency heterozygote. When, as a result of radiation or other mutagenic treatment, one obtains a large number of deficiencies in a particular chromosome, one can construct a so-called *cytological* map for the same genes for which percentages of crossing-over have yielded a *genetic* map. A comparison of these two types of maps shows that the genes are rather evenly distributed over the entire chromosome in the cytological map but not in the genetic map. This is shown in Fig. 7-21. As pointed out in the discussion of crossing-over, physiological factors undoubtedly exist that affect the rates of breakage and reunion of different segments of the chromosomes.

A number of human genetic diseases

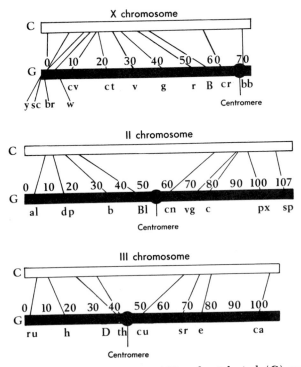

Fig. 7-21. Comparison of crossover or genetic (**G**) and cytological (**C**) maps of *Drosophila melanogaster*.

have been traced to deficiencies in chromosomes. Probably the best known example is the syndrome called *cri du chat*. As its French name imples, the crying of an affected infant resembles that of a suffering kitten. These individuals are found to have a deletion of a portion of the short arm of one of their chromosomes-5 (Fig. 6-4). The children exhibit severe mental and motor retardation and in most cases have to be institutionalized. A different human genetic disease, involving a less severe form of mental retardation which is associated with microcephaly, is found in individuals who have a deletion of part of the long arm of one of their chromosomes-18.

One form of Turner's syndrome involves a deletion of the short arm of one of the female's X chromosomes. These individuals have karyotypes that contain 46 chromosomes. However, one of their X chromosomes is found to be missing its short arm. The clinical manifestations of the affected persons are essentially identical to those of the ordinary XO females. The individuals lacking one of their X chromosome arms are found to be chromatin positive but exhibit an unusually small Barr body. From this evidence, geneticists have concluded that the Turner syndrome phenotype in XO females is actually a result of the lack of a short arm of one of the X chromosomes rather than a result of the loss of an entire X chromosome. It is to be expected that as research in this area progresses, more examples of deficiencies in human chromosomes will be discovered and their effects analyzed.

Deficiencies are not known to impart any beneficial effects on their carriers. In double dose, they are usually lethal, and as mentioned earlier even in single dose they may also be lethal. The occurrence of deficiencies appears to be a hazardous by-product of chromosome breakage and reunion. However, it must be realized that other types of chromosome breakage and reunion can have beneficial effects on their carriers insofar as they cause new combinations and arrangements of genes to occur.

Inversion. As discussed above, a deficiency may result from the union of broken segments of a chromosome. Another possible result of the reunion of two breaks occurring in the same chromosome is for all three chromosomal segments to join together again with the middle piece remaining in the middle but turned end-for-end. This situation is called an *inversion* and is also illustrated in Fig. 7-19. Inversion causes no addition or loss of chromosome material. Only the order of the genes has been changed. The rearrangement of gene order may have important consequences, since contiguous genes are often involved in related steps of the same biochemical pathway. All such alterations of gene function due to a change in location of the gene are called *position effects*. Position effects will actually appear as mutations and may be dominant, recessive, etc. These effects disappear once the gene is returned to its former location. This phenomenon can be substantiated by a cytological study of the altered portion of the chromosome.

Another consequence of inversion occurs in the individual who contains, at least in his gonadal cells, one chromosome with a paracentric inversion (not including the centromere) while its homologue is normal. During gamete formation, synapsis will occur. Since synapsis is a gene-by-gene pairing, one of the homologues will be twisted in the process and a *loop* will be formed. If a crossover occurs within the inverted section, then, on separation of the homologous chromosomes, acentric and dicentric chromosomes will be formed. This situation is illustrated in Fig. 7-22. Under such conditions, gametes will be formed with deficiencies and duplications of chromosome material. These gametes then give rise to chromosomally imbalanced offspring that are usually unable to survive. Since only chromosomes in which no crossing-over occurred survive, inversions appear as dominant mutations that result in the *suppression of crossing-over*.

We thus have, as in the case of deficiencies, both genetic and cytological evidence of

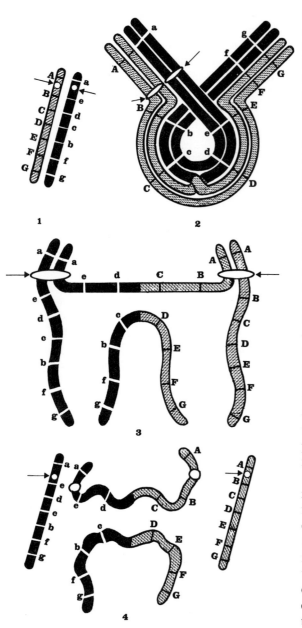

Fig. 7-22. Crossing-over in a heterozygote for a paracentric inversion. **1,** Two chromosomes differing in an inversion; **2,** diplotene stage of meiosis showing a chiasma inside the inverted section; **3,** anaphase of the first meiotic division, showing a chromatid bridge and an acentric fragment; **4,** outcome of meiosis—two noncrossover chromosomes with normal gene complements and (in the middle) dicentric and acentric chromosomes. Arrows indicate the centromeres. (From Sinnott, E. W., Dunn, L. C., and Dobzhansky, T. 1958. Principles of genetics. McGraw-Hill Book Co., New York.)

the occurrence of an inversion. The genetic evidence will be the suppression of crossing-over and possibly the appearance of a mutation-like position effect. The cytological evidence will be the formation of a loop whenever synapsis of chromosomes occurs in an inversion heterozygote.

Before leaving this consideration of inversions, we should ask whether any advantage to the individual and the species results from inversions. The possibility exists that a position effect, since it acts as a mutation, may be beneficial to its carrier. However, this is usually not the case; most mutants tend to be deleterious to their carriers. On the other hand, the apparent suppression of crossing-over in inversion heterozygotes can theoretically have important beneficial effects. Suppression of crossing-over prevents the separation of those genes in the inverted section of the chromosome except to recombine with genes in a similarly inverted section of a homologous chromosome. The net effect is to bind the genes of the inversion together to form a "super-gene." This permits beneficial complexes of genes to form, through mutation, within the inverted section without periodic disruption due to crossing-over at meiosis. The apparent suppression of crossing-over, to be sure, limits the number of new combinations of genes that can be formed in each generation, and this may be considered as an evolutionary drawback. In addition, this apparent crossover suppression is at the cost of 50% of the gametes (i.e., those containing acentric and dicentric chromosomes) formed from cells in which crossing-over has in fact occurred. We, therefore, have a situation that can have both deleterious and beneficial effects. As it turns out, only certain species (e.g., *Drosophila*) have evolved inversion systems to any great extent and then only with certain accompanying changes in the normal pattern of meiosis. The alterations of the normal meiosis pattern have included the suppression of crossing-over in the male (as mentioned earlier) and the segregation of the crossover by-products into the polar bodies of the eggs in the female.

Two breaks in nonhomologous chromosomes

Breaks may occur in two chromosomes that are not homologous. Should the segments from the nonhomologous chromosomes undergo a reciprocal interchange, the result is called a *reciprocal translocation*. In the joining of chromosome segments the formation of acentric and dicentric fragments may again occur, and the situation discussed above will be repeated. However, should the chromosomes heal in such manner that there is a balanced set of chromosomes, with each chromosome containing a single centromere, then we have a situation with important genetic consequences to consider. The condition is pictured in Fig. 7-23. Stage A shows the normal arrangements of two sets of chromosome pairs, pairs arbitrarily called III and IV. Stage B illustrates the same pairs with a single break in one homologue of each pair. Stage C pictures the reciprocal interchange of fragments and healing. The chromosome content of the cell is still complete, and each chromosome contains but one centromere. Should the reciprocal translocation occur in a fertilized egg or, at a minimum, in an embryonic cell whose descendant cells will form the gonads,

we will have the situation in stage D. Stage D shows the possible dispositions of the two chromosomal pairs at gamete formation after homologous chromosomes have separated. The two bottom combinations of chromosomes will contain duplications and deficiencies. Offspring developing from these two bottom combinations will contain genetic imbalances and will die in early embryonic stages if the genetic imbalances are large. The top combination is perfectly normal, and the offspring will develop normally. The gamete containing chromosomes (2) and (4) will contain a balanced set of chromosomes but will still carry the translocation. Offspring developing from a fertilization involving this gamete will be perfectly normal. However, in this and each succeeding generation, the meiotic process outlined above will be repeated, with the production of chromosomally balanced and unbalanced gametes.

The consequences of this situation will be the elimination of 50% of the offspring of any mating in which one of the parents has a reciprocal translocation, involving critical genes, between any two chromosomes. The individual involved is *semisterile*. Of the 50% of the offspring that do survive, half

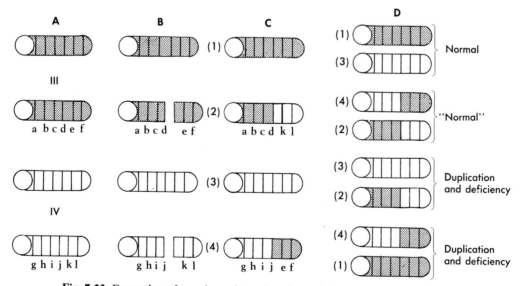

Fig. 7-23. Formation of a reciprocal translocation and its consequences at meiosis.

will contain the translocation and will also be semisterile. Thus a reciprocal translocation occurring in a fertilized egg will appear as a dominant mutation causing semisterility. Its true nature can only be discerned from other evidence to be discussed. Another effect of a translocation is a *change of linkage groups* for those genes located in the translocated segment. Still another possible result of a translocation is the appearance of a mutant due to *position effect* just as was possible with inversions. These three consequences of a translocation represent the genetic evidence that the event has occurred. On the cytological side, we find the formation of a *cross-shaped configuration* at the pachytene stage of meiosis or wherever chromosomes are synapsed. At the subsequent metaphase and anaphase of the first meiotic division, the translocated chromosomes will form a ring. This is shown in Fig. 7-24.

We have thus far considered only reciprocal translocations. It is, of course, possible that a simple translocation will occur that will transfer a segment of one chromosome to a nonhomologous chromosome. Here we get, at meiosis, the formation of some gametes with a deficiency and the formation of other gametes with a duplication. Again, the nature and extent of the abnormality will determine the fate of the offspring that are produced by these gametes.

In our discussion of Down's syndrome in Chapter 6, it was pointed out that these individuals are trisomics for chromosome 21 and their karyotypes contain 47 chromosomes. However, a number of persons have been found who exhibit all the characteristics of Down's syndrome but whose karyotypes contain 46 chromosomes. An analysis of the ideograms of these persons reveals that they have two normal chromosomes 21, one normal chromosome 14 or 15, and one longer than normal chromosome 14 or 15 that appears to represent a fusion of one of these chromosomes with at least a part of a third chromosome 21. This condition is usually referred to as "translocation Down's syndrome-15/21." However, one must be cautioned that some reports indicate that it may be chromosome-14 which is involved

in most cases of this type of translocation chromosome.

A study of the parents of these affected individuals revealed that although the parents exhibited a normal phenotype, one of them did, in some cases, have a karyotype that contained only 45 chromosomes. These persons appeared to be monosomic for chromosome 21. However, one of their chromosomes was the translocation chromosome 15/21, thus giving them a complete or al-

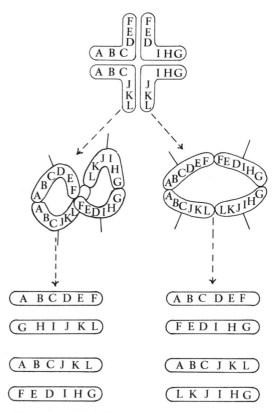

Fig. 7-24. Meiosis in a translocation heterozygote, and gametes formed as a result of it. At the top is the cross-shaped configuration formed at the pachytene stage of meiosis. The second row shows the twisted ring and the open-ring configuration formed by the chromosomes involved at meiotic metaphase. In the two lower rows are the chromosomal complements in the four types of gametes formed; only the two types on the left contain normal sets of genes; each of the other two types contains some genes in duplicate (duplication) and fails to carry some genes (deficiency). (From Sinnott, E. W., Dunn, L. C., and Dobzhansky, T. 1958. Principles of genetics. McGraw-Hill Book Co., New York.)

most complete genome. For reasons that are not at all clear, in most cases it is the mother who carries the aberrant chromosome. In these families, regardless of who carries the atypical chromosome, the production of Down's syndrome children is not mother-age–dependent. In fact, in the 45-chromosome karotype, the translocation chromosome 15/21 acts as the equivalent of a dominant gene for the production of Down's syndrome children.

Any consideration of a chromosomal anomaly must include a discussion of its value to the individual and to the species. The most deleterious effect of a reciprocal translocation is the semisterility it causes. As in the case of inversions, the evolution of reciprocal translocations in a species must carry with it some severe modification of the normal developmental pattern. An example of a species that has evolved reciprocal translocations to a great extent is the evening primrose, *Oenothera*. This was the plant whose variability caused de Vries to propose his mutation theory. The plant is self-fertilizing in nature where it exists in a number of pure breeding strains, each having a characteristic phenotype. The diploid number of its chromosomes is 14, and in some of the strains all the known mutant genes behave as if they were contained in a single linkage group. A study of the meiosis of such strains reveals the formation of a single ring at the metaphase of the first meiotic division, containing 14 chromosomes. This would imply that all 7 paternal chromosomes are united by reciprocal translocations and that the same is true of all 7 maternal chromosomes. In this type of situation, the gametes produced by an individual would be identical to those which united to form it. As stated above, reciprocal translocations can produce semisterility. In the case of *Oenothera,* this is confirmed by an examination of the flowers, which shows that one half of the ovules regularly fail to produce seed when self-fertilized. The evening primrose must therefore be a permanent heterozygote, and those zygotes that are homozygous either for all paternal or all maternal chromosomes must be lethal. Any arrangement whereby both homozygous types die and only the heterozygous type survives is called a *balanced lethal system*. The balanced lethal system is important in population studies and will be considered further in Chapter 13.

Duplication

The final type of chromosomal rearrangement that we shall consider is illustrated in *Drosophila* by the dominant X-linked gene called *Bar*. Its chief effect is to reduce the number of facets in the compound eye, leaving a rather narrow band, or "bar," of ommatidia. Another allele at this locus is *Ultra-Bar*. It is dominant to both the *Bar* and wild type alleles and reduces the number of facets in the compound eye still further. An examination of the salivary gland chromosomes shows that *Bar* is actually a *duplication* of some six disks of the section of the X chromosome labeled as 16A, while *Ultra-Bar* is a *triplication* of this same region. The salivary gland chromosomes for normal, Bar, and Ultra-Bar characteristics are shown in Fig. 7-25.

With the discovery of the chromosomal nature of Bar and Ultra-Bar, it was realized that one could get various arrangements of the duplicated and triplicated regions of section 16A. Some of these are shown in Fig. 7-26. The figures show that the addition of 16A sections results in a progressive reduction in the number of facets in the compound eye of the fly. Illustrations C and D demonstrate the action of position effect. In both C and D, the fly contains four sections of 16A. However, in C the four sections are distributed equally in both X chromosomes; in D, one X chromosome contains three sections and the other X chromosome contains one. The eye is noticeably more reduced in the latter case than in the former.

Karyotypes involving duplications of chromosome material have been found in human beings. One of the more dramatic situations involves the duplication of an entire chromosome arm, with the result that the chromosome is metacentric and contains two identical arms that are mutually homologous. This type of anomalous chromosome

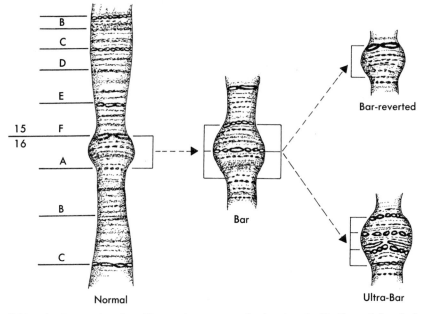

Fig. 7-25. The Bar region in salivary chromosome I, showing duplication of bands in *Bar* *(B/B)* and *Ultra-Bar (BB/BB)*.

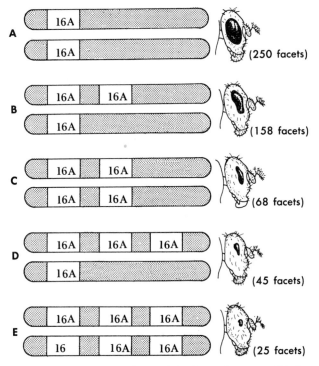

Fig. 7-26. Different arrangements of section 16A in *D. melanogaster* X chromosome and resulting phenotypes. **A,** Wild type female. **B,** Bar female, heterozygous. **C,** Bar female, homozygous. **D,** Ultra-Bar female, heterozygous, showing position effect. **E,** Ultra-Bar female, homozygous.

is called an *isochromosome*. It is assumed to have arisen as a result of a transverse splitting of the centromere with a loss of one arm and a duplication of the remaining arm of the chromosome. An example of this type of chromosomal rearrangement is found in some Turner's syndrome females. These individuals have 46 chromosomes that include one normal X chromosome and one longer than normal X chromosome. On the basis of their identical autoradiographic labeling patterns (Chapter 2), the two arms of the longer than normal X chromosome are considered to be the mutually homologous arms of an isochromosome. These individuals exhibit unusually large Barr bodies and fall into the category of chromatin-positive Turner's syndrome females. The phenotypic characteristics of the Turner's syndrome must be the result of a lack of the short arm of the isochromosome. The above finding supports the conclusion reached from the study of deficiencies, as discussed earlier.

SUMMARY

We have reviewed in this chapter the relation of genes to chromosomes. This has included linkage, crossing-over, and genetic maps. Incorporated into our discussion has been a consideration of the cytological demonstration of crossing-over, the stage at which crossing-over occurs, the nature of recombination, and the factors affecting recombination frequency. There followed a survey of chromosomal rearrangements such as deficiency, inversion, translocation, and duplication. Included was an analysis of the beneficial and deleterious effects of the various chromosomal rearrangements to their carriers and the evolutionary consequences of the chromosomal rearrangements. In the next chapter we shall turn our attention to the cytoplasm and the genetic material contained in it.

Questions and problems

1. Define the following terms:
   a. Linkage group
   b. Parentals
   c. Recombinants
   d. Chiasma
   e. Interference
   f. Coefficient of coincidence
   g. Ascospore
   h. Tetrad analysis
   i. Chromatid interference
   j. Discontinuous pairing

2. What is the relationship between crossing-over and recombination?

3. What are the salient features of the "copy choice" mechanism of recombination? of the "breakage and reunion" mechanism of recombination?

4. Using the radioactive isotope tritiated thymidine, Taylor conducted an experiment designed to shed light on the manner of replication of chromosomes of higher organisms. In one series of autoradiographs, Taylor found that of six chromosome pairs, two displayed segregation of label from nonlabel whereas four chromosome pairs showed evidence of label in segments of both homologues. Explain.

5. The recessive genes for roughoid eyes *(ru)*, curled wings *(cu)*, and ebony body color *(e)* have been calculated as occupying relative map positions of 0.0, 50.0, and 70.7, respectively, on the third chromosome of *Drosophila melanogaster*. A homozygous wild type female is mated to a male, mutant for the above three characters. Assuming no appreciable interference:
   a. Diagram all possible parental gametes.
   b. Diagram all possible F₁ gametes.
   c. Diagram all possible genotypes occurring in the F₂ generation.

6. A male *Drosophila*, having cut wings *(ct)*, vermilion eyes *(v)*, and a yellow body *(y)* is crossed to a female, heterozygous with respect to these three characters. The following classes of phenotypes are observed in the progeny:
   a. Yellow body — 440
   b. Yellow body, vermilion eyes — 50
   c. Wild type — 1780
   d. Vermilion eyes, cut wings — 470
   e. Vermilion eyes — 300
   f. Yellow body, vermilion eyes, cut wings — 1710
   g. Yellow body, cut wings — 270
   h. Cut wings — 55
   (1) Diagram the genotype of each class of flies.
   (2) Which classes represent the parental types?
   (3) Which classes reflect the occurrence of double crossover events?
   (4) Which classes reflect the occurrence of single crossover events?

7. Refer to question 6.
   a. Construct the genetic map of the three loci involved, indicating both map distance and the correct sequence of genes.
   b. Which map distance (of the ones calcu-

lated in a) does not reflect the actual distance as it exists on the *Drosophila melanogaster* chromosome? Why? What is a better estimate of the map distance between the two genes in question?

c. What is the coefficient of coincidence involved here?

d. Interpret your answer in c.

8. Of what value is *Neurospora* in genetic analysis?

9. In *Neurospora crassa* the genes for arginine synthesis *(arg)*, spore pigmentation *(alb)*, and lysine synthesis *(lys)* are located on the first chromosome. Mating is effected between a strain of mating type *a*, which is mutant for the above three characters, and a strain of mating type *A*, which is wild type for the above three characters.

a. Spores isolated in order from one particular ascus type showed the following characteristics when cultured on various media:

Spore position in ascus	Behavior on various media
1 & 2	Colored spores that formed mycelia on media lacking arginine but could not grow when lysine was missing
3 & 4	Colorless spores that formed mycelia on medium lacking arginine but could not grow when lysine was missing
5 & 6	Colored spores that formed mycelia on medium lacking lysine but could not grow when arginine was missing
7 & 8	Colorless spores that formed mycelia on medium lacking lysine but could not grow when arginine was missing.

b. Spores isolated in order from a second ascus type showed the following characteristics when cultured on various media:

1 & 2	Colored spores that grew on minimal medium
3 & 4	Colorless spores that grew on minimal medium
5 & 6	Colored spores that needed both arginine and lysine for growth
7 & 8	Colorless spores that needed both arginine and lysine for growth

c. Spores isolated in order from a third ascus type showed the following characteristics when cultured on various media:

1 & 2	Colored spores that needed lysine but not arginine for growth
3 & 4	Colorless spores that grew on minimal medium
5 & 6	Colored spores that needed arginine but not lysine for growth
7 & 8	Colorless spores that needed both arginine and lysine for growth

d. Spores isolated in order from a fourth ascus type showed the following characteristics when cultured on various media:

1 & 2	Colored spores that grew on minimal medium
3 & 4	Colorless spores that grew on medium lacking arginine but could not grow when lysine was missing
5 & 6	Colored spores that needed both lysine and arginine for growth
7 & 8	Colorless spores that needed arginine but not lysine for growth

Diagram the chromatid exchanges involved in forming each of the four tetrad types described above.

10. Refer to question 9.

a. In which type are no parental types recovered?

b. During which division—first or second—did the genes for spore color segregate in the second type of ascus described?

c. Which ascospore type would be excessively present in the population if positive chromatid interference were operating?

d. Which would be excessively present if negative chromatid interference were operating?

11. In *Neurospora crassa* mapping can be accomplished by assessing how far each gene is from the centromere. If the particular locus is so close to the centromere that no crossover can occur between the centromere and the gene, alleles *A* and *a* will segregate at the first division, and subsequent mitosis will yield ordered tetrads of two types, *AAAAaaaa* or *aaaaAAAA*, depending on how the haploid chromosomes were originally aligned with respect to the spindle. If, however, the distance between the particular gene and the centromere is large, a crossover can occur, delaying segregation of alleles *A* and *a* until the second division. In this case, subsequent mitosis will yield ordered tetrads of four types: *aaAAaaAA* or *AAaaAAaa,* and *AAaaaaAA* or *aaAAAAaa.*

a. What kind of information is necessary for the construction of a linkage map?

b. In order to compare the linkage distances between a gene and its centromere in *Drosophila* and *Neurospora*, the value obtained for *Neurospora* must be divided by 2. Why?

12. A cross is made between a *Z* strain of *Neurospora* and a *z* mutant strain. The following types of asci—with respect to spore order—are recovered among the progeny of this mating:

Spore order				Asci
1 & 2	3 & 4	5 & 6	7 & 8	
z	Z	Z	z	7
z	z	Z	Z	105
Z	z	z	Z	5
Z	z	Z	z	12
Z	Z	z	z	110
z	Z	z	Z	14

How far is gene *Z* from the centromere?

13. Recombination in bacteria can be mediated by transduction. In this process, genes of a donor bacterial strain are incorporated into the bacteriophage during the phage's reorganization within the host. When the phage is liberated from its host, it can introduce the genomic fragments into a second bacterial strain on infection. Homologous genes, so introduced, can pair with alleles of the acceptor strain and undergo recombination events. It is known that the amount of genetic material capable of being contained within the bacteriophage is equal to about 1.5 map units on the bacterial chromosome. How can this be applied to mapping the genes of the bacterial chromosome?

14. Recombination in *E. coli* can be mediated by conjugation. In this process, bacteria of a particular mating type, Hfr, readily transfer their chromosomes to bacteria of an acceptor strain, F⁻. It is known that the chromosomes of Hfr strains are transferred in a characteristically oriented manner. The rate of transfer of the chromosome is constant, and the closer a particular allele is to the end that is transferred first, the more likely it is to be introduced into the recipient strain. Once introduced, it can pair with the recipient's allele and undergo a recombination event. How would you apply the above information toward devising a technique for mapping the genes of the *E. coli* chromosome?

15. Define the following terms:
    a. Acentric fragment   f. Pseudodominance
    b. Dicentric fragment   g. Inversion
    c. Deficiency   h. Reciprocal translocation
    d. Duplication
    e. Balanced lethal   i. Position effect
    system   j. Cytological map

16. Of what aid are cytological maps in the study of chromosomal abberations?

17. *Drosophila* females are treated with x rays and are, as a result, found to contain the following reciprocal translocation involving chromosomes "A" and "B":

Before		After	
ABC DE	LMN OP	ABC NML	ED OP
(A)	(B)	(A)	(B)

This stock is mated to wild type males. Diagram the synaptic configuration assumed by chromosomes A and B.

18. Refer to question 17.
    a. Diagram the different classes of gametes that could be produced by individuals heterozygous for the reciprocal translocation.
    b. What percent of the gametes would be expected to produce viable offspring when mated to normal (i.e., untreated) flies?

19. How could the following aberrations relate to the genetic phenomenon of position effect?
    a. *Bar* eye gene of *Drosophila*.
    b. A reciprocal translocation involving autosomes 7 and 9 in maize.
    c. An inversion in the single linkage group of *E. coli*.

20. In a tester stock of *Drosophila* known as the *Basc* stock, females are homozygous for Bar eyes and apricot eye color. When these females are crossed to wild type males, the F₁ consists of only one class of males, *Basc,* and only one class of females, wide-Bar-eyed and otherwise wild type. When the F₁ females are crossed to their *Basc* brothers, half the sons are wild type and half the sons are *Basc*. Half the daughters are heterozygous Bar-eyed (wide-Bar-eyed) and half are *Basc*.
    a. How would you account for the absence of any recombinants having normal-sized, apricot-colored eyes?
    b. How could the *Basc* stock be of value in detecting an X-linked recessive lethal carried by P₁ sperm?

References

Beadle, G. W. 1946. Genes and the chemistry of the organism. Amer. Sci. **34**:31-53. (Contains a description of the life cycle and genetics of *Neurospora*.)

Belling, J. 1931. Chromomeres in lilaceous plants. Univ. Calif. (Berkeley) Pub. Botany **16**:153. (Original statement of the "copy choice" mechanism of recombination.)

Bridges, C. B. 1936. The Bar "gene," a duplication. Science **83**:210-211. (Discovery of the nature of the Bar gene series.)

Chinnici, J. P. 1971. Modification of recombination frequency in *Drosophila*. II. The polygenic control of crossing over. Genetics **69**:85-96.

(Analysis of the effect of each chromosome on recombination frequency.)

Cleland, R. E. 1962. The cytogenetics of *Oenothera*. Advances Genet. **11**:147-237. (Description of the ring formation seen during synapsis.)

Creighton, H. B., and McClintock, B. 1931. A correlation of cytological and genetical crossing-over in *Zea mays*. Proc. Nat. Acad. Sci. U. S. A. **17**:492-497. (Direct evidence that genetic recombination involves the physical exchange of chromosomal material.)

Ford, C. E., and Clegg, H. M. 1969. Reciprocal translocations. Brit. Med. Bull. **25**:110-114. (Review of the field as it relates to man.)

Jacobs, P. A. 1969. Structural abnormalities of the sex chromosomes. Brit. Med. Bull. **25**:94-98. (Review of the various types of chromosomal rearrangements found in human sex chromosomes.)

Lewis, E. B. 1950. The phenomenon of position effect. Advances Genet. **3**:73-116. (Review of the types of position effects known.)

Meselson, M., and Weigl, J. J. 1961. Chromosome breakage accompanying genetic recombination in bacteriophage. Proc. Nat. Acad. Sci. U. S. A. **47**:857-868. (Direct evidence for the "breakage and reunion" mechanism of recombination.)

Patterson, J. T., Stone, W., Bedichek, S., and Suche, M. 1934. The production of translocations in Drosophila. Amer. Natur. **68**:359-369. (Discussion of the translocation-type of chromosomal aberration.)

Pritchard, R. H. 1955. The linear arrangement of a series of alleles in *Aspergillus nidulans*. Heredity **9**:343. (On the phenomenon of negative interference.)

Stern, C. 1936. Somatic crossing-over and segregation in *Drosophila melanogaster*. Genetics **21**:625-630. (Original description of the phenomenon of somatic crossing-over as it pertains to the "yellow spot" discussion in this chapter.)

Sturtevant, A. H. 1913. The linear arrangement of 6 sex-linked factors in *Drosophila* as shown by their mode of association. J. Exp. Zool. **14**:43-59. (Description of the first successful genetic mapping.)

Whitehouse, H. L. K. 1970. The mechanism of genetic recombination. Biol. Rev. **45**:265-315. (An excellent review of the various theories of crossing-over.)

8 Extrachromosomal inheritance

⚮ ⚮ In our discussion so far, we have
considered only those genes that
⚮ ⚮ are associated with chromosomes.
Chromosomal, or, as it is also called, Mendelian, inheritance is characterized by the fact that genes from the male and female parents contribute equally to the genetic constitution of the progeny. Thus, in Mendelian inheritance, reciprocal crosses between parents of different homozygous genotypes will yield offspring of identical phenotype. In *extrachromosomal inheritance,* male and female parents do not make equal contributions to the genetic constitution of the progeny; hence, reciprocal crosses yield different results. Extrachromosomal inheritance is defined and detected as an exception to Mendelian inheritance. This is a negative type of definition, and in practice the breeding tests for extrachromosomal inheritance are largely step-by-step eliminations of all possible chromosomal explanations of the observed results. As will be detailed below, the study of extrachromosomal inheritance is further complicated by the fact that there are many different types of non-Mendelian inheritance.

INHERITANCE OF COLOR
IN PLASTIDS

The cytoplasmic inheritance of color in *plant plastids* was first described by C. Correns in 1909 in the four-o'clock plant, *Mirabilis palapa*. Plastids are self-reproducing bodies that are distributed throughout the cytoplasm of many plant cells. They may number anywhere from 1 in the marine alga *Micromonas* to 10 in *Euglena* to about 100

in the cells of the advanced plants. There are various types of plastids, each with a separate function. We shall be concerned with two kinds of plastids: the *chloroplast* (green plastid), whose chlorophyll is involved in the photosynthetic process, and the *leukoplast* (white plastid), which results when the chloroplast permanently loses its chlorophyll. The division of the cytoplasm at each mitosis is such that both resultant cells contain plastids. However, the cells need not receive the same number of plastids nor need they receive some of every type of plastid present in the original cell.

An examination of the four-o'clock plant shows that while many of the leaves are completely green, others are mottled with green and white areas. Plastids in the cells of the white areas of the leaves lack chlorophyll, but at least some of the plastids in the cells of the green areas contain chlorophyll. Sometimes whole branches are white, while others are green, and still others are variegated with green and white patches. Although the leukoplasts of the white parts of the plant are incapable of carrying on photosynthesis, the white parts survive by receiving nourishment from the green parts. Correns found that flowers on green branches produce only green offspring, regardless of the phenotype of the pollen parent, while flowers from the white branches produce only white seedlings regardless of the pollen parent. The plants developing from these white seedlings will die because they lack chlorophyll and cannot carry on photosynthesis. Correns further found that flowers from the variegated branches yield a mixed progeny of green, white, and variegated plants in widely varying ratios. His findings illustrate the typical pattern of one type of extrachromosomal inheritance—that in which the offspring resemble the phenotype of the female parent and in which the phenotype can be correlated with a nonchromosomal particle.

A study of the meiosis of the egg cell of the four-o'clock plant reveals that the cytoplasm of the egg cell contains plastids, as do most other cells. If the egg cell is derived from green plant tissue, its cytoplasm will contain colored plastids; if derived from white plant tissue, its cytoplasm will contain white plastids; if derived from variegated tissue, its cytoplasm may contain colored plastids only, white plastids only, or a mixture of colored and white plastids. A study of the meiosis of the pollen cell shows that it contains very little cytoplasm and is, in most cases, completely devoid of plastids. Without plastids, the pollen cannot affect this aspect of the offspring's phenotype. In species with distinct male and female sexes, this type of inheritance is also called *maternal inheritance*. In species where the particle can be definitely located in the cytoplasm, this type of inheritance is also called *cytoplasmic inheritance.*

The above type of inheritance has been shown to occur in many plants, including barley, maize, and rice. In a number of cases, it has been found that chromosomal genes can either produce or stimulate the production of plastid mutations. A well-studied case in maize was reported by M. M. Rhoades in 1943. Certain corn plants have mosaic leaves with stripes of green and white. The inheritance pattern follows that observed in the four-o'clock plants. In one cross of two all-green corn plants, some progeny were found that were green-and-white striped. The striped plants were shown to be homozygous for a recessive chromosomal gene, *iojap (ij),* for which their parents were heterozygous. Since the recessive homozygote was capable of producing both types of plastids, the colorless plastids could

not have been the result of interference by *ij/ij* with the biosynthetic pathway leading to the production of chlorophyll. The simplest hypothesis for this effect is that in the homozygous *ij/ij* individual, an extranuclear gene, located in the plastid and essential for the production of chlorophyll, is induced to mutate to a form that prevents the formation of the green pigment. One further point must be made. The induced mutation must occur independently in a number of plastids of the egg cell before it will result in a green-and-white striped leaf pattern in the corn plant. The crosses that were made are shown in Fig. 8-1.

If plastids are self-duplicating structures that possess their own hereditary factors, it should be possible to isolate them and demonstrate that they contain a complete genetic system. Such a system would, as discussed in Chapter 1, have to provide for DNA replication, DNA-RNA transcription, and RNA-protein translation. This system has, in fact, been found. One of the technical problems was to obtain plastids free from contamination by nuclear material and also by bacteria that can normally be found in the cytoplasm. This was done by Gibor and Izawa in 1963 with the green alga *Acetabularia*. Research in a number of different laboratories subsequently demonstrated that the chloroplasts contain DNA, RNA, and ribosomes. Analysis by CsCl density gradient centrifugation showed the presence of two types of DNA molecules: (1) a molecule of the same weight, and later found to have the same base composition as, nuclear DNA; and (2) a lighter molecule, later shown to have a base composition different from that of nuclear DNA. This lighter DNA has been called *satellite DNA*. The quantity of DNA per chloroplast in this alga was estimated to be about 10^{-16} gram, a quantity that is of the same order of magnitude as that found in rickettsial type organisms and in even-numbered T bacteriophages. This amount of DNA is sufficient to code for over 100 different polypeptides of molecular weight about 15,000. Estimates in different organisms place the total amount of chloro-

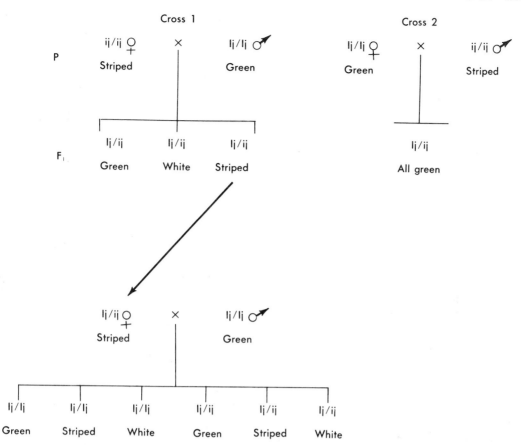

Fig. 8-1. Results of reciprocal crosses between striped (iojap) and normal (green) corn plants and backcross of F₁ striped by normal green.

plast DNA at 1% to 4% of the total cell DNA.

Evidence that *DNA replication* occurs in the plastids is indirect. Our discussion of the *iojap* condition in corn showed that, within a single cell, chloroplasts and leukoplasts will grow and divide, each producing more of its own type. Under such conditions, the DNA of the different plastids could not be produced by nuclear DNA, which would be capable of producing only one form of plastid DNA. Similar evidence has been obtained by Gibor and Granick in 1962 from the green alga *Euglena* in which it is possible to irradiate the cytoplasm with ultraviolet light while shielding the nucleus and vice versa. The ultraviolet (UV) light, when directed at the cytoplasm, causes irreversible bleaching of the plastids, and the irradiated *Euglena* individuals in successive generations never recover the ability to form chloroplasts. The effective UV wavelength was 260 mμ (nm.), suggesting that nucleic acids are involved in the bleaching effect of UV. The fact that one can cause all the chloroplasts (about 10 per *Euglena*) to mutate to leukoplasts indicates that the chloroplasts are highly mutable. Only irradiation of the cytoplasm caused irreversible mutation of the plastids, which then grew and divided as tiny proplastids from generation to generation. Again the reasoning is that if the UV-sensitive DNA units of the cytoplasm had been formed by the nucleus, one might expect that new DNA units, also formed by the nucleus, would replace the damaged DNA units of the cytoplasm and thus prevent bleaching. Since this did not happen, the inference is that the DNA of the plastids is formed from preexisting plastid DNA.

Other evidence that the DNA in plastids replicates to make more DNA is suggested by autoradiographic studies such as were described in Chapter 2. The plastids in *Spirogyra* cells and in *Euglena* have been found to incorporate tritiated thymidine. It is assumed that only DNA involved in replication would incorporate thymidine, since the compound appears to have no other role in cell metabolism.

Experiments on *DNA-RNA transcription* have usually involved isolated chloroplasts. As stated earlier, investigations showed that plastids contain DNA, RNA, and ribosomes. An additional finding was that in *Euglena* a unique RNA appears simultaneously with the differentiation of proplastids into chloroplasts. This unique RNA has a higher ratio of A + U to G + C than does the cytoplasmic RNA. As would be expected from this fact, the DNA of the chloroplasts of *Euglena* is characterized by a higher ratio of A + T to G + C than is nuclear DNA. These findings are compatible with the hypothesis that plastid DNA acts as a template for RNA synthesis. In 1964 J. T. O. Kirk reported on his studies on RNA synthesis in chloroplast preparations of the broad bean. He found that treatment of the chloroplasts with actinomycin D or with DNase inhibited RNA synthesis. He further found that the RNA polymerase of the chloroplast was distinguishable from the RNA polymerase of the nucleus because it was inhibited by Mn^{++} whereas the nuclear RNA polymerase was not.

Investigations of *RNA-protein translation* also have involved isolated chloroplasts. In 1963 Eisenstadt and Brawerman reported the incorporation of radioactive amino acids by isolated chloroplasts of *Euglena* and also by a system containing ribosomes that were isolated from such purified chloroplasts. They further found that the incorporation of the amino acids was inhibited if the chloroplasts were treated by RNase or by actinomycin D. Although protein synthesis in isolated plastids is suggested by the incorporation of amino acids, no net synthesis of a specific protein of the plastids has been demonstrated as yet.

We have reviewed one form of extrachromosomal inheritance. Examples of this type of inheritance are being identified in ever-growing number. We shall now consider other examples of this second genetic system of our bodies.

MITOCHONDRIA

Mitochondria are cytoplasmic organelles that contain the respiratory enzymes whose activities provide the main source of energy for the cell. They may vary in number from 1 in the marine alga *Micromonas* to about 1000 in a human liver cell. Certain strains of the yeast *Saccharomyces cerevisiae* produce tiny colonies when grown on agar. The slow growth of these colonies has been found to be due to the loss of the cell's aerobic respiratory enzymes and the utilization, by the cell, of the less efficient fermentation process. When individuals from a tiny colony are crossed with individuals from a normal-sized colony and the resultant zygote undergoes meiosis, we obtain in the meiotic products a 1:1 ratio of those individuals that will form tiny and normal-sized colonies. Such tiny strains are evidently due to mutant nuclear genes and are called *segregational petites*. In 1952 B. Ephrussi found that when normal yeast cells are treated with an acridine dye (euflavine), many petite colonies arise. These strains are called *vegetative petites*. When vegetative petite cells are crossed with those from normal colonies, the petite phenotype does not segregate normally. This lack of normal segregation, plus the high mutability of normal-colony cells to petite-colony cells, provides good evidence that the vegetative petite phenotype is due to extrachromosomal genes. In crosses of cells from segregational and vegetative petites, some normal offspring are produced. To obtain a normal cell from the fusion of cells of the two types of petite colonies, the segregational petite must contribute the normal extrachromosomal factor that the vegetative petite lacks.

Another mitochondrial enzyme deficiency that appears to be extrachromosomal is found in *Neurospora* and is called *poky*. Poky is a slow-growing strain of the mold. When poky is crossed to a wild strain, no normal segregation of the poky factor occurs. Poky differs from petite in that the two mutants are not deficient for the same enzymes.

As with plastids, investigations have been conducted to discover if mitochondria contain a complete genetic system. In 1964 Luck and Reich were able to isolate DNA from *Neurospora* mitochondrial fractions. Its yield was small and amounted to less than 1% of the total cellular DNA. The mitochondrial DNA was differentiated from the nuclear DNA by its density in a CsCl gradient.

The origin of the mitochondrion (hence, its DNA replication) is less clear than the origin of the plastid. The problem stems from the small size, plasticity, and lack of pigmentation of the mitochondria. A clear example of the origin of a mitochondrion from a preexisting one was reported for a marine alga, *Micromonas*. This organism contains 1 nucleus, 1 chloroplast, and 1 mitochondrion. Electron microscope studies show that the three organelles divide synchronously at the time of cell division. Further evidence that mitochondria arise from preexisting ones has been provided by experiments with a mutant choline-requiring strain of *Neurospora*. This strain incorporates radioactive choline into the lecithin of its mitochondria. Cultures labeled with tritiated choline were transferred to unlabeled medium and permitted to grow for three doubling cycles. The distribution of label among individual organelles was measured by quantitative autoradiography. The results showed that throughout the three doubling cycles, the label was uniformly distributed among all mitochondria. The implication of these observations is that all mitochondria come only from preexisting

ones and that somewhere in the cell cycle the mitochondrial DNA must replicate itself.

Investigations were carried out to discover if RNA- and protein-synthesizing systems were also present in the mitochondria. RNA-synthesizing activity was found in isolated *Neurospora* mitochondria. This RNA-synthesizing activity showed a requirement for all four ribonucleoside triphosphates and was inhibited by actinomycin D. The mitochondrial RNA-synthesizing system thus resembles, in its activity, the enzyme RNA polymerase. As in the case of plastids, it has not yet been possible to demonstrate the synthesis of a specific protein by the genetic system of the mitochondria.

After reviewing the evidence for the genetic systems in plastids and mitochondria, we might well ask whether there is any possible advantage to the organism in having a large number of self-reproducing organelles in the cytoplasm of its cells. The question has even wider significance if we consider that there are other cytoplasmic structures such as centrioles, kinetosomes, kinetoplasts, Golgi apparatus, and lysosomes that may have some or all of the genetic characteristics of plastids and mitochondria.

It is quite clear that if a gene of one organelle mutates in such a way that the organelle multiplies at a slower rate than the cells, that organelle will be eliminated from most of the descendants of the organism. However, there will be no danger of losing all the organelles. Thus the multiplicity of organelles serves to preserve the genic contents of the cytoplasm. Should mutations occur that do not reduce the organelle's rate of multiplication, the mutants may accumulate. The ability to carry along many organelles, each somewhat different in its genetic constitution, may provide the organism with great versatility to survive drastic environmental changes. Thus there may be two advantages in having a large number of self-reproducing organelles in the cell: first, to prevent a total loss of organelles because of a single mutation and, second, to provide a reservoir of cytoplasmic mutations that could become useful under adverse environmental conditions.

ORIGIN OF CHLOROPLASTS AND MITOCHONDRIA

Although our discussion has indicated some possible advantages to an organism in having a number of relatively independent, self-replicating cytoplasmic organelles, we have not yet examined the question of the possible origin of chloroplasts and mitochondria. It seems odd that two genetic systems, each complete with DNA, RNA, and ribosomes, would have evolved simultaneously in an evolutionary line. The need for duplication of enzyme systems and protein-synthesizing machinery appears to be wasteful for species that are under constant competitive conditions and in need of maximum efficiency. The situation becomes all the more perplexing when one contrasts the rigidity of distribution, through mitosis and meiosis, of the genes located in the nuclear genetic system with the almost haphazard distribution of the elements of the cytoplasmic genetic system. An interesting and unanticipated theory on the origin of these cytoplasmic organelles has developed through a study and comparison of the characteristics of the two genetic systems. Some of the data that led to the formulation of the theory are shown in Table 8-1.

It will be seen, from Table 8-1, that the ribosomes of eucaryotic (higher animals and plants) chloroplasts and mitochondria actually resemble those of procaryotic (blue-green algae and bacteria) organisms rather than the ribosomes found in the cytoplasm of eucaryotes themselves. This resemblance of ribosomes includes the sedimentation coefficients of both the entire ribosome and specifically also of its RNA components. A second point of comparison is the ability of various antibiotics to inhibit protein synthesis, which they do by binding to the ribosome and interfering with the normal mRNA-protein translation process. It is apparent that the antibiotics which are effective against the procaryotes will also stop the protein synthesis carried on in chloroplasts and mitochondria. However, these antibiotics will not inhibit the protein synthesis taking place in the cytoplasm of the same eucaryotic cell. In parallel fashion, cycloheximide will stop any protein synthesis involving the cytoplasm of a eucaryote but will be ineffective against the ribosomes of its chloroplasts and mitochondria and also against procaryotes. To the information available in Table 8-1 must be added the fact that the DNA in both chloroplasts and mitochondria are *not* complexed with histone and are often found in ring form. This, as will be recalled from

Table 8-1. Comparison of ribosomes from different sources

	Most eucaryotes			
	Cytoplasm	Chloroplasts	Mitochondria	Procaryotes
Sedimentation coefficient	80s*	70s	70s	70s
RNA components	18s, 25-28s 5s	16s, 23s 5s	16s, 23s 5s	16s, 23s 5s
Protein synthesis inhibition by				
Streptomycin	−	+	+	+
Chloramphenicol	−	+	+	+
Tetracycline	−	+	+	+
Cycloheximide	+	−	−	−

*s = Svedberg unit.

Chapter 2, also resembles the situation found in the procaryotes.

Based on the above and other information, a theory has been developed that not only chloroplasts and mitochondria but possibly many other cell organelles were originally independent and free-living organisms that became obligate endosymbionts of larger cells. Under this hypothesis, chloroplasts are thought to have originally been independent blue-green algae and mitochondria are considered to have originated as bacteria. When one considers the universal distribution of mitochondria in all higher animals and plants and the comparable distribution of chloroplasts in all higher plants, it is obvious that these endosymbiotic associations must have occurred relatively early in the evolution of eucaryotes and must have proved highly advantageous to have become so widespread. It seems also clear that over the countless generations, there have been many alterations through mutation in the functions of the original symbionts, so that today neither chloroplasts nor mitochondria can carry on an independent existence and, comparably, eucaryotes cannot survive without these organelles.

KAPPA PARTICLES IN PARAMECIUM

In 1938 T. M. Sonneborn reported a peculiar phenomenon in *Paramecium aurelia.* He had many stocks of this protozoan, each derived from a different ancestor. Strains of organisms derived from a single ancestor by asexual reproduction are called *clones.* In one experiment he arranged his stocks in pairs, took individuals from the paired clones, and placed them in the same culture vessel. In some combinations, he observed that the individuals from one strain (later called *killers*) survived but individuals from the other strain (later called *sensitives*) died. Further research showed that among the killer stocks there were different kinds of killers. One of the best-studied types is called *hump-killer.* It causes sensitive paramecia to develop a characteristic aboral hump before they die.

Investigation of the genetic control of this phenomenon centered on the results of *conjugation* between killers and sensitives. In *Paramecium* two types of nuclei are present: the *micronucleus,* or germ nucleus, and the *macronucleus,* or somatic nucleus. The micronuclei are two in number and diploid. During conjugation, the macronucleus disintegrates and each micronucleus undergoes meiosis, forming four haploid nuclei. Of the eight haploid nuclei produced by the two micronuclei, seven disintegrate. The remaining nucleus divides mitotically, forming two gametic nuclei. One gametic nucleus is stationary and remains in the animal, while the other is migratory and passes over a cytoplasmic bridge formed by and connecting the conjugants. Thus, there is a reciprocal exchange of migratory nuclei between mates. Each migratory nucleus fuses with the stationary nucleus of the other animal, resulting in a diploid nucleus. The chromosomal content of the conjugants at the end of fertilization is therefore identical. The newly formed diploid nucleus undergoes two mitoses, resulting in four nuclei. Two of these become the new micronuclei of the cell, while the other two become macronuclei, each one going into a separate cell at the next cell division.

The cytoplasmic bridge that connects the conjugants during conjugation may persist only long enough for the exchange of nuclei to occur. Under such conditions, no cytoplasm is exchanged between the animals. However, in some cases the bridge persists beyond the minimum time necessary for nuclear exchange, and an exchange of cytoplasm between the protozoa may also occur. After conjugation, the partners separate and reproduce by fission.

The finding that conjugation could occur between killers and sensitives implies that, at least during conjugation, the sensitives are temporarily resistant to the lethal effects of the killers. Sonneborn conjugated killer animals of stock 51 and sensitives of stock 52. In this mating, the cytoplasmic bridge is formed for only a short period of time. He found that the two exconjugants produced different clones. The clones that derived their cytoplasm from the killer parent were killers, while those that derived it from the sensitive parent were sensitives. Both exconjugants, of course, had identical chromosome complements. These results immediately showed that the killer factor was in the cytoplasm. A further proof of this was afforded by the cross between stock 51 killers and stock 47 sensitives. Here the cytoplasmic bridge may persist after the completion of the exchange of the migratory nuclei, and, as a result, there is also an exchange of cytoplasm. In this experiment both exconjugants produce killer animals, again showing that there is an extrachromosomally transmitted factor in the killer cells that determines their killer trait. This killer factor was called *kappa*.

A killer paramecium can be transformed into a sensitive one by exposure to high temperature, x rays, nitrogen mustard, and Chloromycetin. Such sensitives can then be crossed to killers with the same results as just described. The various crosses are diagrammed in Fig. 8-2. It would appear that the only protection a paramecium can have against the effects of a killer is to contain the killer factor in its own cytoplasm and thus be a killer. Resistance to the hump-killer and possession of the hump-killer trait go together.

From the foregoing, it would seem that the nuclear genes play no role whatsoever in the killer trait. This was found not to be the case. When killer exconjugants from the cross of stock 51 and stock 47 underwent *autogamy,* half of them lost their killer trait and became sensitives. In autogamy, the diploid nucleus undergoes meiosis, after which

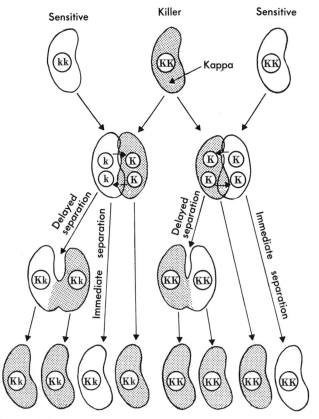

Fig. 8-2. Inheritance of the killer trait in *Paramecium aurelia.* To be a killer, an animal must contain kappa particles and the dominant gene **K.**

all but one haploid nucleus disappears. This remaining nucleus undergoes mitosis, and the two resultant haploid nuclei then fuse to form a new diploid nucleus. The net effect of this process is to make the animal homozygous for all its genes. Thus if paramecia that are heterozygous at a locus *(Kk)* undergo autogamy, they will become homozygous for one or the other of the two alleles at this locus *(KK* or *kk)* with about equal frequency. As a result of this discovery, it was postulated that there must be a nuclear gene that determines whether or not kappa can be maintained in the paramecium cytoplasm. The dominant allele, *K,* of this gene permits the maintenance of kappa, whereas the recessive allele in homozygous condition, *kk,* results in the disappearance of kappa. Stock 51 was assumed to have the genotype *KK,* and stock 47 to have *kk.* The cross between them results in *Kk* individuals, all of whom would be killers if they contained the kappa factor. Autogamy of the *Kk* killers produces *KK* individuals who remain killers, and it also produces *kk* individuals who become sensitives.

Cytological examination, especially with the electron microscope, of the cytoplasm of killers and sensitives reveals that killer paramecia contain characteristic minute particles that are absent in sensitives. This type of particle is about 0.4 μ long, and it has been likened to a rickettsia or a virus. The kappa particle is bounded by a membrane and has a small amount of its own cytoplasm. It contains both DNA and RNA. Kappa would therefore appear to be either a parasite or more possibly a symbiont, since it does not seem to harm its host. Normally, one may find hundreds of kappa particles per killer cell. Approximately half of the total kappa particles are transmitted to each daughter cell at binary fission, and the full population level is restored before the next fission. A killer cell occasionally loses one or more of its kappa particles into the liquid medium. If a lost kappa particle is ingested by a sensitive paramecium, the sensitive paramecium is killed. One kappa particle is capable of killing a sensitive paramecium.

Another characteristic of kappa has to be mentioned. As described above, a *KK* individual will be a sensitive paramecium if no kappa particles are present in its cytoplasm. When such sensitive cells are placed in concentrated suspensions of disintegrated killer animals, some of them acquire kappa from the suspension and are converted into killers. Hence, the extrachromosomal factor of the killer trait is infectious. As can be anticipated, kappa is subject to mutation, and one animal can carry two or more forms of kappa.

As we review our survey of kappa, we come to the conclusion that we are dealing with an extrachromosomal factor whose survival in the paramecium is determined by the genotype of the host animal. Kappa cannot be considered a normal cytoplasmic component of paramecia, since most of them are free of kappa particles. Kappa is probably a foreign intracellular organism. A most interesting question can be raised on this last point. If all paramecia had been found to contain kappa, would we have considered it to be a normal cytoplasmic component? We have already discussed the possibility that chloroplasts and mitochrondria may, in fact, be foreign intracellular organisms that have evolved a symbiotic relationship. This raises the possibility that all organisms, man included, may actually be symbiotic associations of large numbers of viruses, rickettsiae, bacteria, etc. and that the uniqueness of each species may lie in the unique combination of its symbionts.

MATE-KILLER IN PARAMECIUM

Another type of killer trait in paramecia is called *mate-killer.* In 1952 R. W. Siegel first described the mate-killing phenomenon. He found that mate-killers do not kill sensitive animals when the latter are swimming freely in the same culture fluid as the killers. Mate-killers affect sensitives only if there has been conjugation between the two types of animals. Siegel observed that the conjugation process occurred as usual, and the exconjugant that derived its cytoplasm from the mate-killer regularly yielded viable prog-

Table 8-2. Genetic basis of hump-killer, mate-killer, and specific resistance to these killers

Paramecium genotype	Cytoplasmic symbionts	Phenotype
KK or *Kk*	Kappa 51	Hump-killer; resistant to hump-killer
KK or *Kk*	None	Nonkiller; sensitive to hump-killer
kk	None	Nonkiller; sensitive to hump-killer
MM or *Mm*	Mu 540	Mate-killer; resistant to mate-killer
MM or *Mm*	None	Nonkiller; sensitive to mate-killer
mm	None	Nonkiller; sensitive to mate-killer

eny. The other exconjugant either died without dividing at all or produced a few daughter cells that soon died. The basis of the mate-killer trait was postulated to be a cytoplasmic particle called *mu*. As with kappa, large populations of mu multiply in the cytoplasm of the paramecium and are partitioned to daughter cells at fission, after which the full population level is restored. Cells that have populations of mu are mate-killers; cells lacking them are mate-sensitives. Resistance to mate-killers is correlated with possession of mu. Hence, when two similar mate-killers conjugate, neither conjugant is killed. If the two conjugants bear different kinds of mu, one of the partners may kill the other. Mu also parallels kappa in being maintained only in cells with a particular genotype. Mu particles are maintained in cells containing either of the two unlinked dominant genes, M_1 and M_2. When the genotype of a heterozygous mate-killer changes after autogamy, the daughter cells of genotype $m_1m_1m_2m_2$ lose their mu particles and become sensitives. Table 8-2 summarizes the nuclear and cytoplasmic relationships for both kappa and mu. In Table 8-2, only one of the mate-killer genes is indicated. Mu particles are composed of DNA, RNA, protein, and other substances and, like kappa, are considered to be symbionts.

METAGON IN PARAMECIUM

The previous discussion pointed out that the maintenance of the symbiont mu in a paramecium is dependent on the genotype of the paramecium. If a mate-killer animal becomes $m_1m_1m_2m_2$, the mu particles will disappear. However, the symbionts do not disappear immediately. In fact, they maintain their normal number of particles in the paramecium for about seven generations. Then, starting with the eighth generation, the particles suddenly and completely disappear from a small fraction of the cells. The fraction of cells lacking the symbionts increases thereafter until only a small number (7%) of cells containing them can be found by the fifteenth cell generation. This pattern of loss of cytoplasmic particles in mate-killers was investigated by Gibson and Beale. Their findings led them in 1962 to postulate that the maintenance of mu particles in paramecium was due to the presence of another cytoplasmic particle, which they called a *metagon*. The metagon was hypothesized to be a product of one or the other of the genes M_1 and M_2. It was further assumed that when the paramecium became homozygous recessive, the metagons would be passively distributed to the daughter cells. If one assumes an initial number of 1000 metagons per cell, animals lacking metagons would begin to appear after eight fissions, and thereafter the number of paramecia lacking metagons would rise at each fission. Another postulate was that one metagon in a cell is sufficient to maintain the whole mu population of the cell.

Proof of the postulate that only one metagon is necessary to maintain hundreds of mu particles took the form of a painstaking experiment. In this experiment, cells in the fifteenth generation, post autogamy, were analyzed for the mate-killer trait by observing the results of conjugation with known sensitives. Those cells that killed their mates still possessed mu. The question was whether these mate-killers contained only one metagon. Of the two daughter sixteenth-generation cells arising at fission from a fifteenth-generation mu-bearer, both initially contained mu particles. One of the cells retained its mu, but the other cell lost its mu. Loss began within an hour and was completed by the sixth hour of the roughly eight-hour generation time. In another part of this study, some fifteenth-generation cells were allowed to go through three more cell generations. Each cell in the resulting sets of eight eighteenth-generation cells was mated to a known sensitive. It was found that 32 of 38 such sets of eight cells tested included only one mu-bearer. These results agree well with the expectation that fifteenth-generation mu-bearers usually contain only one metagon, which is transmitted to only one descendant in each subsequent generation.

Further analysis of the metagon centered on its chemistry. It was found that mu-bearing cells after exposure to RNase retained mu until after one cell division and then lost mu before the second division. The loss of mu was not due to any detectable defect in mu itself. This was demonstrated in an experiment in which RNase-treated cells, directly after treatment, were mated to mu-less $M_1M_1M_2M_2$ cells. The progeny derived from the RNase-treated exconjugants were mate-killers. It seems evident that during conjugation the RNase-treated cells received functional metagons from the mu-less $M_1M_1M_2M_2$ cells. The effect of RNase treatment on mate-killer cells indicates that the metagon is composed of RNA. This led to the hypothesis that the metagon is in reality the messenger RNA of one or the other of the M genes, since, as pointed out previously, $m_1m_1m_2m_2$ cells eventually lose their mu particles and become sensitives. It follows that the metagon must code for some as yet unknown polypeptide or protein

necessary for the maintenance of mu particles in the cell. The metagon obviously represents a long-lived messenger RNA, since it appears to persist indefinitely through many cell generations.

Later experiments showed that metagons could be absorbed by RNase-treated mate-killers from a nutrient solution that contained the ribosome fraction of mate-killer cells. The treated mate-killers thus retained their mate-killing trait. This was an in vitro duplication of the conjugation experiment outlined earlier. The experiment also indicated that the metagon could be an infective particle. Removal of ribosomal protein by phenol extraction still left metagon activity. DNase treatment did not remove metagon activity, but RNase treatment did. Subsequently, strong confirmation was obtained that the metagon is composed solely of RNA. Material from the ribosome fraction of a mate-killer was subjected to electrophoresis on cellulose acetate paper. One band, apparently composed solely of RNA, contained most of the metagons, as shown by its infective activity.

Another line of research involving the metagon centered on RNA/DNA hybridization. RNA extracted from mate-killers was added separately to DNA extracted from various sources, including mate-killer paramecia, mate-sensitive paramecia, other ciliates such as *Tetrahymena* and *Didinium,* and the bacterium *Aerobacter aerogenes,* which is used as food for the paramecia. In each case, the RNA that had been bound to the DNA was disassociated from the DNA and tested for metagon activity. No metagon activity could be demonstrated in RNA released from any of the DNAs except those from paramecia. Although DNA from both mate-killer and mate-sensitive paramecia released RNA with metagon activity, the RNA released from the DNA of mate-killer paramecia had about 18 times greater metagon activity than the RNA released from the DNA of mate-sensitive paramecia. These results showed that metagon RNA is specifically complementary to part of the DNA of paramecia.

The last aspect of the metagon that we

shall discuss is its transfer from one species to another. Certain ciliates such as *Didinium nasutum* and some species of *Dileptus* are predators that swallow paramecia alive and whole. *Didinium* normally does not contain metagons or mu. Mate-killer paramecia were fed to various stocks of didinia. The didinia were then fed on non-mu-bearing paramecia. It was found that some stocks of didinia still contained mu particles in their cytoplasm after 1000 cell generations, covering a period of more than six months (generation time—four hours). This occurred regardless of whether the paramecia used as food after the initial infection were $M_1M_1M_2M_2$-sensitives or $m_1m_1m_2m_2$-sensitives. It was further found that RNA extracts from the stocks of didinia exhibited metagon activity even if the ancestors of the extracted didinia had been fed only one mate-killer paramecium. Such results suggested that *Didinium* cannot itself produce the metagon, but that once metagons are acquired from ingested paramecia, the metagons replicate indefinitely in didinia. This raises the fascinating possibility that the metagon is a messenger RNA in paramecia and the equivalent of a virus in didinia. It could imply that many of the RNA viruses are actually messenger RNAs from one species in the cells of another species. The metagon may force us to revise our concept of the smallest living particle. Viruses with protein coats have previously been considered as the simplest infectious agents. However, the metagon may be an example of an even simpler infectious unit, capable of replication in a host cell.

SEX RATIO IN DROSOPHILA

In some species of *Drosophila,* an occasional female is discovered that gives rise to progenies that are predominantly or exclusively female. This condition has been called *sex ratio.* In 1957 Malogolowkin and Poulson reported their studies on the sex-ratio condition in *Drosophila willistoni.* They found that females showing the sex-ratio condition produce unisexual, or almost unisexual, progenies when crossed to males

from a variety of strains. The occasional sons of sex-ratio mothers were mated to normal females, and the progenies contained normal numbers of both sexes. These findings indicated that the sex-ratio condition is transmitted through the female line only. It was also shown that the sex-ratio condition was retained even when all the chromosomes of the original sex-ratio strain are replaced, through appropriate crosses, with corresponding chromosomes from lines giving normal progenies.

An examination of the eggs deposited by sex-ratio females revealed that half of them did not hatch. It was concluded that the sex-ratio agent acts by making the eggs fertilized by Y-bearing sperm lethal. Injections were made of the ooplasm of dying male eggs into females of a normal strain. This transformed them to the sex-ratio condition. The same effect could be produced by injections of various tissues from adult sex-ratio females, especially the hemolymph. Investigators also found that exposure of the females to high temperatures (28° C.) would "cure" some of the females of their sex-ratio condition. Such females were crossed to a variety of males, and their progenies showed no restoration of the capacity to produce unisexual progenies. The loss of the sex-ratio condition was permanent. Electron microscopy of infected tissues revealed the presence of a spirochaete of the genus *Treponema,* which was demonstrated to be the infective agent. This type of sex-ratio condition was also found in *D. paulistorum* and *D. equinoxialis.* It was discovered that the spirochaete could even be infectively transferred between *D. willistoni* and *D. equinoxialis.*

The sex-ratio condition in *D. willistoni* and other species follows an extrachromosomal pattern similar to that of the killer traits of paramecium. However, some species of *Drosophila* exhibit a sex-ratio condition that is chromosomally determined. This was first reported in *D. pseudoobscura* by Gershenson in 1928 and later found to occur in other species as well. In the chromosomal sex-ratio condition, the trait is carried through the male line, with sex-ratio males producing only female offspring. A cytological examination of the chromosomes of sex-ratio males revealed that the X chromosome of these males always contains three characteristic inversions. A study of spermatogenesis in sex-ratio males by Novitski and co-workers in 1965 showed that the X and Y chromosomes pair and separate normally in the first division. In the second division, however, the Y chromosome loses its characteristic appearance and appears as a chromatin mass. Counts were made on sperm bundles at spermatid and mature sperm stages in sex-ratio and control males. Sex-ratio males had a full complement of sperm (128) in each bundle. These observations led to the conclusion that the sex-ratio phenomenon in *D. pseudoobscura* was due to a nonfunctioning of two of the products of meiosis. It is assumed that the abnormal Y chromosome is responsible for the non-functioning of the sperm containing it. However, the mechanism involved is unknown. The above explanation does not account for the fact that sperm bearing neither an X nor a Y chromosome are functional and produce viable but sterile XO males.

Up to now, we have discussed two types of sex-ratio conditions, one that is cytoplasmically located and the other that is chromosomally controlled. After our many preceding discussions on the complexities of genetic systems, it should come as no surprise to read that there is a sex-ratio condition that results from a gene-cytoplasm interaction. An example of this type of sex-ratio condition was reported in 1970 by Colaianne and Bell. In *D. melanogaster* there is an X-linked recessive gene called "sonless" *(snl)*. Females that are *snl/snl* produce progeny consisting overwhelmingly of daughters, regardless of the male parent's genotype. The *snl/Y* sons of *snl/snl* mothers die during the embryonic or early larval stages. An unexpected finding was that the progeny of *snl/+* females exhibit a normal 1:1 ratio of males to females and that the *snl/Y* sons of *snl/+* mothers are perfectly viable. The conclusion seems clear that the

snl/Y genotype can be either lethal or viable depending on the genotype of the mother. This would indicate that the egg cell produced by an *snl/snl* female lacks some substance necessary for the normal development of an *snl/Y* zygote. What this substance might be is unknown. However, an *snl/+* female does produce enough of this necessary but as yet unknown substance to permit the normal development of an *snl/Y* individual. A final aspect to be considered in this situation involves those rarely produced sons of *snl/snl* mothers, who, because of nondisjunction of their mothers' X chromosomes, are X/O, having received their X chromosomes from their fathers. If the paternal X chromosome carries the wild-type allele of *snl* (i.e., the sons are +/O), one finds that the X/O sons survive although developing in an egg produced by an *snl/snl* female. These X/O individuals are apparently able to synthesize the substance that is deficient in the eggs of *snl/snl* females. The facts just cited lead to the conclusion that two requirements must be met for this type of sex-ratio condition to be manifested: (1) the female parent must be *snl/snl* and (2) the male offspring must be *snl/Y*. This combination of genotypes results in a sex-ratio condition that is controlled by a gene-cytoplasm interaction.

Our review of the sex-ratio phenomenon in *Drosophila* has considered three different conditions. One, as exemplified by *D. willistoni,* is transmitted through the female line and is caused by an extrachromosomal infective agent that was found to be a spirochaete. The second, as found in *D. pseudoobscura,* is carried through the male line and is chromosomally determined. Its effect is due to the nonfunctioning of the Y-bearing sperm. The third condition, as seen in *D. melanogaster,* involves a gene-cytoplasm interaction. This situation apparently results from the inability of a particular male genotype to produce a substance that is missing from its egg cytoplasm. We thus see the same phenomenon capable of being caused by any one of three quite different mechanisms.

EPISOMES

Our discussions of extrachromosomal inheritance, up to this point, have considered two situations: (1) normal cell constituents (e.g., plastids, mitochondria) and (2) infective particles that may be symbiotic microorganisms (e.g., kappa, mu, metagon, sex ratio). We have also seen that the extrachromosomal genetic system can be influenced by the chromosomal genetic system. Yet at this point in our discussion the two systems would appear to be physically distinct from one another. As one might anticipate, there are particles that can be shown to sometimes be a part of either of the two genetic systems. Such particles, which can participate in a cell either as extrachromosomal or as chromosomal elements, are called *episomes*. The concept of the episome was put forward by Jacob and Wollman in 1958 to explain some of their findings on virus-infected bacteria.

Lysogeny

In Chapter 1 we discussed the life cycle of a T2 virus. The infection of a bacterium by this phage leads typically to the lysing of the bacterium and the liberation of many new virus particles. The T2 virus is said to be *virulent,* and when it is reproducing itself inside the cell, it is said to be in the *vegetative* state. Experiments with other bacteria and their viruses, as discussed in Chapter 2, however, give different results. When the bacterium that causes typhoid in mice, *Salmonella typhimurium,* is infected with the virus P22, one of two patterns may emerge: the virus can either enter the vegetative state, replicate itself, and destroy the host; or it may remain in the cell without killing the host. Among the many descendants of the latter bacterium, from time to time one of the daughter cells will break open and yield a crop of new virus particles. This indicates that the bacterial host did not destroy the invading phage. In some strains, when the descendants of the originally infected bacterium are exposed to UV light, x rays, or active chemicals such as nitrogen mustard or organic peroxides, the whole bacterial population will lyse within an hour, releasing tremendous numbers of viruses. This finding demonstrates that the invading phage particle was transmitted in some inactive form to all the cells of the bacterial culture. Cells that carry inactive viruses and are thus potentially subject to lysis are said to be *lysogenic.* The phages involved are called *temperate;* in their inactive form they are called *proviruses (prophage)* and are said to be in an *integrated* state. In order to explain the above findings, it was postulated that the temperate virus becomes attached to the host bacterial chromosome and from then on behaves as a chromosomal constituent of the host. In this prophage state, the virus is replicated along with the bacterial chromosome and is, as a result, transmitted to all daughter cells. In some of the daughter cells, it enters the extrachromosomal state and becomes virulent. In the extrachromosomal phase, the virus genetic material has two functions, replication and transcription, leading to the production of the protein material of the mature virus. In the chromosomal phase, this latter function is not exercised. In fact, in its integrated state, the virus confers an immunity on the host to new infection *(superinfection)* by identical or closely related phage. Occasionally, lysogenic cells lose their prophage (are cured). The two different patterns of virus life cycle are shown in Fig. 8-3.

Lysogeny has been found to be an important aspect of some human diseases. For an example we can examine a report published in 1972 by M. W. Eklund and his co-workers. They studied one of the strains of the bacterium *Clostridium botulinum,* which is considered to be the causative agent of one type of food poisoning. Previously, investigators had found that toxigenic cultures of the bacterium could be "cured" of their toxigenicity by being grown in a medium containing the dye acridine orange. They had also found that if the newly formed nontoxigenic cultures were incubated in a broth containing the filtrates of toxigenic strains, the nontoxigenic cultures would regain their tox-

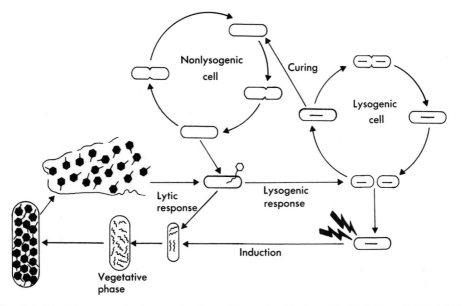

Fig. 8-3. The life cycle of a temperate phage. (From Lwoff, A. 1953. Bact. Rev. **17:**269-337.)

igenicity. It was apparent that toxigenicity, for these bacteria, is infective. Eklund and his co-workers isolated a bacteriophage, which was labeled DEβ, from cells of a toxigenic culture. This virus was then placed in the same food medium as nontoxigenic *C. botulinum.* The nontoxigenic bacteria became toxigenic. On the basis of Eklund's experiments, there seems to be little doubt that the toxigenicity of *C. botulinum* depends on its active and continued association with the bacteriophage DEβ. At this time we are not able to guess how many "bacteria-caused" diseases are the result of the activities of the bacteria and how many are the result of the viruses they contain. However, the discovery of just which organism is causing a particular disease is of immense importance in deciding which therapeutic approach to use. Future research in this area will undoubtedly reveal more examples of the role of lysogeny in microorganism-caused diseases.

Lysogeny may also have great importance in other types of human diseases. It is possible to conceive of viruses entering cells, remaining inactive, and being transmitted to many daughter cells. At some later time, a stimulus such as UV light, x rays, etc. may then cause them to become extrachromosomal and virulent. It might be more than a coincidence that the agents used to activate proviruses are themselves carcinogens and also mutagens. At present, it is not possible to state the mechanism by which these agents cause their effects.

In Chapter 1 we discussed the RNA tumor virus (e.g., Rous sarcoma virus) which on entering a cell of a chicken, mouse, rat, etc., makes a double-helix DNA copy of itself. This DNA copy becomes incorporated into one of the chromosomes of the host cell and is replicated along with the host chromosome. The presence of this "foreign" DNA in a host chromosome raises the question as to what constitutes species-specific DNA. We have already considered the problem posed by cytoplasmic symbionts when we are faced with the task of describing the genetic systems of a species. To this perplexing situation, we must now add lysogeny as a source of further complication in characterizing the genome of a species. We shall now turn our attention to other types of episomes.

Sex factor in Escherichia coli

The sex factor in *E. coli,* F⁺, was discussed in detail in Chapter 5 as an example, in some organisms, of the role of cytoplasm in sex determination. Briefly, conjugation in *E. coli* occurs between a "male" cell which transfers genetic material and a "female" cell which receives it. This difference in the role of the two types of cells during conjugation is ascribed to a sex factor, F, which is present in males (F⁺) but not in females (F⁻). Males may spontaneously give rise to females but not the reverse. The transition from male to female is considered to be due to the irreversible loss of the sex factor. During conjugation, F⁺ males convert F⁻ females into males, presumably by transference of the sex factor. In the F⁺ males, the sex factor appears to exist in an autonomous, extrachromosomal state.

An F⁺ cell can change into another type of male cell, called Hfr. In the Hfr condition, the male cell does not convert F⁻ females into males during conjugation but does transfer its chromosome to the female cell. Should the entire chromosome be transferred, the F⁻ cell becomes Hfr. It is concluded that in the Hfr state the sex factor is integrated into the chromosome. Hfr cells can spontaneously give rise to F⁺ cells. The above facts indicate that the F⁺ and Hfr states of a cell are mutually exclusive; therefore the sex factor in *E. coli* fits the definition of an episome.

One interesting and complicating event can occur when the sex factor, F, leaves the bacterial chromosome and returns to the cytoplasm. The F factor may carry with it a fragment of the bacterial chromosome. This is thought to occur in a manner similar to that discussed in Chapter 2 for transducing phage. The modified sex factor has been called an *F-prime* (F′) or *substituted sex factor.* When the F′ sex factor is transferred to an F⁻ cell during conjugation, the F⁻ cell becomes F′ and also becomes diploid for that portion of its genome which is present in the F′ sex factor. This method of creating partially diploid bacteria has been used to great advantage in many investigations of gene interaction and function.

Colicinogenic factors

Some *E. coli* strains produce proteinaceous substances called *colicins,* which kill related strains of bacteria. A cell capable of producing a colicin is called *colicinogenic.* This characteristic can be transmitted through thousands of cell generations. Although it can be lost spontaneously, spontaneous acquisition of colicinogeny has never been observed. However, cells can become colicinogenic as a result of conjugation with cells possessing this property. This factor thus appears to be extrachromosomal and autonomous.

The evidence that the colicinogenic factor can exist in an integrated chromosomal state is less convincing. It depends on the following reasoning. The frequency, speed, and order with which different chromosomal loci are transferred from Hfr to F⁻ cells vary from one Hfr strain to another. In parallel fashion, the transmission from an Hfr to an F⁻ strain of the ability to produce colicin also varies from one Hfr strain to another. From this it is inferred that the colicinogenic factor can occupy a locus on the Hfr chromosome. Another kind of evidence that the colicinogenic factor may exist in an integrated chromosomal state comes from the following observations. Only a small fraction of the cells of a colicinogenic culture actually produce colicins. However, colicin synthesis can be induced in nearly all cells of the bacterial culture, through exposure to UV light, nitrogen mustard, or hydrogen peroxide. This phenomenon is very similar to what is observed in lysogenic bacteria and may indicate that the colicinogenic factor is a true episome.

Resistance transfer factors

In some strains of the genus *Shigella,* the cells are simultaneously resistant to three or even four therapeutic drugs such as sulfonamide, streptomycin, tetracycline, and chloramphenicol. Experiments have shown that resistance to the different drugs is due in each case to a separable genetic determinant. However, in conjugation, there is a multiple transfer of resistance to sensitive cells. The factor responsible for this is called a *resistance transfer factor* (RTF). The presence of this factor in the cells of a culture is accompanied by the appearance, in a CsCl density gradient, of a DNA fraction with a distinctive G-C content. RTF appears to promote conjugation and drug resistance transfer not only between species within the genus *Shigella,* but also between species from different genera such as *Shigella, Escherichia,* etc. Although originally discovered in 1955 in Japan, the occurrence of multiple-drug–resistant pathogenic bacteria has now become worldwide. The importance of this multiple-drug resistance transfer factor is very great because its distribution to all pathogenic bacteria, which appears to be technically possible, would tend to force medicine back into the preantibiotic era. RTF appears to exist largely in an autonomous state. The evidence that it may exist in an integrated chromosomal state is not at all convincing as yet. Future investigations will determine whether RTF can be considered as a true episome.

EXTRACHROMOSOMAL INHERITANCE IN CHLAMYDOMONAS

In the green alga *Chlamydomonas reinhardi,* an intensive analysis of extrachromosomal inheritance has been pursued by R. Sager and co-workers since 1960. This organism can reproduce asexually, by mitotic cell division, to produce clones. Members from different clones may pair, fuse, and form zygotes. The zygotes undergo meiosis, resulting in four haploid cells, each of which can be isolated to give rise to a separate clone. Investigations have shown that there are two mating types, which have been designated mt^+ and mt^-. The diploid zygote must then be, with respect to mating type, mt^+mt^-. An analysis of the cells resulting from meiosis of the zygote shows a 1:1 ratio of $mt^+:mt^-$. Mating type therefore behaves as a trait that is determined by a nuclear gene. Many other traits similarly determined by nuclear genes are also known.

The wild type *Chlamydomonas* is sensitive to streptomycin and will not grow in a food medium containing the drug. These cells are called streptomycin-sensitive *(ss)*. In contrast to the wild type cell, mutants have been discovered that actually require the presence of streptomycin in the food medium in order for them to grow. These mutants are designated streptomycin-dependent *(sd)*. It was found that when $sd\ mt^+$ cells fused with $ss\ mt^-$ cells, the resultant zygote underwent meiosis and yielded the expected 1:1 ratio of mt^+ and mt^- cells, but that almost all the cells were *sd*. The reciprocal cross of $ss\ mt^+$ and $sd\ mt^-$ cells yielded the same results as the above for the mating type trait, but produced almost all *ss* cells. It was quite evident from these experiments that the reaction of the cell to streptomycin is caused by a gene that does not comply with the usual rules of transmission for chromosomal genes. The inheritance is obviously uniparental and associated with the mt^+ individual. It is as if all extrachromosomal factors brought to the zygote by the mt^- parent are lost. Crosses illustrating this type of inheritance are diagrammed in Fig. 8-4.

As stated previously, in both types of crosses, not all zygotes produce haploid cells that follow the uniparental rule. There are a small number of exceptional zygote colonies, about 0.07%, that show the extrachromosomal traits of both parents. These colonies have been analyzed through a number of cell divisions, and they eventually segregated out into pure *ss* and *sd* cells. This would imply that both extrachromosomal factors were present in the four haploid cells formed by these exceptional zygotes and that the segregation of *ss* and *sd* occurred in the postmeiotic (i.e., mitotic) divisions. When two sets of extrachromosomal factors were studied in exceptional zygotes, the two sets of gene-pairs were found to segregate independently of one another in the postmeiotic divisions.

The inheritance of extrachromosomal factors in *Chlamydomonas* presents a number of intriguing facets. Normally, the mating type locus, mt^+, which is chromosomal, appears to control the selective elimination of the extrachromosomal genes from the mt^- parent at some time between zygote formation and meiosis. This selective elimination does not occur in all zygotes, as seen by the rare haploid colonies that exhibit the traits of both parents. The number of such exceptional colonies can be increased tenfold by administering a heat treatment to the zygotes immediately after they are formed. This could imply that the selective elimination of extrachromosomal genes is under enzymatic control. When the haploid cells do contain extrachromosomal genes from both parents, these genes eventually segregate from each other in one of the mitotic divisions. Should such a system be found to operate in all organisms, it will force a reconsideration of our views on the stability of the organism's genome.

MATERNAL PREDETERMINATION

Maternal predetermination considers transitional phenotypic effects that normally wear off in one or more generations. It usually involves the imposition of characteristics on the cytoplasm of the egg by an

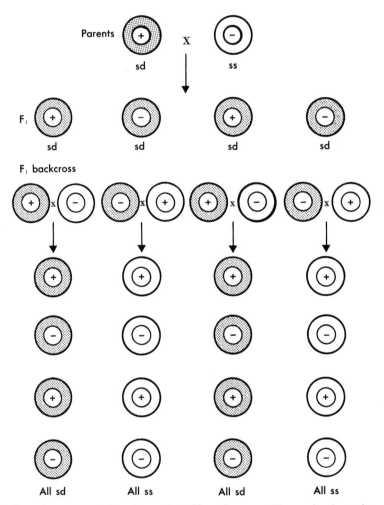

Fig. 8-4. Extrachromosomal inheritance in *Chlamydomonas*. Plus and minus signs refer to mating type (mt). The cross *sd mt⁺* × *ss mt⁻* gives rise exclusively to *sd* offspring, which segregate 2:2 for mating type and other chromosomal genes. F₁ clones of plus mating type, backcrossed to *ss mt⁻*, produce all *sd* offspring (4:0); but F₁ clones of *minus* mating type, backcrossed to *ss mt⁺*, produce only *ss* progeny (0:4). (From Sager, R. 1954. Proc. Nat. Acad. Sci. U. S. A. **40:**356-363.)

outside force. Even though such phenomena are not, strictly speaking, modes of inheritance but rather developmental processes, they are generally included in discussions of extrachromosomal inheritance so as to distinguish them from what are true examples of such inheritance.

Shell-coiling in snails

An example of maternal predetermination, involving shell-coiling in snails, was reported by Boycott and colleagues in 1930. Many species of snails are known in which the shell aways coils to the right *(dextral),* whereas in other species the shell always coils to the left *(sinistral).* In a few species, both dextral and sinistral individuals occur. In the species *Limnaea peregra,* dextrality

is dominant to sinistrality. However, the coiling of a snail's shell is not determined by the individual's own genes but by those of its mother. Offspring whose mothers are either homozygous or heterozygous for dextrality are also dextral, including those offspring that are homozygous for sinistrality. Correspondingly, offspring whose mothers are homozygous for sinistrality are also sinistral even if these offspring carry the dominant gene for dextrality. The crosses illustrating maternal predetermination of shell-coiling in snails are shown in Fig. 8-5.

Experiments have demonstrated that the direction of coiling in a snail's shell is determined by the orientation of the spindle at the second cell division after fertilization. If the spindle is tipped toward the left of the

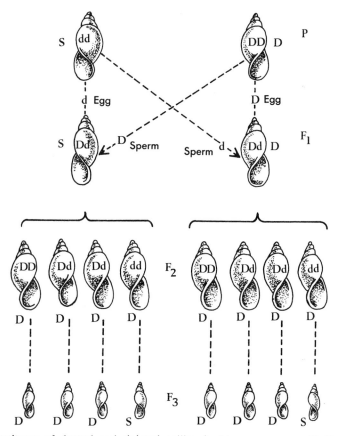

Fig. 8-5. Inheritance of dextral and sinistral coiling in *Limnaea peregra.* **D,** Dextral phenotype; **S,** sinistral phenotype; **D,** gene for dextrality; **d,** gene for sinistrality. This species is hermaphroditic and may reproduce either by crossing (F₁) or by self-fertilization (F₂ and F₃).

median line of the cell, the sinistral pattern will develop. Conversely, if the spindle is tipped toward the right of the median line of the cell, the dextral pattern will develop. By some unknown mechanism, the genotype of the mother must affect the eggs developing in the ovary in such a way as to predetermine the direction of the second cell division after fertilization.

SUMMARY

In this chapter we have considered various types of extrachromosomal inheritance. These have included plastids and mitochondria as examples of normal cell elements that have independent non-Mendelian genetic systems. Also discussed were the kappa, mate-killer, and metagon characteristics in *Paramecium* and the sex-ratio condition in *Drosophila*, as examples of infective particles that become part of the inheritable characteristics of the individual. Our review of episomes considered that type of hereditary particle that can alternately be part of the cell's Mendelian and non-Mendelian genetic systems. The discussion of episomes included lysogeny, sex factor, colicinogenic factors, and resistance transfer factors. We completed our consideration of extrachromosomal inheritance with a discussion of this type of inheritance in *Chlamydomonas*. The chapter was concluded with an example of maternal predetermination, as seen in shell-coiling in snails, which is sometimes confused with extrachromosomal inheritance. In the next chapter, we shall turn our attention once more to the structure and functions of the gene. We shall look at the gene as a mutational unit, as a recombinational unit, and as a functional unit.

Questions and problems

1. What factors distinguish extrachromosomal inheritance from Mendelian inheritance?
2. What are the advantages of self-replicating organelles to a cell?
3. For each of the following, cite the evidence that allowed investigators to say non-Mendelian inheritance was operating:
   a. Iojap mutation in maize
   b. Petite yeast
   c. "Poky" *Neurospora*
   d. Kappa *Paramecium*
   e. Sex ratio in *Drosophila*
   f. Streptomycin dependence and sensitivity in *Chlamydomonas*
4. Define the following terms:
   a. Maternal inheritance
   b. Cytoplasmic inheritance
   c. Satellite DNA
   d. Macronucleus
   e. Micronucleus
   f. Autogamy
5. In each of the following examples of conjugation involving *Paramecium,* what will be the phenotypes of the exconjugants with reference to the killer trait?
   a. Stock 51 killers × Stock 52 sensitives
   b. Stock 51 sensitives × Stock 52 killers
   c. Stock 51 killers × Stock 47 sensitives
   d. Stock 51 sensitives × Stock 47 killers
6. Two paramecia of genotype *Kk* conjugate. With what frequencies would you expect each of the following combinations of nuclear genotypes to occur?

	Exconjugant 1	Exconjugant 2
a.	*KK*	*Kk*
b.	*KK*	*KK*
c.	*kk*	*kk*
d.	*Kk*	*Kk*

7. Define the following terms:
   a. Lysogeny
   b. Temperate phage
   c. Lysogenic bacteria
   d. Prophage
8. How is the sex factor of bacteria relevant in a discussion of extrachromosomal inheritance?
9. In each of the following types of matings involving *Chlamydomonas,* what will be the phenotypes of most of the derived haploid clones with reference to streptomycin dependence?
   a. *sd mt⁺ × ss mt⁻*
   b. *ss mt⁺ × sd mt⁻*
   c. *sd mt⁺ × sd mt⁻*
   d. *ss mt⁺ × ss mt⁻*
10. In wild type *Drosophila,* adults can be exposed to pure CO_2 for as long as fifteen minutes with no ill effects. In certain strains, however, adult flies are invariably killed on exposure to CO_2. By performing appropriate crosses, one can replace each of the chromosomes in a sensitive strain with the corresponding chromosome derived from a CO_2-resistant strain. This procedure has no effect in changing the CO_2 response of a sensitive strain; the progeny are still sensitive. In crosses between resistant and sensitive strains, the CO_2 sensitivity trait does not segregate among the progeny of hybrids. What do you think is the genetic basis for the CO_2 sensitivity in *Drosophila?* Explain!
11. Phenotype is determined in some cases chro-

mosomally, in other cases extrachromosomally, and in yet other instances as a result of interaction between nuclear and cytoplasmic factors. Amplify and illustrate this statement in light of information presented in this chapter.

12. What will be the direction of shell-coiling in the progeny of the following crosses between *Limnaea* snails:
    a. Female *Dd* × male *dd*
    b. Female *DD* × male *Dd*
    c. Female *dd* × male *DD*
    d. Female *Dd* × male *DD*
    e. Female *DD* × male *dd*
    f. Female *dd* × male *Dd*

13. Refer to question 12.
    a. What kind of inheritance is operating?
    b. How does it differ from extrachromosomal inheritance?

14. Diagram a cross between an inbred *dd* male snail with sinistral coiling and a female snail, *DD*, whose shell coils to the right. Carry the cross to the F_2 and represent the expected results from inbreeding each of the F_2 snails.

15. Define the following terms:
    a. Colicinogenic factor   d. Hump-killer
    b. Resistance transfer factor e. Mate-killer
    c. Metagon   f. Sex factor

16. How do you explain the fact that in corn the *Ij/Ij* genotype can result in three different leaf phenotypes—green, white, and striped?

17. A particular phenotype can be brought about as a result of two vastly different genetic mechanisms. Discuss this statement with reference to:
    a. Sex ratio in *Drosophila*
    b. Mitochondria mutations in yeast
    c. Leaf color in plants

18. Certain strains of mice have a strong tendency toward developing mammary tumors. How could you determine whether (a) nuclear genes, (b) extranuclear factors, or (c) environmental factors were involved?

19. What are the special advantages of *Paramecium* as an experimental organism for genetic investigation?

20. When a portion of black skin derived from a black and white spotted guinea pig is grafted onto a white area on the same animal, the pigment grows outward from the margin of the graft into the white area. How do you account for this pigment spread?

References

American Society of Naturalists. 1965. Symposium on Cytoplasmic Units of Inheritance. Amer. Natur. **99:**193-334. (A series of papers discussing various aspects of extrachromosomal inheritance.)

Beale, G. H. 1954. The genetics of *Paramecium aurelia.* The University Press, Cambridge. (An excellent short book on the earlier genetic studies of this organism.)

Campbell, A. M. 1969. Episomes. Harper & Row, Publishers, New York. (An authoritative review of our knowledge of episomes.)

Colaianne, J. J., and Bell, A. E. 1970. Sonless, a sex-ratio anomaly in *Drosophila melanogaster* resulting from a gene-cytoplasm interaction. Genetics **65:**619-625. (Demonstration of one of the patterns that results in the non-Mendelian inheritance of the sex-ratio phenomenon.)

Edelman, M., Cowan, C. H., Epstein, H. T., and Schiff, J. A. 1964. Studies of chloroplast development in *Euglena.* VIII. Chloroplast-associated DNA. Proc. Nat. Acad. Sci. U. S. A. **52:**1214-1219. (Evidence for the existence of extranuclear DNA.)

Eklund, M. W., Poysky, F. T., and Reed, S. M. 1972. Bacteriophage and toxigenicity of *Clostridium botulinum* type D. Nature New Biol. **235:**16-17. (Demonstration of the role of lysogeny in "bacteria-caused" disease.)

Gibor, A., and Granick, S. 1964. Plastids and mitochondria: inheritable systems. Science **145:** 890-897. (Excellent review of the genetics of these cytoplasmic organelles.)

Jacob, F., and Wollman, E. L. 1961. Sexuality and the genetics of bacteria. Academic Press, Inc., New York. (An excellent general treatise on the subject; see Chapter 16 for an account of episomes.)

Luck, D. J. L. 1963. Genesis of mitochondria in *Neurospora crassa.* Proc. Nat. Acad. Sci. U. S. A. **49:**233-240. (A report on experiments designed to determine the origin of mitochondria.)

Margulis, L. 1970. Origin of eukaryotic cells. Yale University Press, New Haven, Conn. (An authoritative review of the evidence on the primary role of symbiosis in the evolution of higher plants and animals.)

Mitchell, M. B., and Mitchell, H. K. 1952. A case of "maternal" inheritance in *Neurospora crassa.* Proc. Nat. Acad. Sci. U. S. A. **38:**442-449. (A description of the "poky" mutation in *Neurospora.*)

Nanney, D. L. 1958. Epigenetic control systems. Proc. Nat. Acad. Sci. U. S. A. **44:**712-717. (A general consideration of the nature of extrachromosomal systems.)

Novick, R. P. 1969. Extrachromosomal inheritance in bacteria. Bacteriol. Rev. **33:**210-235. (A thorough review of this very complex aspect of bacterial genetics.)

Poulson, D. F., and Sakaguchi, B. 1961. Nature of the "sex ratio" agent in *Drosophila.* Science **133:**1489-1490. (A discussion of the phenomenon responsible for the non-Mendelian inheritance of the sex-ratio factor.)

Sager, R., and Ramanis, Z. 1963. The particulate nature of nonchromosomal genes in *Chlamydomonas*. Proc. Nat. Acad. Sci. U. S. A. **50:** 260-268. (Experimental evidence for the existence of nonchromosomal mechanisms of heredity in *Chlamydomonas*.)

Sonneborn, T. M. 1959. Kappa and related particles in *Paramecium*. Advances Virus Res. **6:** 229-356. (A discussion of factors involved in nonchromosomal inheritance in *Paramecium*.)

Sturtevant, A. H. 1923. Inheritance of the direction of coiling in *Limnaea*. Science **58:**261-270. (A discussion of the manner of coil direction inheritance in snails.)

9 Gene concepts

In our considerations of the gene thus far, we have reviewed a large body of information about genes, including their various possible locations and actions. Our discussions have raised a number of problems that we have not yet analyzed. The statement that genes may exist in differing allelic forms left unspecified how the various alleles could have arisen. Our consideration of the genetic code and protein synthesis did not indicate how one determines exactly what constitutes a functional gene. Our study of linkage was limited to the recombinations that occur between genes and left open the possibility of crossing-over's taking place within a gene. Finally, the phenomenon of position effect clearly showed that genes do not necessarily act in autonomous fashion but that their activities may be controlled by adjacent loci. We shall now consider these problems and see what are the limits of our understanding of them. It seems evident that we may examine the gene from three points of view: (1) as a mutational unit, (2) as a functional unit, and (3) as a recombinational unit. In addition, we can consider the overall question of how gene action is regulated. We ought to anticipate that the gene may not show the same characteristics when considered in these different aspects.

THE GENE AS A MUTATIONAL UNIT

Our discussion of the genetic code revealed that, with the exception of three triplets (UAA, UAG, and UGA), all triplet combinations resulted in a code word for a particular amino acid. The change from one code word to another can, under these conditions, be achieved in a number of ways. A nucleotide can be *added* to the message, a nucleotide can be *deleted* from the message, or one nucleotide of the message can be *replaced* by another. The first two types of nucleotide change will usually not only alter the particular code word but will also cause radical changes in the rest of the message because every code word after the affected triplet will also be modified. This can easily result in the production of an inactive protein and can lead to the death of the cell. The third type of nucleotide change, the replacement of one nucleotide of a triplet by another, will affect only that particular codon. Such an altered code word may designate a different amino acid, resulting in the production of a protein with a single amino acid substitution. These three types of nucleotide change set the lower limit of the gene as a mutational unit at a single nucleotide. This does not necessarily mean that all mutants are the result of single-nucleotide changes. It does indicate that a mutant may be produced by that small a change in the nucleotide sequence of the DNA. Stating that the lower limit of the gene as a mutational unit is a single nucleotide does not indicate how the change may take place. This aspect of mutation must now be considered.

Chemical basis of mutations
Transitions

Our review of the structural arrangement of the DNA molecule pointed out that the purine *adenine,* A, is normally linked to the pyrimidine *thymine,* T, by two hydrogen bonds, while the purine *guanine,* G, is normally linked to the pyrimidine *cytosine,* C, by three hydrogen bonds. However, each of the four bases in DNA can exist in alternative states—the so-called *rare states* of each of the bases. Rare states occur as a result of rearrangements in the distribution of hydrogen atoms in the molecule. These rearrangements are called *tautomeric shifts* and are shown in Fig. 9-1. When a base is in its rare state, it cannot be linked to its normal partner. However, a purine such as adenine can in its rare state form a bond with cytosine, provided the cytosine is in its normal state. The four pairings that can occur when one member of a base pair has

undergone a tautomeric shift are shown in Fig. 9-2.

Watson and Crick in 1953 hypothesized that the occurrence of the bases in their rare states provides a mechanism for mutation during DNA replication. If, for example, adenine in an old chain is in its rare state at the moment that the complementary new chain reaches it, cytosine can pair with it and be added to the growing end of the new chain. The result of this type of pairing is the formation of a DNA molecule that contains an exceptional base pair. This situation is not stable, and at the next replication adenine is expected to return to its common state and to pair with thymine. Cytosine introduced into the complementary strand due to the tauto-

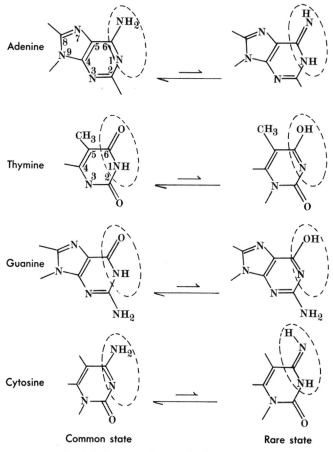

Fig. 9-1. Tautomerism of the bases of DNA.

meric shift in adenine would then pair with guanine. Thus, two kinds of DNA molecules would be formed, one that is identical to the original DNA and another that has undergone a base-pair substitution of G-C for A-T. Such a change—involving the replacement of one purine in a polynucleotide chain by another purine and, correspondingly, in the complementary chain the replacement of one pyrimidine by another pyrimidine— is called a *transition*. When transitions occur, the two kinds of DNA molecules will eventually be separated into different cells that will form different clones. In our ex-

ample, the cell line containing the transition will have a G-C pair in place of an A-T pair. This could result in the coding for a different amino acid at that point of the DNA and would presumably result in a recognizable mutation. Mutations formed during DNA replication are called *copy error mutations*. Should the mutation code for one of the terminator or nonsense triplets (UAA, UAG, or UGA), the reading of the mRNA would stop at that point. This would prevent a complete polypeptide chain from being formed and could result in a lethal mutation. The steps involved in a base-pair transition following tautomerism are diagrammed in Fig. 9-3. As can be seen in the illustration, a transition can also result if the base of a nucleotide, in its rare form, is added to the growing end of a new chain during DNA replication. Using the rare state of adenine again as our example, we find that the transition will result in the base-pair substitution of A-T for G-C. It is to be noted that any base can become involved in a transition, provided it is in its rare state.

Transitions may also follow the *ionization* of a base at the time of DNA replication. Ionization involves the loss of the hydrogen from the number 1 nitrogen of a base. For example, in its ionized state, thymine can pair with guanine if the guanine is in its common form. In similar fashion, guanine in its ionized state can pair with thymine in its common form. From any such unstable base pair, a transition will result following the steps outlined above for tautomeric shifts. The formation of T-G and G-T base pairs after ionization of these two bases is shown in Fig. 9-4.

Experimental evidence for transitions. The experimental evidence for transitions is best seen in the studies on *rII* mutants of the *Escherichia coli* phage T4. When phage and bacteria are mixed together in melted nutrient agar and poured into a petri dish, the agar will solidify. The bacteria will multiply to form a dense sheet of cells, causing the agar to be cloudy. Each of the virus particles will have absorbed to a bacterium when the phage and bacteria were first

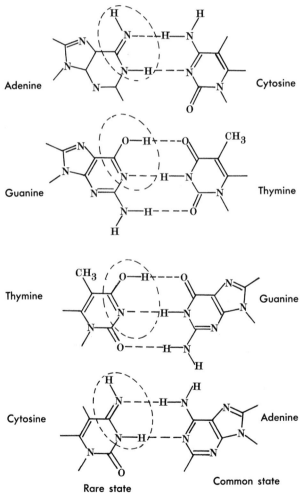

Fig. 9-2. "Forbidden base pairs" resulting from tautomerization.

mixed together. The phage will multiply within the bacterium and will eventually lyse the cell. On their release from the cell, the virus particles will absorb to neighboring cells and will repeat the process. Within a few hours, clear areas *(plaques)* will form in the otherwise continuous sheet of bacterial cells. The number of plaques will correspond to the number of virus particles mixed with the bacteria initially, and the areas of the plaques will represent the millions of bacteria lysed by the successive waves of phage particles released from previously infected adjacent cells. T4rII particles, when grown on *E. coli* strain B, make plaques that are larger and have a sharper edge than do T4r⁺. Thousands of *rII* mutations have been isolated from independent occurrences of mutation. On the basis of their different rates of back mutation to *r⁺*, most of these mutants are distinguishable from one another, although this is not the only criterion of their uniqueness. The tremendous number of such mutants within a restricted

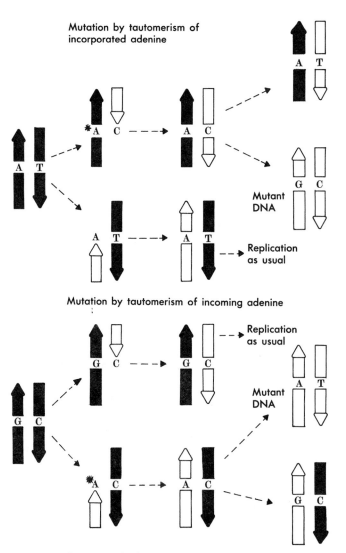

Fig. 9-3. Consequences of tautomerization of an adenine residue (*) at the time of DNA duplication.

region of the T4 chromosome led to the hypothesis that most of these mutants must represent changes of single base pairs.

In addition to the above observational evidence, the ability to induce forward and back mutations at the *r* locus has been used as a test of the Watson-Crick hypothesis of the transition type of mutation. The study of the induction of base-pair transitions has centered to a large extent on the use of chemical mutagens such as *5-bromouracil* and *nitrous acid*.

5-BROMOURACIL. 5-Bromouracil (5BU) is a structural analogue of thymine (5-methyl uracil) (see structural formulae in Chapter 1). When virus-infected bacteria are placed in a medium containing 5BU, the phage can incorporate some of the chemical, in place of thymine, into their newly replicated chains of DNA. The 5BU can, like thymine, be found sometimes in a rare tautomeric state and sometimes in an ionized state. However, the frequency of 5BU found in these uncommon states is much higher than for thymine. As a result, phage grown under conditions that permit 5BU incorporation exhibit a higher than normal mutation rate. The common, tautomeric, and the ionized forms of thymine and its analogue

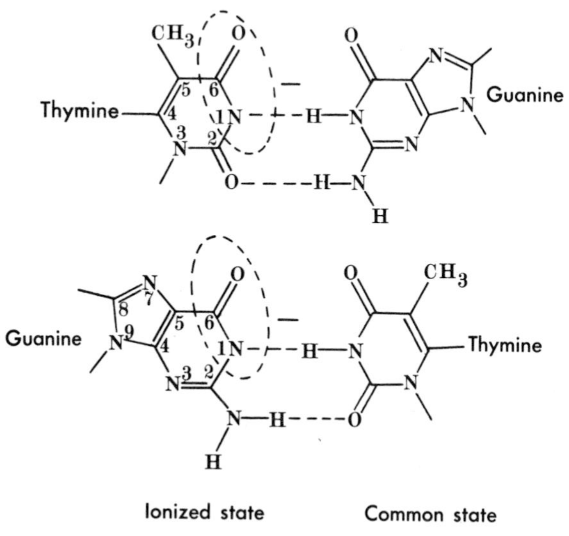

Ionized state   Common state

Fig. 9-4. "Forbidden base pairs" resulting from ionization of the number 1 nitrogen.

5BU are shown in Fig. 9-5. As can be seen there, tautomerization and ionization both involve the loss of hydrogen from the number 1 nitrogen. This common structural feature results in a base-pair transition when either tautomerization or ionization occurs.

Experimental evidence that 5BU produces its mutations through transitions rests on the following observations. Mutations induced in phage through the use of 5BU can, in turn, be induced to back mutate (revert) to their original form through the use of 5BU. This is explainable only if 5BU can effect transitions both from A-T to G-C and also from G-C to A-T. In the case of the tautomerism of adenine, either type of transition can occur, depending on whether the rare state of the adenine is in the old DNA chain or in the newly formed chain. In similar fashion, 5BU, in place of thymine, can in its rare form cause either type of transition to occur. Further evidence for the hypothesis that 5BU causes base-pair substitutions to occur comes from the observation that some 5BU-induced mutants can revert only when the 5BU is present in the old DNA chain, while other mutants can revert only when 5BU is present in the nucleotide forming the new DNA chain. This too implies the reversibility of 5BU effects and fits the model for transitions very nicely.

NITROUS ACID. Mutations, as discussed previously, can occur as copy errors during DNA replication. However, DNA synthesis is not a requirement for the production of all mutants. Spontaneous forms can occur in nondividing cells. It has also been found that mutants can be produced in nondividing cells by various mutagenic agents. A good example of this type of mutagen is *nitrous acid*. Phage can be treated with nitrous acid, removed from it, and then allowed to multiply in bacteria. A small percentage of the treated virus particles will give rise to mutant offspring. However, all of these affected phage will also give rise to wild type offspring. This result is expected if the chemical change is induced in only one of the two DNA strands. On subsequent replication, the modified strand produces mutant progeny, and the unmodified strand produces normal phage.

Nitrous acid treatment of amino-substituted purines and pyrimidines results in the replacement of the amino group by a hydroxyl group. The deamination of cytosine leads to the formation of uracil, whose hydrogen-bonding properties are essentially those of thymine, while the deamination of adenine results in the formation of hypoxanthine, whose hydrogen-bonding properties are similar to those of guanine. The

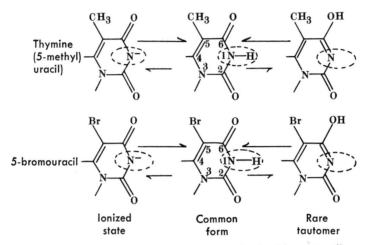

Fig. 9-5. Molecular basis of mutation induction by 5-bromouracil.

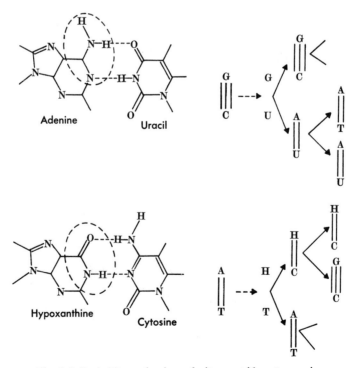

Fig. 9-6. Probable mechanism of nitrous acid mutagenesis.

net result of these changes should be, after a cycle of DNA synthesis, the substitution of A-T for G-C when cytosine is deaminated, and the substitution of G-C for A-T when adenine is deaminated. The transitions involved are shown in Fig. 9-6.

As in the case of 5BU, nitrous acid can be used to cause the reversion of nitrous acid–induced mutants back to their previous state. In fact, each of these two mutagens can be used to revert the other's induced mutants. It has been further shown that any base analogue (5BU, hydroxylamine, 2-amino-purine, etc.) can be used to revert any transition type mutant regardless of the mutagen that induced it.

• • •

We may now ask whether all known T4rII mutants are reversible by base analogues. Their being so would mean that all known mutants at this locus were caused by transitions. The answer to the question is in the negative. There are a large number of *rII* mutants that cannot be reverted by base analogues. Among them are most (85%) of the spontaneous mutants, and those mutants induced by acridine dyes, x rays, UV irradiation, and other mutagens. Such mutants must be other than transition type mutations. We shall now consider some nontransition type mutations.

Additions and deletions

A group of mutants of a nontransition type are induced by the acridine dye *proflavin*. A proflavin molecule can, it is believed, insert between two successive bases of a DNA strand, thereby stretching the strand lengthwise. At replication, this situation would allow the insertion of an extra nucleotide in the complementary chain at the position occupied by the proflavin molecule. After another cycle of DNA synthesis, there would emerge a double helix with an *added* nucleotide pair. In corresponding fashion, it is believed that during DNA replication a proflavin molecule can insert temporarily in the newly formed strand, thus excluding the normal base at that position.

The new DNA strand will then have one less nucleotide than the template. Should the proflavin molecule subsequently be removed from the strand, a double helix with a *deleted* nucleotide pair would emerge after the next cycle of DNA synthesis.

The intercalation of a proflavin molecule between two nucleotides can produce both an addition-mutant and a deletion-mutant if it occurs at a time when DNA molecules are pairing and crossing-over with one another. The stretching of the strand by the molecule of the mutagen could shift one of the bases in such a fashion that an unequal crossing-over would occur. This would result in one chain's having one less nucleotide than it should and its complementary chain's having an extra nucleotide. After the next cycle of DNA synthesis, one DNA double helix would appear with a *deleted* nucleotide pair, and the other with an *added* nucleotide pair. Recent evidence in favor of such a hypothesis is afforded by the demonstration that acridine orange is mutagenic for bacteria only if it is added when the bacteria are undergoing recombination.

Through one or both of the above-mentioned mechanisms, proflavin is thought to induce the addition or deletion of one or even, possibly, several adjacent nucleotide pairs. Proflavin-induced mutants can be induced to revert at high frequency with additional proflavin treatment. They do not revert after treatment with mutagens that cause transitions. Conversely, proflavin does not cause base-analogue–induced mutants to revert. It is interesting to note that most *rII* spontaneous mutants can be induced to revert by proflavin. This would imply that *rII* spontaneous mutants and proflavin mutants involve the same kind of change. However, the addition-deletion type of spontaneous mutation is not characteristic of all organisms. In other organisms most spontaneous mutants are not of the addition-deletion type but, rather, of the transition type. This can be seen in the fact that mutations have not been produced as readily by acridine dyes in most organisms as they have been in bacteriophage.

Earlier, we discussed the enormous effect of the addition or deletion of nucleotides, in other than multiples of three, on the protein subsequently produced. If the genetic message is read as a series of triplets starting from a fixed point, the addition or deletion of nucleotides would result in a completely altered protein distal to that point. Therefore, it is not surprising to find that the proflavin mutants that survive occur only in genes that control what appear to be dispensable functions. The loci that control such essential proteins as the head protein and the tail fibers of phage T4 cannot be successfully mutated by acridine dyes, although these loci can be successfully mutated by base analogues.

We shall consider one more aspect of addition and deletion mutants: their reverse mutations to wild type. Only those mutants involving very small numbers of nucleotides revert to wild type at a measurable rate (greater than about 10^{-9}). The inability of mutants involving large numbers of nucleotides to revert is easily understood, since gross changes in the nucleotide sequence of a DNA molecule are not likely to be repaired by the chance process of mutation. Another interesting feature of reverse mutation in these mutants is the finding that most of the wild type revertants are actually double mutants in which a second mutation has occurred that cancels the effect of the primary mutation. Such reversions are called *suppressor mutations*. It is easy to see that the effect of an addition would be reversed by a deletion and vice versa, if the two mutations occurred reasonably close to one another. The reading of the message would be unaltered, except for the region between the two mutations. If the segment of the polypeptide coded for by this altered region is not critical, the resultant protein could still function. In 1961 F. H. C. Crick and coworkers reported on the experimental evidence they had gathered on the T4rII mutants. They found that most proflavin-induced mutations fell into two mutually exclusive classes, arbitrarily called *plus* and *minus*. When plus and minus mutants were put into the same phage by recombination, they could reverse one another and produce the wild type (T4r⁺) phenotype. Neither two plus mutants nor two minus mutants could do this. Presumably, one class represented addition mutants, and the other contained deletion mutants. Thus, the experimental evidence fitted the model proposed for the acridine dye–induced mutants. We shall now consider another kind of base-substitution type mutation, called *transversion*.

Transversions

Base-pair substitution mutants need not always involve transitions. A different type of mutation caused by the replacement of one base by another is called a *transversion*. The transversion, whose existence was first postulated by E. Freese in 1959, involves the substitution, in a base pair, of a purine for a pyrimidine, and vice versa. Evidence for the occurrence of a transversion is indirect. As an example of the reasoning involved, we can consider the different human hemoglobins. Analyses of normal human hemoglobin and sickle cell hemoglobin show that the two differ by a single amino acid. Normal hemoglobin has glutamic acid at a given point in the protein, while sickle cell hemoglobin has valine at the same position. The genetic code for glutamic acid is GAA or GAG, while that for valine is GUU or GUC or GUA or GUG. If one makes the assumption that sickle cell hemoglobin is a mutant form derived from normal hemoglobin, the substitution of the "U" in GUA or GUG for the middle "A" in GAA or GAG would represent a transversion. This kind of reasoning represents the simplest hypothesis that will explain a mutation from normal to sickle cell hemoglobin. Obviously, other models of the steps involved are possible, but they are more complicated and involve less probable events.

Experimental evidence for the occurrence of transversions comes from studies of such mutagenic treatments as (1) the combination of acid and heat and (2) the use of an alkylating agent, ethyl ethane-sulfonate

(EES). These mutagenic treatments have the common property of depurinating DNA. The removal of a purine from a strand of DNA leaves a gap at that point. At the time of replication, it would be possible for any of the four bases to insert in the complementary newly formed strand. If the inserted nucleotide contained a purine, the complementary strand would contain a transversion. The next cycle of DNA synthesis would then yield a DNA molecule containing the complete transversion. The steps hypothesized for a transversion are shown in Fig. 9-7. Transversions would not be expected to be reversible either by mutagens that cause only transitions (5BU, nitrous acid, etc.) or by mutagens that cause only additions and deletions of nucleotides

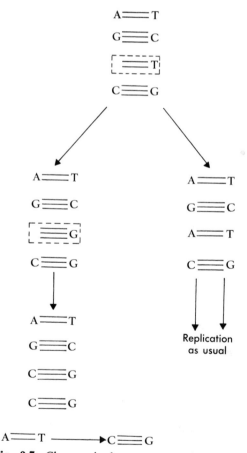

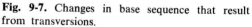

Fig. 9-7. Changes in base sequence that result from transversions.

(acridine dyes). Investigators have found that about one third of the mutants produced by EES in T4 appear to fall into the category of transversions. The other two thirds of the mutants appear to be transitions, indicating that ethylated bases undergo tautomeric shifts.

Radiation-induced mutations

We now turn our attention to the mutagenic effects of radiations. We shall ask whether all radiations produce mutations in the same manner and also whether the damage caused by radiations as well as other mutagens can be repaired.

The radiations that are important in mutagenesis fall into two categories, *ionizing radiations* and *nonionizing radiations*. The former include x rays and gamma rays; alpha and beta rays; electrons, neutrons, protons, and other fast-moving particles. Nonionizing radiations include ultraviolet and visible light. Of these latter two, only ultraviolet light is important in causing mutations. Both ionizing and nonionizing radiations, in sufficiently high doses, will kill exposed organisms and will also induce mutations among the survivors.

Ionizing radiation

Relatively little is known about the mechanism by which ionizing radiations produce mutations. When ionizing radiations pass through matter, they hit and eject electrons from the outer shells of atoms. This results in the atoms' becoming positively charged ions. In addition, the ejected electrons move at high speed and, in turn, knock other electrons free from their respective atoms. Eventually, when their energy is dissipated, the free electrons attached to other atoms which then become negatively charged ions. Since each electron lost from one atom is eventually gained by another, ions that are formed by radiations occur as pairs. Since each particle travels but a short distance, there results a cluster of ion pairs along the path of the radiation. Ions must undergo chemical reactions in order to neutralize their charge and reach a stable configuration. The mutagenic effects of ionizing radiations are thought to take place during these chemical reactions.

One of the known effects of ionizing radiation is the breaking of chromosomes and chromatids. Such breaks must involve the sugar-phosphate backbone of the polynucleotide strands. When both strands of the double helix are broken, one can get a loss of the distal end of the chromosome with the possible development of a breakage-fusion-bridge cycle after its duplication. Two simultaneous breaks in the same chromosome can lead to a deletion or inversion, while simultaneous breaks in nonhomologous chromosomes can lead to translocations. The characteristics of these various types of chromosomal aberrations were discussed in Chapter 7. Breaks can also occur in only one of the two strands of the double helix. Two simultaneous breaks in the same strand of the double helix might, at DNA replication, result in the loss of the intermediate segment. This could conceivably involve a single nucleotide and would produce a deletion mutant indistinguishable from one produced by proflavin.

A seemingly anomalous role is played by oxygen in chromosome breakage during irradiation and in chromosome restitution after irradiation. X-ray–induced chromosomal aberrations are more frequent at high oxygen tensions. This observation has led to a theory that x rays may act in an indirect way. Oxygen is important in the formation of H_2O_2 and HO_2 in irradiated water, and these products may be the factors that induce the breaks. Investigators have also found that, after irradiation, reducing the oxygen tension increases the time necessary for chromosome breaks to fuse. It is quite possible that oxygen is important in very different biochemical reactions, promoting both chromosome breakage and restitution.

Nonionizing radiation

When ultraviolet light (UV) is absorbed by nucleic acids, the absorbed energy can cause alterations in the bond characteristics of the purines and pyrimidines. Bases so altered are called *photoproducts*. Pyrimidines are found to be more liable to such changes than purines. One of the consequences of the altered bond characteristics is the formation of covalent bonds between adjacent pyrimidines of the same DNA strand. The linked pyrimidines are called *dimers*. Of the three possible types of pyrimidine dimers that can be formed in DNA, the thymine dimer is the one that forms most readily. Dimerization interferes with the proper base-pairing of thymine with adenine and may result in thymine's pairing with guanine. This will produce a T-A to C-G transition. Other photoproducts such as cytosine dimers are formed to a lesser extent and cause mutations by becoming deaminated to uracil dimers. Since uracil acts like thymine, this will eventually produce a G-C to A-T transition.

Studies of the mutagenic effects of UV led to the answer of a question about spontaneous mutation rates that had nagged geneticists for some time. They had noted that spontaneous mutation rates were extremely low, being in the order of one in 100 million replications for any particular gene. It seemed unlikely that each gene was able to replicate itself with such remarkable accuracy without the help of an error-correcting mechanism. The first indication of the operation of an error-correcting mechanism came in 1949, when A. Kelner reported that the number of bacteria surviving large doses of UV radiation could be increased by a factor of several hundred thousand if, subsequent to irradiation, the organisms were exposed to an intense source of visible light. The phenomenon was labeled *photoreactivation*. Other investigators subsequently showed that during UV radiation a particular enzyme is selectively bound to the bacterial DNA. During the photoreactivation process, the enzyme is activated by the visible light, which serves as an energy source. The enzyme then cleaves the pyrimidine dimers, thereby restoring them to their original form. This repair process is enzyme mediated and light dependent.

Another type of UV repair mechanism

was separately discovered, which is not light dependent and which has been termed *dark-reactivation.* In 1946 E. Witkin discovered a mutant of *E. coli* strain B that was resistant to UV radiation. It was called "B/r." This strain resisted not only the killing effect of UV but also its mutagenic effect. The strain did not require light to manifest its radiation resistance. In 1958 R. Hill discovered a mutant of strain B that was extremely sensitive to radiation, and this was called "B_{s-1}." B_{s-1} was sensitive to both the killing and the mutagenic effects of UV. Conjugation experiments between B/r and B_{s-1} revealed the presence of at least three genes that can be transferred from B/r to B_{s-1}, thereby making the latter radiation-resistant. The question arose whether the UV resistance of B/r depended on the splitting of the thymine dimers. The answer was provided in an experiment reported by Setlow and Carrier in 1964. Cultures of UV-resistant and UV-sensitive bacteria were grown separately in broth containing radioactive thymine. After sufficient time had elapsed so that the radioactive thymine had become incorporated into the DNA, the organisms were exposed to UV and then placed in the dark. Thirty minutes later the bacteria were broken open and the location of the labeled thymine was determined. In the UV-sensitive strain, all the radioactive thymine that had been incorporated into DNA was found in the bacterial chromosome. However, in the UV-resistant strain, some radioactive thymine was found in small fragments of DNA that were separated from the bacterial chromosome. These fragments were analyzed by paper *chromatography,* a technique by which substances are separated according to their characteristic rates of travel along a piece of paper that has been wetted with a solvent. With this technique, thymine dimers can easily be distinguished from single bases or, even, combinations of bases that are not in a dimer configuration. It was found that the fragments of DNA contained thymine dimers. The experiment provided strong evidence that dark-reactivation of UV-damaged DNA does not involve the splitting of dimers but rather involves their actual removal from the DNA molecule. This hypothesis was confirmed and extended by later experiments.

The process of dark-reactivation, as pictured by Howard-Flanders and Boyce in 1964, appears to be as follows: (1) an unidentified enzyme *(endonuclease)* makes a cut in the polynucleotide strand on either side of the dimer and excises a short, single-strand segment of the DNA; (2) according to evidence that we have not reviewed, another enzyme *(exonuclease)* is believed to widen the gap produced by the action of the endonuclease; (3) DNA-*polymerase* resynthesizes the missing segment, using the remaining opposite strand as a template; and (4) the final gap is closed by some enzymatic rejoining process, probably mediated by a ligase-type enzyme. Thus *E. coli* has two UV repair mechanisms, one light dependent and acting through the splitting of thymine dimers, and the other light independent and acting through the excising of the thymine dimer from the DNA strand. It is interesting to note that the enzymes involved in dark-reactivation appear to also be involved in recombination. *E. coli* mutants that are Rec⁻ have also been found to be UV-sensitive.

These discoveries of repair processes provide an explanation for still another phenomenon associated with UV-induced mutants. It has been found that one can affect the yield of mutants by various postirradiation treatments. Cells that are prevented from replicating DNA immediately after UV-irradiation undergo a decline in frequency of mutation. Cells that are allowed to replicate DNA normally during the postirradiation period show a maximum number of mutations. It has been hypothesized that the ability of a given photoproduct to cause a mutation depends on which enzyme system acts on the photo-product first. If the enzymes of DNA replication act first, a mutation may result. If the enzymes of DNA repair act first, the photoproduct is removed and the potential mutation is lost. Any post-

UV treatment that affects the relative rates of DNA replication and DNA repair will affect the number of mutations resulting from the radiation.

This discussion of repair processes has broad implications, as it has been discovered that DNA structural defects other than pyrimidine dimers can be repaired. DNA repair has been shown to occur in *E. coli* treated with nitrogen mustard, a chemical that reacts with guanine bases and causes a guanine-guanine cross-linkage between bases in opposite strands of the double helix. It has also been found that DNA damage caused by such diverse agents as x rays and the antibiotic Mitomycin C can all be repaired in radiation-resistant strains of *E. coli*. Another important fact about repair phenomena is that they occur in organisms other than *E. coli*.

The inability of cells to carry out DNA repair may have considerable medical importance for certain human beings. One case in point involves the genetic disease *xeroderma pigmentosum*. This is a rare disease, inherited as an autosomal recessive, in which the skin of the affected individual is unusually sensitive to UV light. Exposure of these individuals to the sun, even in moderate amounts, results in the development of an extreme degree of "freckling." This sensitivity to UV radiation is associated with the formation of skin cancers, usually within the first few years of life. When cultures of xeroderma pigmentosum cells are experimentally exposed to UV, it is found that there is no excision of pyrimidine dimers nor is there any other type of DNA repair. Investigations have shown that these cells lack a functional form of the endonuclease enzyme. Cultures of cells from persons free of this genetic disease do excise pyrimidine dimers after exposure to UV and do contain a functional endonuclease enzyme. On the basis of these findings, one cannot assume a direct relationship between DNA repair and cancer formation. However, as pointed out earlier, a relationship does exist between delayed DNA repair and the increased occurrence of mutations, after

exposure of bacterial cells to UV. Presumably this type of relationship would also exist for human cells that completely lack a DNA-repair capability. When considered together, all the above information would indicate that further studies of DNA repair should increase our understanding of the molecular basis of some human diseases.

Crossing-over and mutation

In our discussion of meiosis in Chapter 7, we considered a molecular model of crossing-over (Fig. 7-15). It was pointed out that "errors" in the process might occur. We shall now look at the mutational consequences of some of the faulty crossover events that can take place. As was stressed in our earlier discussion, the only known attractive force at work on a DNA molecule involves hydrogen bond formation between complementary bases. This situation, however, allows the possibility that a mismatching will occur between DNA strands that contain sequences of identical complementary bases. As an example we may consider two DNA double helixes (DNA strands 1 and 2 versus DNA strands 3 and 4), each coming, of course, from a different nonsister chromatid, as follows:

```
(1)    ←——————————
            CAAAAAC
(2)         GTTTTTG
       ——————————→

(3)    ←——————————
            CAAAAAC
(4)         GTTTTTG
       ——————————→
```

Let us assume that a break occurs in strand 2 immediately after the last T-containing nucleotide and in the complementary nucleotide chain of opposite polarity, strand 3, immediately before the first A-containing nucleotide. A number of possibilities for mismatching of bases exist. One of these is as follows:

```
(2)    ——————————
            GTTTTT
(3)         AAAAAC
       ——————————
```

Toward completion of the process that results in the formation of a crossover molecule, one would get the following:

```
              ------------------>
(2)        GTTTTT  G
(3)          C  AAAAAC
              <------------------
```

At this stage of the crossover process, the cell's DNA repair mechanism can insert a T-containing nucleotide in the upper strand and an A-containing nucleotide in the lower-strand. This would result in an addition of a nucleotide pair to the DNA molecule.

As pointed out earlier, mismatching of bases can occur in different ways. Let us assume that the breaks in the DNA strands have occurred as described above and that the following type of mismatching has taken place:

```
         ------------------
(2)        GTTTTT
(3)        AAAAAC
         ------------------
```

If, somewhere in the crossover process, the cell's DNA-repair system "recognizes" the mispaired bases (G-A and T-C) and uses the upper strand as the template for "correction" of the lower strand, one would get the following double helix:

```
         ------------------>
(2)        GTTTTT
(3)        CAAAAA
         <------------------
```

If the lower strand is used as the template, one would get:

```
         ------------------>
(2)        TTTTTG
(3)        AAAAAC
         <------------------
```

It is interesting to note that in both cases just mentioned, there will be deletion of a G-C nucleotide pair. In the double helix derived from the upper strand, the G-C pair from the right end of the sequence was deleted, whereas in the DNA molecule derived from the lower strand, the G-C pair from the left was missing.

All the types of addition and deletion mutations we have discussed will subject the cells containing them to the various hazards that such mutations cause in the reading of the DNA code.

Although we shall not review it in detail, it should be recognized that mispairing of nucleotides can, as a consequence of DNA repair, result in the production of a replacement-type mutation (transition or transversion). It should also be realized that mismatching of nucleotide bases can involve very extensive regions of a chromosome. In some cases these have been found to include several thousand nucleotides. The effect of these kinds of "errors" in crossing-over is the production of mutations. Some regions of certain chromosomes are very mutable and are known as "hot spots." Investigations have shown that these are regions of extensive repetitions of particular nucleotide bases and, as such, are regions with high mismatching frequencies. From all that we have said, the most striking thing about crossing-over is that it is "without error" so much of the time.

• • •

In this portion of the chapter we have examined the gene as a mutational unit and have found that a single-nucleotide change is sometimes sufficient to cause a mutation. We have also reviewed some of the mechanisms by which mutations are produced. We shall now turn our attention to the gene as a functional unit.

THE GENE AS A FUNCTIONAL UNIT

In the beginning of this chapter we pointed out that our earlier consideration of crossing-over indicated that a gene must have a point on the chromosome at which it begins and another point at which it ends. We now want to analyze the gene as a functional unit and review the evidence for our present ideas on its spatial limits. As part of our analysis of the gene as a functional unit, we shall have to consider the type of genetic test that demonstrates the functional identity of, or the difference between, two genes. This test is called the *complementation test*. To carry out this test, it must be possible to make the organism both diploid and heterozygous for the gene to be studied.

The first requirement of the complementation test presents no difficulties for normally diploid organisms. We can even solve the problem for haploid species through conjugation, transduction, and transformation. The second requirement restricts us to the study of those genes that are known to exist in at least two allelic forms.

The rII locus of phage T4

Earlier in this chapter it was stated that when a bacterial virus is mixed with a culture of bacteria and poured into an agar plate, most of the agar will become cloudy due to the dense growth of bacteria. Where the virus has infected a bacterium, a clear *plaque* will be present, due to the successive invasions and lyses of bacteria adjacent to the point of infection. The morphology of the plaque (size, nature of its edges, etc.) is characteristic for each type of phage. Mutants can occur in the viruses that affect plaque morphology. In T4, such mutants have been found and are called *r* mutants. Most of the *r* mutants lyse the bacterial cell rapidly and produce a larger plaque with sharper edges than do the phage containing the wild type allele, r^+. There are three regions of the T4 chromosome that can mutate to form *r* mutants. Depending on the region of their origin, the mutants have been labeled *rI, rII,* or *rIII.* The *r* mutants in all three regions produce the characteristic large plaques when *E. coli* strain B is used as a host. However, when *E. coli* strain K12(λ) is used as a host, only mutants *rI* and *rIII* form plaques. Although the mutant *rII* can enter K12 (λ) cells, it cannot reproduce and lyse the cells. Phage with the wild type alleles, r^+, form their smaller, fuzzy-edge plaques with both strain B and strain K12(λ). The interactions of the various viruses and the different bacterial strains are shown in Table 9-1.

One of the techniques that has been extremely important in analyzing the fine structure of the T4*rII* locus is that of mixed infections. It has been found that if K12(λ) bacteria are infected simultaneously

Table 9-1. Types of plaques formed by T4 phages in *E. coli* strains B and K12(λ)

Phage T4 genotype	Plaques formed on E. coli strain	
	B	K12 (λ)
rI	r	r
rII	r	None
rIII	r	r^+
r^+	r^+	r^+

with both *rII* and r^+ viruses, phage of both genotypes will enter the cells and reproduce. As noted above, *rII* mutants by themselves cannot reproduce in K12(λ) bacteria, although they can enter the cells. It is apparent that the r^+ virus produces some substance that the *rII* phage can utilize for its own reproduction, with the result that both phages can reproduce. The mixed infection provides the equivalent of the diploid state for the virus genome with the genes located in the *trans* position (i.e., in different homologous chromosomes). Using this technique, we can test for both functional and structural allelism between different *rII* mutants. If the mutants are functional alleles, no growth will occur. If the mutants are located in different genes, they will complement each other, and both mutants will be able to reproduce.

Complementation within the rII locus

Much of our knowledge of the fine structure of the T4*rII* locus comes from the work of S. Benzer, as reported in 1955. Thousands of *rII* mutants have been isolated. They have been tested, two at a time, by mixedly infecting K12(λ) bacteria with them. On the basis of their behavior after mixed infection, the *rII* mutants have been divided into two classes, arbitrarily called A and B. When a mutant from class A and a mutant from class B are used, an area of bacterial lysis is obtained. This implies that the *rII* locus is composed of two subunits, A and B, and that the products of both are required to produce bacterial lysis. In the mixed infection of an A and a B mutant, the A mutant presumably produces a wild type B product, and the B mutant in turn produces a wild type A product. However, if two class A mutants or two class B mutants are used, the virus will not reproduce and the K12(λ) cell will not be lysed. The subunits of the *rII* region and similar subunits of other loci have been called *cistrons*. A cistron denotes a functional unit that has been recognized as the result of *cis-trans* tests of pairs of mutants. *Cis-trans* tests are complementation tests involving mutations that have been placed in the same *(cis)* versus different *(trans)* chromosomes. A cistron is, in effect, the gene when it is viewed as a functional unit. The *rII* region of phage T4 is considered as containing gene *rIIA* and gene *rIIB*. In terms of DNA, a cistron is that section of the DNA molecule which specifies the composition of a particular polypeptide chain. The average cistron (gene) consists of 1000 to 1500 nucleotide pairs.

One tangential point has to be made on the mixed infections involving *rIIA* and *rIIB* mutants. The phage emerging from the K12(λ) cells are mainly *rIIA* and *rIIB*. A few r^+ and *rIIArIIB* genotypes are also found. The latter two genotypes are the result of the occasional recombinants that occur. We shall now consider in detail the gene as a unit of recombination.

THE GENE AS A RECOMBINATIONAL UNIT

We now want to analyze the gene as a unit of recombination and review the evidence for our present ideas on this aspect of its unity. Involved in this study will be the question of whether crossing-over can occur within the gene, and if so, to what extent is the gene structurally subdivisible.

Pseudoalleles

We have defined alleles in Chapter 3 as homologous genes that yield, in homozygous conditions, different phenotypes. In heterozygous condition, the phenotype will reflect the interaction of the alleles involved. Under this definition of alleles, one would not expect to observe any recombinants among the offspring of heterozygous individuals. The absence of recombinants would equate functional allelism with structural allelism. A study demonstrating that these two aspects of allelism need not be identical was reported by E. B. Lewis in 1952. He investigated the locus known as "white" in *Drosophila melanogaster*. The locus, located on the X chromosome, affects eye color and represents a multiple-allelic series. Among its alleles are w^a and w which, in homozygous condition, result respectively in apricot- and white-colored eyes. The compound w^a/w, which can occur only in females, has very light–colored eyes. If many offspring of w^a/w females are examined, one finds an occasional red-eyed (wild type) fly. The wild type flies occur too frequently to be mutations. Genetic analysis involving other genes on both sides of the white locus demonstrates that the wild type progeny are the result of crossing-over between these two alleles. The male recombinants have one of the following genotypes: $w^a w/Y$ or $w^{a+}w^+/Y$. The female parents must now be considered as double heterozygotes with the genotype $w^a w^+/w^{a+}w$.

The frequency of crossing-over in the experiment just described was very small (less than 1:1000), indicating that the alleles are closely linked. However, the fact that the alleles can be separated demonstrates that functional allelism cannot be taken as proof of structural allelism. The term *pseudoallelism* is used to designate situations in which these two aspects of allelism are not identical. It has been shown that the white locus is composed of at least five separable sites, all of which function as alleles in compounds with one another. Pseudoallelic systems have been found to be quite common in diploid organisms.

Another important observation was made in Lewis's experiment. It was possible to form a doubly heterozygous female whose genotype is $w^a w/w^{a+} w^+$. Such a fly has a wild type phenotype. It will be recalled that the genotype $w^a w^+/w^{a+} w$ resulted in very light–colored eyes. Thus the phenotype of the heterozygote depends on the location of the wild type and mutant alleles. If both wild type alleles are together in one chromosome and both mutant alleles in the other, the organism is wild type. The alleles involved are said to be in the *cis* position. If one wild type allele and one mutant allele are on each chromosome, the *trans* position, the organism is mutant. The difference in results is explained as follows. In the *cis* configuration, the chromosome containing the wild type alleles is producing mRNA that codes for an enzyme that contributes to the production of red eye pigment. Its homologous chromosome is also producing mRNA but for a much lighter colored pigment, which is masked by the red. In the *trans* configuration, both chromosomes carry mutant alleles, with the result that only light-colored pigments are produced. The *cis-trans* configuration will occupy us a great deal in our discussion of the fine structure of the gene.

Our study of pseudoalleles has shown that the recombinational gene can be much smaller than the functional gene. However, the experiment involved did not give us any indication of the fine structure of any of the genes involved in eye pigment production. In order to get information on the nature of intragenic structure, we shall have to turn to studies on the genes of microorganisms where the necessary experiments have been performed.

Recombination within rII cistrons

Our understanding of the fine structure of the T4*rII* locus did not end with the realization that it consists of two complementary cistrons. It has been possible to analyze the structure of both the A and B cistrons. To do this, we must be able to demonstrate which of the many A mutants are alleles of one another and which of the many B mutants are alleles. This necessitates the detection of recombinants that would occur if two

A or two B mutants were reproducing in the same organism. The experimental procedure is relatively simple. We shall take as our example two A mutants, with the understanding that similar results are obtained when two B mutants are chosen. Two *rIIA* mutants are mixed together and are used to infect strain B bacteria. Both mutants can reproduce in strain B cells and will normally produce a large, sharp-edged plaque. The phage that are released from the lysed bacteria are used to infect strain K12(λ) cells. If no recombinants have occurred, no plaque will be formed. If recombination involving the two A mutants has occurred, one of the possible recombinants will have the wild type genotype and will produce an r^+ plaque (small with fuzzy edges). The double-mutant recombinant will not be able to reproduce in K12(λ) cells. By proper dilution techniques and plating on strain B cells, we can determine the total population size of phage used on the K12(λ) cells. By this experimental procedure, recombinants occurring at frequencies as low as 0.00001% (1 recombinant in 10 million) can be detected.

Deletion mutants of the rII locus

A very large number of different *rIIA* and *rIIB* mutants were analyzed in the above fashion. From the results, a linear map of the *rII* locus was constructed. All the A mutations were found to be located in one section of the map, and all the B mutations were found to be located in an adjacent section, with no overlap between A and B mutations. In carrying out the above analysis, it was found that about 10% of the *rII* mutants did not act as point mutations. For example, mutants A-1 and A-2 were found to recombine at some given frequency. However, mutant A-3 would not recombine with either A-1 or A-2, although A-3 would recombine with other A mutants. It was concluded that A-3 must represent a deletion that includes the mutant sites A-1 and A-2. The assumption that the unusual mutations were in fact deletions was supported by the discovery that these mutations did not revert to wild type, as did the normal point mutations. These deletion mutants can also be used to study recombination, through the same technique as was used for point mutations. A deletion mutant will only give r^+ recombinants in crosses with another deletion mutant if the two do not lack a common segment. An example of this type of situation is shown below, where each line represents the relative map location and size of a deleted section of the *rII* region in a different virus.

No wild type recombinants can be formed between a T4 phage that contains deletion 3 and one that contains deletion 1 or 2, since in each case the phage lack a common chromosomal segment (i.e., their deletion lines overlap). For the same reason, no wild type recombinants can be formed between a virus that contains deletion 4 and one that contains deletion 1. However, wild type recombinants can be produced if the two viruses involved contain, respectively, deletions 1 and 2, or deletions 4 and 2, or deletions 4 and 3. By means of the technique used to study recombination within *rII* cistrons, deletion mutants were tested in pairs and their positions in the *rII* locus were determined. Some deletions were found to include portions of both A and B cistrons, but the majority of them were restricted to one or the other cistron.

Utilizing the recombination data both from point mutations and deletion mutations, investigators were able to subdivide the *rII* region into more than 400 recombinable sites. From other data, which we have not reviewed, it has been estimated that the *rII* locus consists of about 2000 nucleotide pairs. This would place the lower limit of recombination at about five nucleotides. However, since there is no structural feature of the DNA molecule that occurs only with every fifth nucleotide and since the 400 recombinable sites represent the minimum number known to date, we may reasonably assume that the smallest recombinational unit in phage will eventually be shown to be one nucleotide. This would mean that the smallest mutational unit (as discussed earlier in this chapter) and the smallest recombinational unit were equal in size. We shall now turn our attention to the question of how gene action is regulated.

REGULATION OF GENE ACTION

Studies on the regulation of gene action have centered largely on *E. coli* and its enzymes. Most important have been those enzymes that are produced only when there is need for them. As an example, we can consider the enzymes associated with the "lactose" *(lac)* region of the *E. coli* chromosome. These enzymes are involved in the hydrolysis of lactose into galactose and glucose and are synthesized only when lactose is present in the medium. Enzymes that are produced only when their substrates are available are called *inducible* enzymes. Through the study of such enzymes has come our basic concept of the nature of gene regulation.

Lac region of the E. coli chromosome

The *lac* region of the *E. coli* chromosome controls the synthesis of three enzymes. These are β-galactosidase, which catalyzes the hydrolysis of lactose into galactose and glucose; β-galactoside permease, which is responsible for the uptake and concentration of lactose in the cell; and galactoside transacetylase, an enzyme whose function in vivo is not known. Mutations affecting the structure of these enzymes have been found. Those affecting β-galactosidase (*z* gene), permease (*y* gene), and transacetylase (*ac* gene) map in adjacent loci of the bacterial chromosome. In addition, the mutations complement one another in diploids, formed by conjugation, thus demonstrating that they are located within three different cistrons.

Studies have shown that when neolactose is used in the food medium instead of lactose, the wild type *E. coli* cannot utilize the neolactose for energy; hence, it cannot grow in a medium containing neolactose as the sole energy source. Eventually a mutant bacterium was discovered that could utilize this sugar as an energy source. The mutant was found to produce β-galactosidase continuously *(constitutively)*, and the enzyme was found to be perfectly capable of hydrolyzing neolactose. It was concluded that neolactose is a substrate for, but not an inducer of, β-galactosidase. It was also apparent that what had mutated was not the gene that determines the structure of the enzyme but rather a gene that determines the conditions necessary for the enzyme's synthesis. The wild type form of this *regulator* gene was designated i^+ (inducible) and the mutant gene was designated i^- (constitutive). Subsequent genetic analysis showed that the i gene was located close to, but was distinct from, the gene controlling the synthesis of β-galactosidase.

In another experiment, bacterial cells were made diploid for the *lac* region. One DNA section contained the genes i^+ z^+ y^+ ac^+, which would be expected to produce its enzymes inducibly. The other chromosomal *lac* region contained i^- z^- y^+ ac^+, which would be expected to produce a structurally modified form of β-galactosidase and the normal form of its other enzymes constitutively. It was found that i^+ (inducible) is dominant to i^- (constitutive) and that both types of β-galactosidase and the other two enzymes were all produced only under induction, that is, when lactose was present. It seemed evident that whatever controlled the structural genes was transmissible through the cytoplasm and that the i^+ gene had produced a *repressor* substance that prevented the z^- y^+ ac^+ genes from functioning despite their association with the i^- gene. It also seemed clear that the repressor substance was not binding the structural genes individually but rather must be affecting a receptor site that controlled the synthesis of all the enzymes. This idea was supported by the following observation.

A *lac* region was studied that we shall designate as i^x z^+ y^+ ac^+. All the enzymes were formed constitutively, which would lead one to believe that i^x was in fact i^-.

However, when a diploid was formed with a *lac* region known to be i^- z^- y^+ ac^+, no structurally modified form of β-galactosidase was produced constitutively. This meant that i^x was actually i^+ and was producing a repressor substance. However, this repressor substance was ineffective on its own structural genes. To explain the above observations, we must hypothesize another locus called the *operator* locus *(o)*, which determines the action of $z, y,$ and ac genes. The operator locus must then be the receptor site to which the repressor substance binds when enzyme synthesis is stopped. The operator gene itself can be found in different allelic forms. The wild type operator gene is designated as o^+ and is the receptor site for the repressor substance. Another allele of this gene is o^c, which is insensitive to the repressor substance. When o^c is present, the $z, y,$ and ac genes act constitutively regardless of the form of the i gene. We may now designate the genotype of i^x z^+ y^+ ac^+ as i^+ o^c z^+ y^+ ac^+ and the genotype of i^- z^- y^+ ac^+ as i^- o^+ z^- y^+ ac^+. Since the z^- gene did not function in the diploid formed by these two *lac* regions, we may conclude that the operator gene can affect only those genes with which it is in the *cis* position.

All the experiments just discussed led to the formulation by Jacob and Monod in 1961 of the *operon* hypothesis that forms the basis of our concept of the nature of gene regulation. An operon consists of a series of structural genes whose expression is controlled by an operator locus situated at one end of the group. The operator responds to a cytoplasmic repressor substance produced by a regulator gene, so that in the presence of the repressor the operon is inactive. The function of the inducer (in our example, lactose) must then be to inactivate the repressor. A map of the *lac* region is shown in Fig. 9-8, and the mechanics of operon function is shown in Fig. 9-9. We may summarize our discussion by stating that the gene may be part of a coordinated complex consisting of an operator plus one or more structural genes.

Although the concept of the operon has

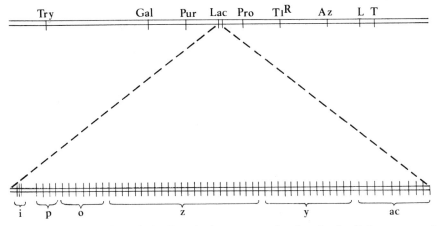

Fig. 9-8. Map of a portion of the *E. coli* chromosome showing the detailed structure of the lactose region.

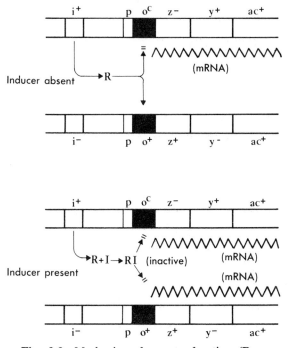

Fig. 9-9. Mechanics of operon function (**R** = repressor substance and **I** = inducer substance).

and *o* genes (Figs. 9-8 and 9-9). Its existence was hypothesized in 1964 by Jacob and co-workers to explain the occurrence of *lac* mutants that produce reduced amounts of enzymes despite the fact that they are shown to be o^c (constitutive). This is an especially powerful line of reasoning when we realize that most, if not all, o^c mutations are found to actually be deletions for that portion of the chromosome. It therefore seemed inconsistent to ascribe a mutable rate-regulating function to the operator locus. Support for the promoter hypothesis came from a study of overlapping deletions that were distributed over the portion of the chromosome between the *i* and *o* sites. Investigators found that the greater the size of the deletion between these points, the greater was the reduction in enzyme synthesis. In some fashion the promoter site was regulating the efficiency of transcription of the structural *(z, y, ac)* genes. The conclusion reached was that the promoter site is that portion of the DNA molecule at which RNA polymerase molecules attach themselves before initiating transcription. Promoter mutants, resulting from deletions or point mutations, reduce the site's affinity for RNA polymerase. Under this extension of our concept of gene regulation, it is believed that when a repressor molecule binds to an operator locus, it functions to block the pro-

been amply supported by experimental evidence over the years, the overall picture of gene control has been complicated by the discovery of yet another "regulatory" site. It is called the *promoter (p)* and, in the *lac* region of *E. coli,* is located between the *i* gress of the RNA-polymerase molecule to the structural genes of the operon. We thus have two mechanisms involved in the regulation of gene transcription. One mechanism provides for initiating and terminating transcription (i.e., the *o* locus), and the second mechanism provides for determining the rate of transcription (i.e., the *p* locus). Promoter genes have been found in many organisms.

SUMMARY

In this chapter we have considered three concepts of the gene: a mutational unit, a recombinational unit, a functional unit. Viewing it as a mutational unit, we found that the addition, deletion, or substitution of a single nucleotide can result in a muta- tion. Our discussion of the gene as a recombinational unit found that genes that were not structural alleles could still be functional alleles. We also found that, most probably, the smallest unit of recombination was a single nucleotide. As a functional unit, the average gene (cistron) consists of 1000 to 1500 nucleotide pairs. Our consideration of gene regulation revealed that a gene may be part of a coordinated complex called an operon. The operon was found to consist of an operator plus one or more structural genes. Our concepts of the gene are quite different from the classic view that presumed each gene to be functioning as a single entity and responding as a whole to mutation and recombination.

Questions and problems

1. Define and illustrate the following:
   a. Addition   c. Transition
   b. Deletion   d. Transversion
2. Which type of induced mutation—acridine dye or base analogue—do you think is more deleterious to an organism? Why?
3. Define the following terms:
   a. Tautomeric   c. Suppressor
      shift   mutation
   b. Ionization   d. Depurinating
                        agent
4. How do the base analogues act to increase the base-level rate of spontaneously occurring mutations?
5. Discuss the role of ionizing radiation in mutagenesis.
6. What mechanisms exist for the "repair" of

ultraviolet-induced mutations in *E. coli?* What are the salient features of each type of mechanism?

7. Illustrate and discuss the following statement: A single nucleotide change is sometimes sufficient to cause a mutation.
8. Define the following terms:
   a. Complementation test
   b. *Cis* configuration
   c. *Trans* configuration
   d. *r* mutants of T4
   e. Cistron
   f. Functional allelism
9. Is it possible for two genes to be functionally but not structurally allelic? Explain.
10. How would you go about determining whether the genes in the following cases were allelic:
    a. Two *rII* mutants in T4?
    b. Two galactose mutants in *E. coli?*
11. How do you explain the occasional wild type and double-mutant plaques that are observed in mixed infections of *rIIA* and *rIIB* mutants in *E. coli* K12(λ)?
12. Five *rIIA* deletion mutants of phage T4 were tested in all possible pair-combinations for wild type recombinants. The following results were obtained (+ = wild type recombinants; 0 = no recombination):

	1	2	3	4	5
5	+	+	+	+	0
4	0	0	0	0	
3	0	0	0		
2	+	0			
1	0				

Construct a topological map for the five deletions in question, illustrating the spatial relationships among them.

13. Define the following terms:
    a. Inducible   c. Regulator gene
       enzyme   d. Operator gene
    b. Constitutive   e. Repressor
       enzyme   f. Inducer
14. Five point mutations were tested for wild type recombinants with each of five deletion mutants. The following results were obtained (+ = wild type recombinants; 0 = no recombination):

	1	2	3	4	5
a	0	+	0	0	+
b	+	+	+	0	+
c	+	+	+	+	0
d	0	+	+	0	+
e	+	0	0	0	+

Determine the sequence of the point mutations and their location with reference to the deletion mutants.

15. Why are operator mutations effective only in the *cis* configuration? Why are regulator gene mutations active in both *cis* and *trans* positions?

16. Five independently isolated mutations of a locus of *Neurospora* were tested with one another for complementation. The following results were obtained (+ = some functional allelism; 0 = no functional allelism):

	1	2	3	4	5
5	+	0	0	0	0
4	0	0	+	0	
3	+	+	0		
2	+	0			
1	0				

Construct a topological map for the five genes in question, illustrating the spatial relationships among them.

17. With a variety of base analogues, alkylating agents, and acridine dyes at your disposal, how would you go about determining the nature of the base change involved in a particular *rII* mutant?

18. What type of mutations do the following base substitutions reflect?

a. $AT \rightarrow GC$ b. $GC \rightarrow AT$ c. $GC \rightarrow TA$
d. $GC \rightarrow CG$ e. $CG \rightarrow GC$ f. $AT \rightarrow CG$

19. Explain each of the following experimental findings.

a. A specific *rII* mutant can be induced by both 5BU and nitrous acid.
b. The above mutant can be induced to revert to wild type by 2-aminopurine.
c. The above mutant cannot be induced to revert by proflavin.

20. State whether active β-galactosidase will be produced inducibly, constitutively, or not at all in each of the following cases:

a. $i^+o^+z^+$ e. $i^-o^cz^+$
b. $i^+o^+z^-$ f. $i^+o^cz^-$
c. $i^+o^cz^+$ g. $i^+o^+z^-/i^-o^+z^+$
d. $i^-o^+z^+$ h. $i^+o^cz^-/i^-o^+z^+$

References

Ames, B. N., and Hartman, P. E. 1963. The histidine operon. Cold Spring Harbor Symp. Quant. Biol. **28**:349-356. (A discussion of the operon hypothesis as applied to the histidine gene cluster in *Salmonella*.)

Bautz, E., and Freese, E. 1960. On the mutagenic effect of alkylating agents. Proc. Nat. Acad. Sci. U. S. A. **46**:1585-1593. (On the mechanism of the mutagenic action of alkylating agents.)

Beckwith, J. R. 1967. Regulation of the *lac* operon. Science **156**:597-604. (Discussion of the regulatory genes controlling the *lac* operon in *E. coli*.)

Benzer, S. 1955. Fine structure of a genetic region in bacteriophage. Proc. Nat. Acad. Sci. U. S. A. **41**:344-354. (A brilliant study of rII phage mutants' map positions, designed to determine the limitation of the gene as a unit of crossover.)

Champe, S., and Benzer, S. 1962. Reversal of mutant phenotype by 5-fluorouracil: an approach to nucleotide sequences in mRNA. Proc. Nat. Acad. Sci. U. S. A. **48**:532-546. (A discussion of the mode of action of an RNA-specific mutagen.)

Crick, F. H. C., Barnett, L., Brenner, S., and Watts-Tobin, R. J. 1961. General nature of the genetic code for proteins. Nature **192**:1227-1232. (The use of acridine-induced mutants in phage as a tool for determining the triplet nature of the genetic code.)

Freese, E. 1959*a*. On the molecular explanation of spontaneous and induced mutations. Brookhaven Symp. Biol. **12**:63-73. (A report on the molecular basis of mutation explained as a result of electron shifts and altered hydrogen bonding properties.)

Freese, E. 1959*b*. The specific mutagenic effect of base analogs on phage T4. J. Mol. Biol. **1**:87-105. (On the mechanism of the mutagenic action of base analogues.)

Hill, R. F. 1958. A radiation sensitive mutant of *Escherichia coli*. Biochim. Biophys. Acta **30**:636-637. (A description of a mutant *E. coli* strain possessing altered sensitivity to UV light.)

Ippen, K., Miller, J. H., Scaife, J., and Beckwith, J. 1968. New controlling element in the *lac* operon of *E. coli*. Nature **217**:825-827. (Discussion of the promoter locus and the experimental evidence for determining its location within the *lac* region.)

Jacob, F., and Monod, J. 1961. Genetic regulatory mechanisms in the synthesis of proteins. J. Mol. Biol. **3**:318-356. (The classic, original description of the operon hypothesis, based on the lactose gene cluster in *E. coli*.)

Kelner, A. 1949. Photoreactivation of ultraviolet irradiated *E. coli* with special reference to the dose reduction principle and to ultraviolet-induced mutations. J. Bacteriol. **58**:511-522. (Original report on photoreactivation phenomenon.)

Martin, R. G. 1969. Control of gene expression. Ann. Rev. Genet. **3**:181-216. (Critical review of the experimental evidence for the operon model.)

Setlow, R. B., and Carrier, W. L. 1964. The disappearance of thymine dimers from DNA: an error-correcting mechanism. Proc. Nat. Acad. Sci. U. S. A. **51**:226-231. (A proposal for the mechanism of removal of thymine dimers.)

Watson, J. D., and Crick, F. H. C. 1953. The structure of DNA, p. 269-276. *In* Taylor, H. (ed.). 1965. Selected papers on molecular genetics. Academic Press, Inc., New York. (A hypothesis for spontaneous mutations based on the occurrence of tautomeric shifts.)

Yanofsky, C. 1960. The tryptophan synthetase system. Bacteriol. Rev. **24**:221-242. (A review of the experimental evidence for the existence of a tryptophan operon in *E. coli* and in *Neurospora*.)

Yanofsky, C. 1963. Amino acid replacements associated with mutation and recombination in the A gene and their relationship to in vitro coding data. Cold Spring Harbor Symp. Quant. Biol. **28**:581-588. (Review of the experimental evidence that some recombinational events appear to occur between adjacent nucleotides in the same coding unit.)

10 Genes and metabolism

⚰ ⚰ In earlier chapters we studied the
⚰ ⚰ gene's composition, construction,
and functions. We also considered
various types of gene interactions as they
are reflected in the individual's phenotype.
However, we have not yet analyzed the gene
as the controlling element of metabolism and
development. The genetic control of metabolism will be the object of our study in this
chapter. We shall consider the genetic control of development in the next chapter.
Since most genes influence metabolism
through the production of structural and
enzymatic proteins, our attention will be
directed to the process of transferring genetic
information from DNA to protein and to
the effects of the proteins produced on cell
metabolism. Our study will include a consideration of the factors that can affect
transcription and translation and of the consequences to the organism of changes in the
types and amounts of different proteins produced. Inevitably, some of our findings will
increase our understanding of previously discussed genetic phenomena, since all activities
of the cell are interrelated.

FACTORS AFFECTING TRANSCRIPTION

Transcription is the process by which
RNA polymerase catalyzes the synthesis of
an mRNA from a DNA template. A number
of agents have been found that affect transcription. One of these is *actinomycin D*.
This antibiotic preferentially inhibits the
RNA polymerase reaction only when DNA
containing *guanine* is involved. No inhibition occurs when a synthetic DNA polymer
that contains only adenine, thymine, and
cytosine is used. Other agents that affect
transcription are proteins called *histones*. It
is found, for example, that thymus histone
leads to an inhibition of mRNA production
but exerts a more extensive inhibition on
A-T–rich DNA than on G-C–rich DNA.
Thus we can see that the presence of certain
molecules in the cell can lead to a preferential inhibition of transcription, depending on
the nucleotide composition of the section of
DNA being transcribed. We can visualize a
situation in which some proteins are not produced at all while others are formed in normal or slightly diminished amounts. Any
radical changes in the amounts of different
proteins in the cell will usually be lethal.
Thus far, investigations of factors affecting
transcription have revealed only agents that
inhibit the process. Another possibility is the
existence of agents that cause a misreading
of the DNA and thereby produce an altered
message. This would, in turn, lead to a
modified protein, which could alter cell
metabolism.

FACTORS AFFECTING TRANSLATION

Translation is the process by which amino
acids are assembled into a polypeptide chain
in an order specified by the sequence of
codons in an mRNA. Errors in translation
(misreading) can result in a protein with
properties different from those of the protein
encoded in the mRNA. Conceivably, such
changes in protein characteristics could be
beneficial to the organism. However, in most
cases, alterations of the protein structure

lead to defective molecules that are detrimental to their carriers. We may anticipate that some of the factors affecting translation will be genetic and will themselves be inherited. Other conditions will be environmental and will affect the individuals only while the environment is so altered. Theoretically, factors affecting the translation process could include the amino acids themselves, the tRNAs involved in the transportation of the amino acids, and the ribosomes on which the polypeptide chains are formed. We shall examine some of the experimental evidence bearing on these possibilities and determine what the consequences to the organism might be.

Role of amino acids in translation

An experiment designed to test the role of amino acids in translation was reported by F. Chapeville and co-workers in 1962. A mixture of the various kinds of tRNA was extracted from *Escherichia coli*. The tRNAs were incubated with all the amino-acyl synthetases, ATP, and the amino acid cysteine that contained radioactive carbon. The cysteine became attached to its specific tRNA, and the resulting compound was isolated. The compound was then exposed to a nickel catalyst, called Raney nickel, that removed the sulfur from the cysteine, thereby converting it to alanine. The radioactive cysteine thus became radioactive alanine yet remained attached to the cysteine-tRNA. It was found that the compound replaced cysteine with alanine in a subsequently formed polypeptide. This experiment demonstrated that an amino acid does not determine its own location in the polypeptide. The amino acid appears to lose its specificity while combined with a tRNA and is carried along to the codon recognized by the tRNA. We shall now consider the role of tRNA in translation.

Role of tRNA in translation

In considering tRNA as a possible factor affecting translation, we must remember that a tRNA molecule has two recognition sites, one for a specific amino-acyl synthetase

(activating enzyme) that is carrying a particular amino acid and the other for a specific codon in an mRNA. Each recognition site is subject to alteration by the occurrence of a mutation in that portion of the organism's DNA from which the particular kind of tRNA molecule is transcribed. A mutation that causes a "wrong" amino acid to be incorporated into a protein is called a *missense mutation*. This type of mutation can produce its effects through a change in either of the two recognition sites of the tRNA. Experiments were reported in 1966 by Gupta and Khorana who worked with a cell-free amino acid–incorporating system prepared from mutant *E. coli*. The system they used was designed to produce polypeptide chains consisting solely of the amino acids valine and cysteine. They observed an incorporation of glycine instead of cysteine into the polypeptide chains formed in their experiment. This was found to be caused by a change in the amino-acyl synthetase recognition site of the cysteine-tRNA molecules produced by their mutant bacteria. In the same year Carbon and co-workers reported experiments that also involved the in vitro synthesis of polypeptide chains. Their system was designed to produce polypeptide chains consisting solely of glutamic acid and arginine. They obtained evidence that an incorporation of glycine instead of arginine was the result of a change in the anticodon of the glycine-tRNA molecules produced by their particular strain of *E. coli*. Although a missense mutation causes the production of a different protein from that specified by the organism's DNA, the mutation does not, strictly speaking, result in an alteration of the translation process.

In addition to missense mutations, one can have mutations that produce a terminator-type codon. This type of triplet does not code for any amino acid, and the mutation that results in a terminator-type codon is called a *nonsense mutation*. This type of mutation directly affects the translation process, since it results in the premature termination of a growing polypeptide chain. In many cases, a missense or a nonsense mutation can be "corrected" by the occurrence of a second mutation in a different gene of the organism. This second mutation results in the production of a functional (wild type) protein by the mutant that previously had produced only an inactive protein. Those genes which cause the suppression of the effects of mutations in other genes are called *suppressors*. Much of the research on the question of the role of tRNA in translation has centered on *nonsense suppressor mutants*. In *E. coli*, the triplet UAG (amber), when located within a message, normally results in the termination of translation of the mRNA with the release of an uncompleted polypeptide. Several amber suppressor genes have been found that permit UAG to be read as an amino acid codon and thus allow continued translation of the message. The amino acid inserted is specific for each suppressor (for example, su_I^+ inserts serine; su_{II}^+ inserts glutamine; su_{III}^+ inserts tyrosine). In 1965 Capecchi and Gussin reported on their findings that *E. coli* su_I^+ cells contain a serine-accepting tRNA that in vitro suppresses the UAG codon in the mRNA of other organisms (e.g., phage). This observation made it clear that the *su* gene somehow controls the functioning of the tRNA rather than any other part of the translation machinery of the organism.

In 1966 Smith and co-workers reported on their investigations of su_{III}^+ (tyrosine). They were able through the use of a transducing phage to transfer su_{III}^+ from one *E. coli* to another. In the new cell, the su_{III}^+ gene product increases as the transducing phage multiplies. Under these conditions, there results a considerable increase of tyrosine-accepting tRNA in the cell. The tyrosine-tRNA so formed recognizes UAG and appears not to recognize the two normal tyrosine codons, UAU and UAC. When the transducing phage is used to transfer the wild type su_{III}^- gene, a similar increase in tyrosine-accepting tRNA occurs. However, this latter tRNA does not recognize UAG but codes for both UAU and UAC. These findings demonstrate that the mutation from wild type su^- to su^+ is accompanied by a

change in the codon recognition site of the tRNA, resulting in an exclusive recognition of the UAG codon. From the experiment just described, one could hypothesize that the *su* gene is the structural gene for tyrosine-tRNA. However, proof of this hypothesis must await a demonstration of a difference in the nucleotide sequence of su^+ and su^- tRNAs.

A number of other suppressor genes have been analyzed. In each case, the codon recognition site of the tRNA appears to be implicated in the changed action of the molecule. This does not preclude the existence of other mutations, which would affect translation through an alteration of the amino-acyl synthetase recognition site of the tRNA. Let us now review the role of the ribosome in translation.

Role of the ribosome in translation

Another factor that could affect the translation process would be any ribosomal change that resulted in a misreading of an mRNA codon. An example of this phenomenon was reported by Gorini and Kataja in 1964. They were working with strains of *E. coli*, each of which was deficient in a different metabolic function. One such strain was unable to manufacture a necessary growth factor, the amino acid arginine. From this strain some streptomycin-resistant mutants were obtained. A few of these mutants which were both arginine-deficient and streptomycin-resistant could grow on a minimal medium containing streptomycin. This type of mutant was called "conditionally streptomycin dependent" (CSD). It was known that the arginine deficiency was due to a defect in an enzyme, ornithine transcarbamylase (OTC), which in its normal form converts ornithine to citrulline. This conversion, as shown in Fig. 10-1, is a necessary step in the formation of arginine. Active OTC is found only in cells grown in the presence of streptomycin. If the antibiotic is removed, enzyme production ceases. In addition, streptomycin is effective only on whole cells and does not restore OTC activity in cell extracts. From this information,

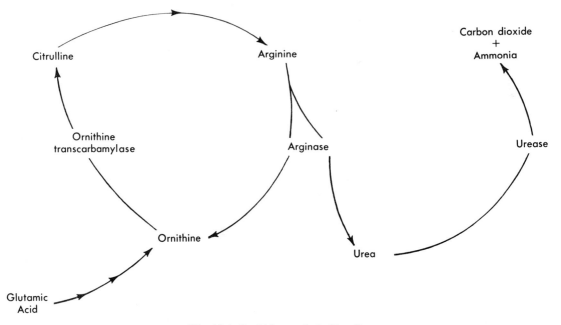

Fig. 10-1. Ornithine cycle in *E. coli.*

it was concluded that streptomycin must be acting somewhere between DNA and protein formation rather than on the functioning of completed protein molecules.

This discussion has indicated that only some of the cells that were both arginine deficient and streptomycin resistant responded to a streptomycin-containing minimal medium with the production of active OTC. The other doubly mutant cells did not grow on the medium. Both these types of streptomycin-resistant mutants map at the same position in the streptomycin locus of the *E. coli* chromosome and seem to be alleles of one another. The action of the antibiotic appears therefore to depend on which allele is present in the bacterial chromosome. From other experiments, which we shall not review here, it was found that the streptomycin locus is one of the genetic determinants of the 30s ribosomal subunit. It was therefore conjectured that the action of the streptomycin takes place at the translation step. Support for this hypothesis came from in vitro experiments which took advantage of the fact that it is possible to separate the 30s and 50s subunits of ribo-

somes from different cells and rejoin them in new combinations. The experiments also involved the use of a synthetic mRNA, poly-U, which codes for the amino acid phenylalanine. Into each of a series of test tubes were placed streptomycin, one of the types of recombined ribosomes, the synthetic mRNA, all twenty amino acids, all the different tRNAs, ATP, etc. In the test tube containing the 30s ribosomal subunits from CSD cells there occurred both a decreased incorporation of phenylalanine and a misincorporation of substantial amounts of the amino acids isoleucine, serine, tyrosine, and leucine. Strong evidence was thus obtained for a role of the ribosome in controlling the accuracy of codon reading. This demonstration that the 30s subunit of a ribosome can play an active role in translation raised the question as to whether it was the RNA or the protein of the 30s subunit that was important in streptomycin-induced mRNA misreading. It had been shown, mainly by M. Nomura and co-workers, that one could dissociate a 30s ribosomal subunit into its 16s RNA and its 20 different proteins and that, under the proper experimental conditions, these components would reassemble themselves into a perfectly functional 30s particle. This technique permitted the substitution of single proteins, one at a time, from 30s subunits of a streptomycin-resistant strain of *E. coli* for their counterparts from a susceptible strain. When this was done for each of the 20 proteins, it was found that the protein labeled P10 determines the translational role of the entire 30s ribosomal subunit in a streptomycin-containing medium.

The observed effect of streptomycin on codon reading indicates that translation can be affected by other molecules present in the cell. A somewhat unrelated conclusion may be drawn from the observed effect of streptomycin on translation. Misreading of mRNA may be the reason why streptomycin acts as an antibiotic. If enough defective proteins are produced, the effect of the streptomycin will be bactericidal. Various in vitro experiments have demonstrated that other antibiotics (e.g., neomycin and kanamycin) also produce misreading of synthetic mRNAs. The possibility exists that many antibiotics produce their effects by interfering with the translation process in the cell.

GENE END PRODUCTS

The end products of gene action are mostly proteins. These may function either as structural or as enzymatic components of the cell. We shall consider examples of both types—hemoglobin and tryptophan synthetase—and indicate how the genes control cell metabolism by specifying the amino acid sequence of the cell's proteins.

Hemoglobins and the anemias

Hemoglobin is a conjugated protein whose prosthetic portion (nonamino acid) consists of four iron-containing heme groups and whose protein portion consists of four polypeptide chains. One heme group is associated with each polypeptide chain. The heme group is the "working section" of the molecule, in that it contains at its center an iron atom to which oxygen atoms are loosely bound. However, it is the protein portion

that actually determines the functional efficiency of the heme groups, and we shall now consider the protein portion in more detail. There are four known types of polypeptide chains: alpha (α), beta (β), gamma (γ), and delta (δ). The chains differ from one another in the sequence of their amino acids, and some chains contain more amino acids than others. This indicates that there are at least four genes controlling hemoglobin polypeptide chain synthesis. Although in rare cases it has been found that all four polypeptide chains of a hemoglobin molecule are identical, it is more usual to find that there are two pairs of identical chains. Most (98%) of the hemoglobin of normal persons consists of two identical alpha chains and two identical beta chains. Hemoglobin of this composition is called "A" type hemoglobin and by conventional nomenclature has the following constitution:

$$\alpha_2{}^A\beta_2{}^A$$

The remaining 2% of normal adult hemoglobin has delta chains in place of beta chains. This hemoglobin is designated "A₂" type and has this formula:

$$\alpha_2{}^A\delta_2{}^{A_2}$$

The other chain, gamma, is manufactured during fetal life and can also serve in place of the beta chain. The result is designated as type "F" hemoglobin and has the constitution shown below:

$$\alpha_2{}^A\gamma_2{}^F$$

Any reduction in the oxygen-carrying capacity of an individual's erythrocytes results in the pathological condition called *anemia*. An anemia may result from (1) an inadequate number of red blood cells, (2) an insufficient number of hemoglobin molecules per erythrocyte, or (3) any alteration of the hemoglobin molecules reducing the oxygen-carrying efficiency of the heme groups. The type of anemia due to an inadequate number of red blood cells is usually caused by a widespread infection or some metabolic disturbance. A severe and often fatal expression of this kind of condition, called *perni-*

cious anemia, is associated with an inadequate absorption of vitamin B_{12} from the digestive tract.

Thalassemia. An example of an anemia brought about by an insufficient number of hemoglobin molecules per erythrocyte is *thalassemia.* The disease is genetically determined and is caused by a gene exhibiting incomplete dominance. Heterozygotes exhibit a mild anemia (thalassemia minor) but usually survive to the age of reproduction. Homozygotes are characterized by a severe anemia (thalassemia major) and a lack of ability to cope with infections. Transfusions of whole blood are essential to maintain an adequate hemoglobin concentration. Despite transfusions, most homozygotes die before the age of puberty.

Attempts have been made to discover the mechanism by which thalassemia is produced. In most patients, the amount of A_2 hemoglobin (i.e., alpha and delta chains) increases from the usual 2% to 3% to a level of 4% to 7%. In these patients a large amount of fetal hemoglobin F (i.e., alpha and gamma chains) also appears but in widely varying amounts. Type A hemoglobin (i.e., alpha and beta chains) continues to be produced, although in much smaller than usual amounts. An analysis of the A hemoglobin from thalassemic and normal persons shows no differences in amino acid composition or sequence. The above findings led to the hypothesis that the gene for this kind of anemia reduces the hemoglobin content of the erythrocytes by curtailing the production of the beta chains below the level that can be compensated with delta and gamma chain formation. Experimental support for this theory was provided by Bank and Marks in 1966. They obtained reticulocytes from nonthalassemic patients with hemolytic anemias (i.e., pernicious anemia and sickle cell anemia), from patients with thalassemia major, and from those with thalassemia minor. The reticulocytes of different origins were incubated separately in a medium containing ^{14}C–amino acids, in order to label the newly synthesized hemoglobin molecules. The cells were then lysed, and each

of the various types of polypeptide chains of the cells' hemoglobin was isolated and its amount determined. The investigators were especially interested in the relative amounts of alpha and beta chains (α/β ratio) present in the hemoglobin. The α/β ratio in cells of nonthalassemic subjects averaged 1.0. The α/β ratio in cells from patients with thalassemia major averaged 5.2, while that of patients with thalassemia minor averaged 2.3. The amount of hemoglobin A in the blood of patients with thalassemia major was found to be such as could account for only 40% of the alpha chains found in their red blood cells, but 90% of the beta chains. These findings indicated that whereas almost all the beta chains synthesized are present in hemoglobin A, less than half the alpha chains are in the molecule. Since the amounts of A_2 and F type hemoglobins are too small to account for the tremendous excess of alpha chains, it was hypothesized that the reduced production of beta chains led to a relative excess of alpha chains, which were released as free chains in the red blood cells. This type of thalassemia that is the result of a drastic reduction of beta chain synthesis is called *beta thalassemia.*

If the gene controlling beta chain production can mutate to a form that reduces the rate of beta chain formation, it is to be expected that the gene controlling alpha chain production can mutate in a corresponding fashion. A second type of thalassemia, called *alpha thalassemia,* has been found in a group of patients in whom there is a reduced synthesis of alpha chains. In this condition, the rate of production of A and A_2 types of hemoglobin is severely retarded because of the lack of alpha chains. However, beta, gamma, and delta chains are produced in great excess over the available alpha forms, and one finds hemoglobin molecules consisting of four beta or delta or gamma chains.

Our discussion of a beta and an alpha thalassemia leads to the conjecture that thalassemia represents mutation of a regulator gene rather than of a structural gene. Any model of a regulator gene in connection with thalassemia would also have to account for the increased production of delta chains and the persistent formation of gamma chains in afflicted persons. No clear-cut evidence has so far been found to support the popular hypothesis that thalassemia is a regulator gene mutation.

Sickle cell anemia. The last type of anemia that we shall discuss is the kind due to an alteration of the hemoglobin molecule, which reduces the oxygen-carrying efficiency of the heme groups. Examples of this type of anemia include *sickle cell anemia* and *hemoglobin C anemia.* Patients with sickle cell anemia normally show only a moderate anemia. However, after acute infections there is usually a hemolytic crisis characterized by a sharp drop in red blood cells. Death may occur from a rapidly developing severe anemia or from an acute infection. It is interesting to note that in most cases there is no reduction of the hemoglobin content within the cells. The erythrocytes of persons with sickle cell anemia undergo a characteristic change from their normal biconcave form to elongated, fila-

mentous, or crescentric sickle forms when the oxygen tension around the cells is reduced. A relatively benign condition called *sickle cell trait* is known, which is characterized by the ability of the erythrocytes to sickle under reduced oxygen tension. People possessing the sickle cell trait appear otherwise to be healthy. A diagram of normal and sickled erythrocytes is shown in Fig. 10-2.

In 1949 J. V. Neel and E. A. Beet independently demonstrated that the sickle cell trait was the heterozygous manifestation of a gene that, when homozygous, resulted in sickle cell anemia. In that same year Pauling and colleagues reported their discovery that hemoglobins from persons with sickle cell anemia, those with sickle cell trait, and normal individuals behave differently on electrophoresis. As shown in Fig. 10-3, normal and sickle cell anemia hemoglobins appear quite different from one another, while the heterozygote has the electrophoretic charac-

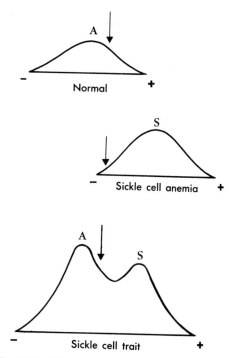

Fig. 10-3. Electrophoretic pattern of the hemoglobin of a homozygous normal individual, **A,** an individual with sickle cell anemia, **S,** and one with sickle cell trait, **AS.**

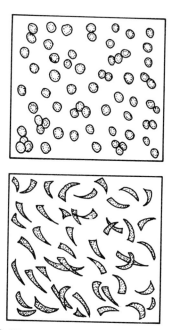

Fig. 10-2. Diagram of normal and sickled erythrocytes.

teristics of both types of hemoglobin. These findings provided the first evidence that adult human hemoglobin exists in more than one molecular form.

The discovery that there are different molecular forms of hemoglobin stimulated research to determine the amino acid sequences of the different forms. This work was greatly aided by a technique described by V. M. Ingram in 1956 and called "fingerprinting." In the fingerprinting procedure, the polypeptide chains are broken up into peptide fragments by means of the enzyme trypsin, which specifically cleaves the peptide bonds between the carboxyl group of lysine or arginine and the amino group of other amino acids. The peptide fragments are then spread out on paper by electrophoresis followed by chromatography at right angles to the direction of electrophoresis. Each peptide fragment is given a different number. The separated peptide fragments then constitute the "fingerprint" of the protein. A diagram of the results obtained with normal and sickle cell anemia hemoglobin is shown in Fig. 10-4. The peptide fragments in both patterns are identical in position except for fragment 4, which is one of the fragments derived from the beta polypep-

tide chain. An analysis for amino acid sequence of fragment 4 from normal and sickle cell anemia hemoglobins revealed a difference in the sixth amino acid. The results, shown in Fig. 10-5, indicate that normal hemoglobin has glutamic acid in the sixth position while sickle cell anemia hemoglobin has valine. Sickle cell hemoglobin is labeled as "S" type hemoglobin and has the following constitution:

$$\alpha_2{}^A\beta_2{}^S$$

These findings clearly demonstrated that the cause of the reduced efficiency of the oxygen-carrying capacity of the sickle cell anemia hemoglobin is a substitution in its beta chain of one amino acid by another. In Chapter 9 we explained that the substitution of valine for glutamic acid is best interpreted as a transversion type mutation in the genetic code. A number of individuals have been discovered who are heterozygous both for thalassemia and for sickle cell anemia. These persons exhibit a severe form of anemia. A study of their hemoglobin reveals a complex mixture of S, F, and A types.

Hemoglobin C anemia. In 1950 Itano and Neel, studying the parents of some children who exhibited an atypical form of

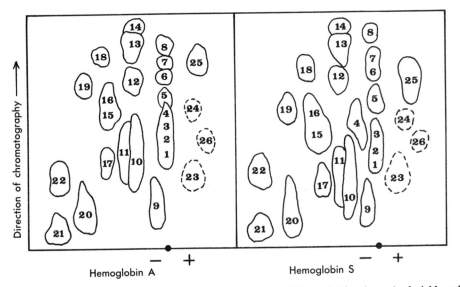

Fig. 10-4. Diagram of the "fingerprint" pattern of normal hemoglobin, **A,** and of sickle cell anemia hemoglobin, **S.**

sickle cell anemia, found that only one of the parents had blood that could be induced to sickle. The erythrocytes of the other parent appeared to contain normal hemoglobin. When, in 1958, Hunt and Ingram subjected the hemoglobin of the "normal" parent to electrophoresis and amino acid analysis, they found that peptide fragment 4 contained lysine as its sixth amino acid but was otherwise identical with A and S type hemoglobins (Fig. 10-5). This new type of hemoglobin was labeled "C" and has the polypeptide chain composition shown below:

$$\alpha_2{}^A \beta_2{}^C$$

The genes determining C, A, and S type hemoglobins are alleles, since their amino acid differences occur within the same cistron. Hemoglobin C heterozygotes appear to be unaffected by the gene. However, homozygotes suffer from a mild anemia that is accentuated by infection. Individuals who are genetic compounds of both thalassemia and hemoglobin C suffer from a severe anemia, much as do the people who are genetic compounds of sickle cell anemia and hemoglobin C.

• • •

Our discussion of alterations in the amino acid sequence of hemoglobin has revolved about peptide fragment 4 of the beta chain. However, other variations in amino acid sequence have been found, which involve other peptide fragments of the beta chain or peptide fragments of the alpha, delta, or gamma chains. Theoretically, mutations in the genetic code for hemoglobin should be possible in every one of its codons.

We have considered the hemoglobin molecule as an example of how genes control cell metabolism through the production of structural proteins. We shall now consider the genetic control of cell metabolism through the synthesis of enzymes.

Tryptophan synthetase

Our discussion of the genetic control of cell metabolism through enzyme synthesis will take as its example *tryptophan synthetase,* one of the enzymes that mediate the production of the amino acid *tryptophan.* Tryptophan synthetase has been studied mostly in *E. coli,* and our review will emphasize the knowledge obtained from those investigations. The amino acid tryptophan is important in many biochemical pathways, one of which is shown in Fig. 10-6. In the diagram, three "reactions" are numbered. All of these are catalyzed by tryptophan synthetase. Reaction 1 is the normal physiological one, while the others are accessory reactions.

In *E. coli,* tryptophan synthetase is easily dissociable into two protein subunits called components A and B. Mutants have been discovered that produce inactive forms of one or the other of these protein subunits. Investigators have found that mutants lacking the A component or containing it in an inactive form cannot carry out reactions 1 or 3 in Fig. 10-6 but that if they contain a B component that is normal, they can catalyze reaction 2 (indole to tryptophan). Thus the A mutants can grow on either indole or tryptophan. Mutants lacking the B component or having it in an inactive form cannot catalyze reaction 1 or 2; but if their A pro-

Kind of hemoglobin	Amino acids numbered in order							
	1	2	3	4	5	6	7	8
A	valine	histidine	leucine	threonine	proline	glutamic acid	glutamic acid	lysine
S	valine	histidine	leucine	threonine	proline	valine	glutamic acid	lysine
C	valine	histidine	leucine	threonine	proline	lysine	glutamic acid	lysine

Fig. 10-5. Results of analyses of amino acid sequences of peptide 4 of the beta chain of various hemoglobins.

tein is normal, they can catalyze reaction 3. Therefore B mutants can grow only if tryptophan is present in the medium. One may summarize these findings by pointing out that protein A alone exhibits activity for reaction 3 and B alone exhibits activity for reaction 2 but that the combination AB is necessary for the reactions 1, 2, and 3 to occur. It is of importance to note that the physiologically significant reaction, indoleglycerol phosphate to tryptophan, requires the combination of both protein subunits.

Complementation tests between A and B mutants, in the *trans* configuration, show that all A mutants complement all B mutants whereas no complementation is found between pairs of A mutants or between pairs of B mutants. These results demonstrate unequivocally that two independent genes, or cistrons, are involved in the production of the two components of bacterial tryptophan synthetase. In 1967 Yanofsky and associates reported on their analysis of the A protein subunit of tryptophan synthetase. They first subjected the protein to proteolytic enzyme activity, which resulted in the formation of polypeptide fragments. These fragments were then separated from one another by a combination of chromatography and electrophoresis. Next, each fragment was analyzed for its amino acid

content and sequence. The geneticists were able to determine the complete sequence of the 267 amino acids in the A protein. Similar studies on mutants of the A protein demonstrated the amino acid substitutions that had occurred in each case. Their findings of the amino acid sequence of the normal tryptophan synthetase A protein of *E. coli* are shown in Fig. 10-7, while the genetic map and the corresponding acid substitutions of some of the known mutants are shown in Fig. 10-8. Notice that mutants A38 and A96 do not produce detectable altered A proteins. They are located at the extreme ends of the *A* gene and may affect its function in some other manner. Notice also that at positions 48, 210, and 233 more than one amino acid substitution has been discovered. The occurrence of crossing-over between the different mutants of position 210 and also between the different mutants of position 233 reflects the probability that, in both instances, the mutants that are found at the same position represent single changes in adjacent nucleotides of the original codon. The lack of crossing-over between the mutants of position 48 undoubtedly reflects the complicated codon changes that are involved in the substitution of methionine (AUG) for glutamic acid (GAA or GAG) as compared to theoretically simple changes that

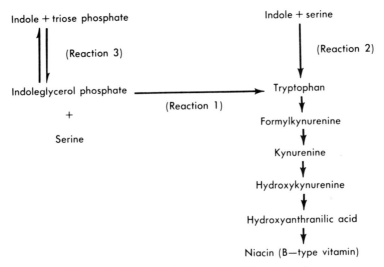

Fig. 10-6. Biochemical pathway involving tryptophan.

Met — Gln — Arg — Tyr — Glu — Ser — Leu — Phe — Ala — Gln — Leu — Lys — Glu — Arg — Lys — Glu — Gly — Ala — Phe — Val —
10 20

Pro — Phe — Val — Thr — Leu — Gly — Asp — Pro — Gly — Ile — Glu — Gln — Ser — Leu — Lys — Ile — Asp — Thr — Leu — Ile —
30 40

Glu — Ala — Gly — Ala — Asp — Ala — Leu — Glu — Leu — Gly — Ile — Pro — Phe — Ser — Asp — Pro — Leu — Ala — Asp — Gly —
50 60

Pro — Thr — Ile — Gln — Asn — Ala — Thr — Leu — Arg — Ala — Phe — Ala — Ala — Gly — Val — Thr — Pro — Ala — Gln — Cys —
70 80

Phe — Glu — Met — Leu — Ala — Leu — Ile — Arg — Gln — Lys — His — Pro — Thr — Ile — Pro — Ile — Gly — Leu — Leu — Met —
90 100

Tyr — Ala — Asn — Leu — Val — Phe — Asn — Lys — Gly — Ile — Asp — Glu — Phe — Tyr — Ala — Gln — Cys — Glu — Lys — Val —
110 120

Gly — Val — Asp — Ser — Val — Leu — Val — Ala — Asp — Val — Pro — Val — Gln — Glu — Ser — Ala — Pro — Phe — Arg — Gln —
130 140

Ala — Ala — Leu — Arg — His — Asn — Val — Ala — Pro — Ile — Phe — Ile — Cys — Pro — Pro — Asn — Ala — Asp — Asp — Asp —
150 160

Leu — Leu — Arg — Gln — Ile — Ala — Ser — Tyr — Gly — Arg — Gly — Tyr — Thr — Tyr — Leu — Leu — Ser — Arg — Ala — Gly —
170 180

Val — Thr — Gly — Ala — Glu — Asn — Arg — Ala — Ala — Leu — Pro — Leu — Asn — His — Leu — Val — Ala — Lys — Leu — Lys —
190 200

Glu — Tyr — Asn — Ala — Ala — Pro — Pro — Leu — Gln — Gly — Phe — Gly — Ile — Ser — Ala — Pro — Asp — Gln — Val — Lys —
210 220

Ala — Ala — Ile — Asp — Ala — Gly — Ala — Ala — Gly — Ala — Ile — Ser — Gly — Ser — Ala — Ile — Val — Lys — Ile — Ile —
230 240

Glu — Gln — His — Asn — Ile — Glu — Pro — Glu — Lys — Met — Leu — Ala — Ala — Leu — Lys — Val — Phe — Val — Gln — Pro —
250 260

Met — Lys — Ala — Ala — Thr — Arg — Ser

Fig. 10-7. Amino acid sequence of the tryptophan synthetase A protein of *E. coli.* The underlined residues are at the positions in the protein at which amino acid changes have occurred in mutants. (From Yanofsky, C., Drapeau, G. R., Guest, J. R., and Carlton, B. C. 1967. Proc. Nat. Acad. Sci. U. S. A. **57:**296-298.)

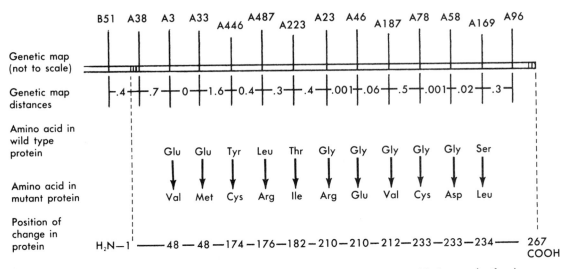

Fig. 10-8. Genetic map of the *A* gene and the corresponding amino acid changes in the A protein. The positions of these changes in the amino acid sequence are also indicated. (From Yanofsky, C., Drapeau, G. R., Guest, J. R., and Carlton, B. C. 1967. Proc. Nat. Acad. Sci. U. S. A. **57:**296-298.)

could result in the substitution of valine (GUU, GUC, GUA, GUG) for glutamic acid.

METABOLIC PATHWAYS AND ENZYME BLOCKS

As was indicated in Chapter 1, findings such as those on hemoglobin and tryptophan synthetase support the hypothesis of the colinearity of gene structure and protein structure. In addition, and more important for the purposes of our present discussion, these analyses demonstrate that the genes control cell metabolism through specification of the amino acid sequence of the cell's enzymes. Mutants that affect enzyme activity have in most cases been shown to alter the amino acid sequence of the enzyme. Exceptions to this generality are mutants that involve the ends of the gene and probably affect other functional characteristics of the enzymes. We shall now consider some metabolic pathways and indicate the kinds of control that genes exert over cell function.

Eye pigment production in Drosophila

Pigment systems in plants and animals have received special attention in genetics, since mutations affecting them are so easily observable. The red eye color of the wild type *Drosophila* is a result of the deposition of two different classes of pigments in each facet of the fly's compound eyes. There are the brown pigments (*ommachromes*) that are deposited at the periphery of each ommatidium and the red pigments (*pteridines*) that are deposited in the center of each facet. A good deal is known of the biochemical steps leading to the brown pigments, but unfortunately very little is known about the formation of the red pigments. The brown pigments have their origin in tryptophan. In ommachrome production tryptophan is degraded along the niacin pathway to hydroxykynurenine (Fig. 10-6), from which the various brown pigments are formed through a series of complicated steps. The red eye pigments have their origin in some unknown precursor, which could possibly be a purine and sugar phosphate.

A number of mutants in *Drosophila melanogaster* are known that affect ommachrome synthesis. In each case there is a buildup of the precursor just before the enzyme block. The mutants and their points of actions are shown in Table 10-1. In contrast to the number of known enzyme blocks

Table 10-1. Mutations in *Drosophila* affecting ommachrome synthesis

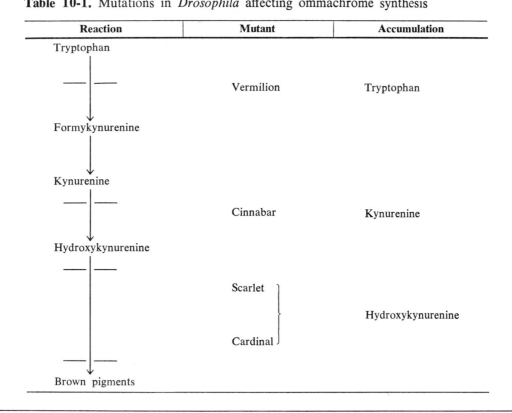

Reaction	Mutant	Accumulation
Tryptophan		
Vermilion		Tryptophan
Formykynurenine		
Kynurenine		
Cinnabar		Kynurenine
Hydroxykynurenine		
Scarlet ⎫		
⎬		Hydroxykynurenine
Cardinal ⎭		
Brown pigments		

in ommachrome synthesis, so far only one mutational enzyme deficiency is known in pteridine synthesis. The deficiency is for the enzyme *xanthine dehydrogenase,* which mediates a number of reactions in both purine and pteridine metabolism. At least three loci affect the enzyme's activity, and these genes are believed to code for three different polypeptide chains, all of which must be present in the wild type for the enzyme to be active.

Cyanogenesis in white clover

White clover, *Trifolium repens,* is grown as a forage crop throughout the temperate zones. The leaves of some of these plants start to release easily detectable amounts of hydrogen cyanide (HCN) as soon as they are removed from the plant. This acts to deter complete predation of the area by her-

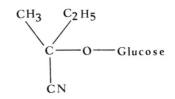

Fig. 10-9. Chemical structure of lotaustralin.

bivores. In contrast to this, the leaves of other white clover plants do not release HCN when similarly treated. Cyanogenesis depends primarily on the presence of two "cyanogenic glucosides," *linamarin* and *lotaustralin.* The chemical structure of lotaustralin is shown in Fig. 10-9. In the other glucoside, linamarin, a methyl group replaces the ethyl group. In the production of HCN from lotaustralin, hydrolysis removes the

Table 10-2. Tests of leaf extracts for HCN production

Genotype	Leaf extract alone	Leaf extract and glucosides	Leaf extract and linamarase
Ac—Li—	+	+	+
Ac—li li	–	–	+
ac ac Li—	–	+	–
ac ac li li	–	–	–

+ = HCN production; – = no HCN production.

glucose, and the remainder of the molecule breaks down to give HCN and a methyl-ethyl ketone.

Two independent enzymes are known to be involved in the production of HCN. One enzyme mediates the production of the glucosides from some precursor, while the second enzyme mediates the breakdown of the glucosides. The gene that controls the first enzyme is called *Acyanogenic (Ac),* and its dominant allele permits the formation of the glucosides. The gene that controls the second enzyme (linamarase) is called *Linamarin (Li),* and its dominant allele permits the release of HCN. The overall diagram can be pictured as follows:

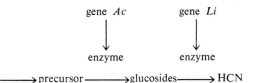

Plants that cannot produce HCN (are acyanogenic) may be homozygous for either or both recessive alleles of the above two genes. One may distinguish which condition exists for a particular plant by testing whether extracts of its leaves require the glucosides, the enzyme linamarase, or both for HCN production. The expected results are shown in Table 10-2.

Phenylalanine and tyrosine metabolism in man

One of the better studied metabolic pathways is that of phenylalanine and tyrosine metabolism in man. Human beings, in con-trast to *E. coli,* cannot manufacture phenylal-anine or tyrosine, and these must be sup-plied in the diet. A complete lack of these amino acids in the diet will result in the death of the individual. Phenylalanine and tyrosine, as will be seen in Fig. 10-10, are involved in a widely diverging series of bio-chemical reactions. We shall examine a number of these reactions and the effects on the individual of defective enzymes at these points.

Alcaptonuria. One of the end points of phenylalanine and tyrosine metabolism is the breakdown of homogentisic acid to car-bon dioxide and water. This reaction is accomplished under the influence of an enzyme that is present in the liver. In 1902 A. E. Garrod described the heritable disease known as *alcaptonuria.* The disorder results from a defect in an enzyme, homogentisic acid oxidase, that normally mediates the breakdown of homogentisic acid. When the enzyme is defective, large amounts of homo-gentisic acid are excreted in the urine, which turns black on exposure to the air. In ad-dition, quantities of homogentisic acid ac-cumulate in the body and become attached to the collagen of cartilage and other connec-tive tissues. As a result of these accumula-tions, the cartilage of the ears and the sclerae are stained black. In the joints of the body, the accumulations can lead to arthritis. When alcaptonurics are fed in-creased quantities of phenylalanine or tyro-sine, there is a corresponding increase of homogentisic acid excreted in their urine. An increase of phenylalanine or tyrosine in

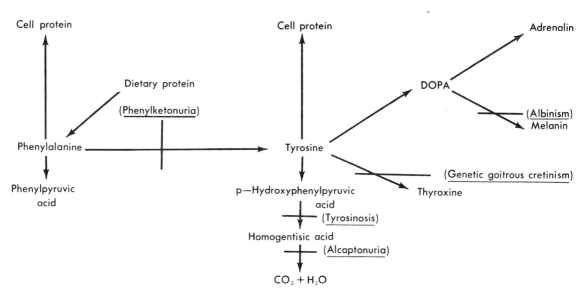

Fig. 10-10. Phenylalanine and tyrosine metabolism in man.

the diet of normal persons is not followed by the appearance of homogentisic acid in their urine. Alcaptonuria is inherited as an autosomal recessive trait and is an example of a genetic enzyme block in which the phenotypic features are due to the accumulation of a substance just proximal to the block.

Tyrosinosis. Another point at which a genetic block is known to occur in phenylalanine and tyrosine metabolism is at the oxidation of *p*-hydroxyphenylpyruvic acid to homogentisic acid. A defective enzyme at this juncture results in the accumulation of *p*-hydroxyphenylpyruvic acid and tyrosine in the urine of the affected person. This condition is called *tyrosinosis*. This rare human abnormality has been carefully studied only once, and its hereditary nature is unknown. In this disorder there is a backup of the biochemical precursors of homogentisic acid such that the accumulated substances include one that is actually one step removed from the point of enzymatic block.

Phenylketonuria. Another abnormality resulting from a block in phenylalanine and tyrosine metabolism results in a condition called *phenylketonuria*. It is due to a defect in the enzyme phenylalanine hydroxylase, which, in the liver of normal individuals,

mediates the oxidation of phenylalanine to tyrosine. The gene involved is an autosomal recessive, and in homozygotes the phenylalanine accumulates and is converted to phenylpyruvic acid. Some of the phenylalanine and phenylpyruvic acid become concentrated in the cerebral spinal fluid, while the rest is excreted in the urine. Phenylketonuria will be discussed in some detail in Chapter 12. For our present purposes, it is important to note that persons with this enzyme deficiency are feebleminded and have light pigmentation. The feeblemindedness is thought to be due to an impairment of the brain tissue by the phenylalanine in the cerebral spinal fluid. The light pigmentation is due to a decreased formation of tyrosine, which is one of the precursors of melanin. Here we have one enzyme block that affects two widely divergent end products of this metabolic system.

Genetic goitrous cretinism. One of the pathways of phenylalanine and tyrosine metabolism leads to thyroxine. An enzyme block at this point results in *genetic goitrous cretinism* with its accompanying physiological and mental retardation. A recessive gene in homozygous condition causes the enzyme block.

Albinism. The last biochemical block in

phenylalanine and tyrosine metabolism that we shall consider is that which can occur between 3,4-dihydroxyphenylalanine (DOPA) and melanin. This biochemical transformation is mediated by the enzyme tyrosinase. A defective enzyme, produced by a recessive gene in homozygous condition, results in the absence of melanin in the skin, hair, and eyes of an affected individual *(albinism)*.

• • •

We have reviewed a number of examples of enzyme blocks within a given metabolic system. It is apparent that any interference with the normal biochemical pathways of the cell may result in far-reaching consequences to the organism. Another situation in which the genetic control of cell metabolism has been important is in the individual's sensitivity to drugs.

DRUG METABOLISM IN MAN

With the widespread use of therapeutic drugs in modern medicine, an increasing number of discoveries have been made of individuals whose metabolism is adversely affected by the drugs. In many instances the

sensitivity to the administered drug has been found to be inherited. From studies of these situations has developed the field of *pharmacogenetics*. We shall examine some examples of drug sensitivity in man and attempt to identify the physiological bases for the disturbances to cell metabolism.

Primaquine sensitivity

Primaquine, one of the 8-aminoquinoline compounds, has been used in the treatment of malaria since 1926. In that same year W. Cordes reported the occasional occurrence of hemolytic anemia associated with the administration of this drug. In 1954 R. J. Dern and co-workers demonstrated that the hemolysis was due to an intrinsic characteristic of the erythrocytes of an affected individual. They transfused ^{51}Cr-labeled erythrocytes from a primaquine-sensitive person into nonsensitive recipients. The survival of the transfused cells was normal until primaquine was administered to the nonsensitive individuals. Rapid destruction of the transfused erythrocytes then occurred. However, when ^{51}Cr-labeled red blood cells from a normal individual were transfused into a

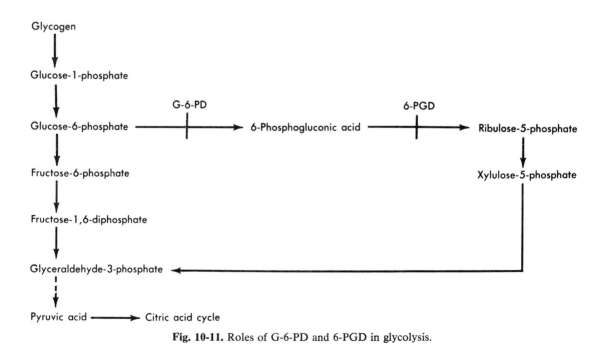

Fig. 10-11. Roles of G-6-PD and 6-PGD in glycolysis.

person known to be primaquine sensitive, primaquine administration had no effect on the survival of the transfused cells, although the primaquine-sensitive recipient underwent a typical acute hemolytic crisis. Subsequently, in 1956 P. E. Carson and co-workers demonstrated that primaquine-sensitive individuals have a deficiency of the enzyme glucose-6-phosphate dehydrogenase (G-6-PD) whose function in glycolysis is shown in Fig. 10-11. The deficiency of G-6-PD results in an abnormality in the direct oxidation of glucose in the erythrocytes of the affected individuals. The relationship of reduced glycolysis efficiency and erythrocyte fragility has not been satisfactorily explained.

Primaquine sensitivity has been found to be inherited as an X-linked recessive trait. It has also been shown that a hemolytic crisis in persons with G-6-PD deficiency follows the ingestion of fava beans, naphthalene mothballs (e.g., accidentally by children), sulfanilamide, sulfoxone (antileprosy drug), nitrofurantoin (urinary antiseptic), and others. G-6-PD deficiency thus appears to be the basis for a number of drug-induced hemolytic anemias. In the absence of these drugs, the affected individuals appear to have normal health. We shall discuss G-6-PD deficiency in Chapter 11, in connection with a different genetic phenomenon.

Isoniazid metabolism

Isoniazid (INH) was first used in the treatment of tuberculosis in 1952. It was found that the drug is inactivated in the body through acetylation, which is mediated by the liver cell enzyme *acetyltransferase.* It soon became apparent that persons taking the drug vary in the speed with which they inactivate INH. Measurements were taken of the time necessary for the blood plasma level of the drug to fall to one half its level shortly after injection (half-life). The half-life of INH in the plasma of rapid inactivators ranges from 45 to 80 minutes, while in that of slow inactivators it ranges from 140 to 200 minutes. The amount of active acetyltransferase was found to be greater in rapid

inactivators of INH than in slow activators. Through family studies, Evans and co-workers in 1960 were able to demonstrate that slow inactivators were homozygous for a recessive gene. In INH inactivation, as in primaquine sensitivity, there is no indication of a metabolic difference between individuals in the absence of the drug.

Barbiturate reaction

A sometimes fatal drug reaction has come to light with the introduction of barbiturates in modern medicine and particularly with their use in anesthesia. This violent reaction is found in people who suffer from a disturbance in the porphyrin metabolism of the body. Porphyrins are ringed compounds that are the precursors of the heme-containing portions of such proteins as hemoglobin, myoglobin, catalase, peroxidase, and the cytochromes. The excessive formation and liberation of porphyrins in the body, called *porphyria,* results in an individual whose skin is extremely photosensitive. Prolonged and chronic exposure to sunlight results in vesicular eruptions of the skin, which heal only with great difficulty. Liver disorders may also be present.

A number of different types of porphyria are known. Our consideration will limit itself to the type found in some 8000 people in South Africa, all of whom, as reported by G. Dean in 1963, are descendants of the Dutch settler Gerrit Jansz and his wife Ariaantje Jacobs who were married in 1688. Family studies of these 8000 individuals indicate that the trait is inherited as an autosomal dominant. The condition usually is not serious. The skin of affected people is sensitive both to sunlight and to mechanical trauma, and it abrades easily with occasional blister formation. Acute attacks of abdominal pain may also occur. When these persons are given anesthetic doses of barbiturates to relieve their abdominal pain or for any other purpose, they become severely paralyzed. They either die or take months to recover. The biochemical mechanism by which barbiturates exert their effects on porphyric patients is not yet fully understood. In the case of barbiturate reaction, we have a known genetic disease (porphyria) whose usually mild effects become severe and sometimes even fatal when the drug is administered to susceptible individuals.

SUMMARY

In this chapter we have studied the gene as the controlling element of metabolism. This has involved a consideration of the factors that affect transcription as well as those that affect translation. We found that although the amino acids play no role in reading mRNA codons, tRNA and the 30s subunit of the ribosome are involved in this function. Our attention then focused on the different kinds of gene end products, which can be either structural proteins such as hemoglobin or enzyme proteins such as tryptophan synthetase. In considering hemoglobin, we reviewed thalassemia, a hereditary disease that results from a diminished number of hemoglobin molecules per erythrocyte, and two hereditary diseases, sickle cell anemia and hemoglobin C anemia, that involve alterations of the amino acid sequences of the normal hemoglobin molecule. There followed a review of some metabolic pathways and the enzyme blocks that could occur. The pathways studied included eye pigment production in *Drosophila,* cyanogenesis in white clover, and phenylalanine and tyrosine metabolism in man. In the last-named metabolic pathway, we discussed certain hereditary diseases: alcaptonuria, tyrosinosis, phenylketonuria, genetic goitrous cretinism, and albinism. The chapter closed with a consideration of drug metabolism in man and the genetic bases of primaquine sensitivity, isoniazid metabolism, and barbiturate reaction.

Questions and problems

1. Define the following terms:
   a. Histone
   b. Amber mutant
   c. Amber suppressor mutant
   d. Fingerprinting
   e. Hemoglobin C
   f. Conditionally streptomycin-dependent mutant
2. How do nonsense mutations act to alter the normal translation process?
3. a. How can ribosomes affect normal translation?

b. What is the relationship between ribosomal structure, antibiotics such as streptomycin, kanamycin, etc., and translation?
4. From your knowledge of the mechanism of the transcription process, what factors, do you think, can lead to alterations in normal transcription?
5. The genetic basis for thalassemia is not conclusively known. Offer a plausible mechanism for this genetic disease.
6. Diagram the results you would expect on electrophoresis of the following:
   a. Blood isolated from a normal individual
   b. Blood isolated from an individual heterozygous for beta thalassemia
   c. Blood isolated from an individual homozygous for beta thalassemia
   d. Blood isolated from an individual heterozygous for sickle cell anemia
   e. Blood isolated from an individual homozygous for sickle cell anemia
   f. Blood isolated from an individual who is a genetic compound for HbC and thalassemia diseases
7. What experimental evidence exists for the belief that two cistrons are involved in the production of *E. coli* tryptophan synthetase?
8. Assume that four genes, A_1, B_1, C_1, and D_1, are involved in the synthesis of enzymes a, b, c, and d, respectively, and that the enzymes are needed in the metabolic pathway leading to end product E. Six strains of *E. coli,* each defective in one of the enzymes, are cultured on media containing each of the precursors—A, B, C, and D—that are involved in the pathway. The following results are obtained, in which "+" indicates the presence of the end product E and "-" indicates the inability to synthesize the end product:

| Mutant | Precursor added | | | |
	A	B	C	D
1	+	+	+	-
2	-	-	-	-
3	-	+	+	-
4	-	+	-	-
5	-	-	-	-
6	-	+	+	-

a. For each of the six mutants tell where the mutation occurs.
b. What is the order of the metabolic pathway involved?
9. Refer to question 8.
   a. How would you be able to detect the presence of a mutant strain?
   b. What possibilities exist as to why the enzymatic block occurs?
10. Discuss the genetic control of metabolism with respect to the following:
    a. Eye pigmentation in *Drosophila*
    b. HCN production in white clover

11. What is responsible for the phenotypic features (dark urine on exposure to air, darkly stained cartilage, arthritis) characteristic of individuals afflicted with alcaptonuria?
12. a. What do you think is the role of the v^+ gene in *D. melanogaster* eye pigmentation?
    b. Of the cn^+ gene?
13. Define the following and tell how they apply in discussions of the genetic control of metabolism:
    a. Tyrosinosis
    b. Phenylketonuria
    c. Albinism
    d. Genetic goitrous cretinism
14. a. What is the genetic basis of primaquine sensitivity?
    b. Of isoniazid inactivation?
    c. Of barbiturate sensitivity in porphyria sufferers?
15. What is the action of each of the following enzymes?
    a. Acetyltransferase
    b. Ornithine transcarbamylase
    c. Linamarase
    d. Xanthine dehydrogenase
    e. Tryptophan synthetase
    f. Glucose-6-phosphate dehydrogenase
16. What conclusive experimental evidence exists for the statement that an amino acid does not determine its own location in the polypeptide chain?
17. How is the change from a wild type su^- gene to an su^+ gene believed to be reflected in the components of the protein-synthesizing machinery?
18. In what sense can inborn errors of metabolism, such as phenylketonuria, be rectified?
19. How would you explain the observation that, in *Neurospora,* inactivation of a single gene often results in the organism's exhibiting a double growth factor requirement?
20. Discuss the role of tRNA in translation.

References

Anfinsen, C. B. 1959. The molecular basis of evolution. John Wiley & Sons, Inc., New York. (An excellent book, available in paperback editions, discussing the relationship between gene and protein.)

Baglioni, C. 1963. Correlations between the genetics and chemistry of human hemoglobins, Chapter 9, p. 405-475. *In* Taylor, J. H. (ed.). Molecular genetics. Part I. Academic Press, Inc., New York. (A very good discussion of gene-protein relationships in terms of the human hemoglobin system.)

Beadle, G. W., and Tatum, E. L. 1941. Genetic control of biochemical reactions in *Neurospora*. Proc. Nat. Acad. Sci. U. S. A. **27**:449-506. (The classic paper in which the one-gene–one-enzyme theory was suggested.)

Capecchi, M. R., and Gussin, G. N. 1965. Suppression in vitro: identification of a serine s-RNA as a nonsense suppressor. Science **149**: 417-422. (A report of experiments designed to examine a particular class of *su* genes and their effects on translation.)

Davis, J., Gilbert, W., and Gorini, L. 1964. Suppression and the code. Proc. Nat. Acad. Sci. U. S. A. **51**:883. (A report on experiments demonstrating that streptomycin causes misreading of the genetic code.)

Garrod, A. E. 1909. Inborn errors of metabolism. Oxford University Press, Inc., New York. (The classic book heralding the beginnings of biochemical genetics.)

Haldane, J. B. S. 1954. The biochemistry of genetics. The Macmillan Co., New York. (An excellent book analyzing gene action in man, animals, and plants.)

Harris, H. 1970. The principles of human biochemical genetics. American Elsevier Publishing Co., New York. (An excellent book on genes and metabolism, as related to human beings.)

Ingram, V. M. 1956. A specific chemical difference between the globins of normal human and sickle cell anemia hemoglobins. Nature **178**: 792-794. (A report on the technique of fingerprinting and its application in the study of abnormal hemoglobins.)

Kurland, C. G. 1970. Ribosome structure and function emergent. Science **169**:1171-1177. (A detailed discussion of the various 30s ribosomal subunit proteins and the functions of each.)

Ozaki, M., Mizushima, S., and Nomura, M. 1969. Identification and functional characterization of the protein controlled by the streptomycin-resistant locus in *E. coli*. Nature **222**:333-339. (Original announcement of the role of the 30s ribosomal subunit P10 protein in mRNA translation.)

Pauling, L., Itano, H. A., Singer, S. J., and Wells, I. C. 1949. Sickle cell anemia, a molecular disease. Science **110**:543-548. (One of the first examples of how a gene controls polypeptide synthesis.)

Pestka, S. 1971. Inhibitors of ribosome functions. Ann. Rev. Microbiol. **25**:487-562. (Detailed discussion of inhibitors that affect the functioning of either the 30s or the 50s ribosomal subunit.)

Yanofsky, C., Helinski, D. R., and Maling, B. D. 1961. The effects of mutation on the composition of the A protein of *E. coli* tryptophan synthetase. Cold Spring Harbor Symp. Quant. Biol. **26**:11-24. (A report on the relationship between the *A* gene and the A protein of the *E. coli* tryptophan synthetase system, including a section on the characteristics of the metabolic pathway in question and a discussion of suppressor mutations.)

11 Genes and development

🐟 🐟 In previous chapters, we have seen
that genes determine the morpho-
🐟 🐟 logical and physiological characteristics of the organism. This is achieved
through a great number of interacting
metabolic pathways, each of which is composed of gene-produced structural and/or
enzymatic proteins. However, for integrated
systems to develop and function, a high degree of control over gene action is necessary.
This would include regulation of DNA
replication, DNA transcription, mRNA
translation, and all subsequent cell metabolism. It is well known that an organism's
form and activities are not constant but are
continuously changing in response to alterations in the environment and in accordance
with the organism's life cycle. It is further
apparent that an organism's life cycle is
itself genetically determined and whatever
may be its "normal" life cycle is subject to
change by environmental alterations and
mutational modifications. We may consider an organism's *development* as including all morphological and physiological
changes that contribute to the course of its
life cycle. For unicellular forms, the changes
are obviously restricted to those that occur
within the single cell. However, for multicellular forms, the changes would include
all those activities that result in cell, tissue,
organ, and system differentiation and integration. Throughout our discussions, great
emphasis will be placed on the factors
that control gene action, for it is only
through a rigid regulation of gene
action that complex, integrated metabolic
pathways can develop and be maintained.

REGULATION OF DNA REPLICATION

The normal pattern of a cell cycle includes the growth of the cell, the replication
of its DNA, and the subsequent division of
the cell into two equivalent cells. The most
important of the events of the cell cycle is
the replication of its DNA, for without the
proper genetic material, information for the
regulated construction of cell material cannot be transferred to future cell generations.
It is interesting to note, in this connection,
that inhibition of DNA synthesis blocks cell
division, while inhibition of cell division
does not necessarily block DNA synthesis.
A continuing study of the regulation of
Escherichia coli chromosome replication has
been made by K. G. Lark and co-workers,
starting in 1956. In their experiments the
strain used was *E. coli* 15T$^-$, a quadruple
auxotroph that requires thymine, tryptophan,
methionine, and arginine.

DNA synthesis and growth rates

One of the investigations dealt with the
relationship of DNA synthesis to growth
rates. The bacteria were grown in various
media that differed in their energy-supplying compounds. Bacteria do not obtain the
same amount of energy from each of their
energy-supplying compounds, and in the
present experiment this resulted in the bacteria's having a different generation time in
each medium. The findings that show the
relation of cellular DNA content to bacterial
growth rate are given in Table 11-1. These
data suggest that at rapid growth rates each
cell may possess two or more chromosomes,
whereas at slower growth rates each cell contains only one.

Table 11-1. DNA content of *E. coli* cells grown in different media

Energy source	Generation time (minutes)	DNA content (μg per 10^7 bacteria)
Glucose	40	0.138
Succinate	70	0.103
Proline	180	0.051
Acetate	270	0.055

From Lark, C. 1966. Regulation of chromosome replication and segregation in bacteria. Biochim. Biophys. Acta **119**:517-525.

A question raised by the above data is whether, in the different media, chromosome replication occurs continuously or whether it occurs only during a portion of the cell cycle, as observed in the cells of higher organisms (Chapter 2). This was tested by allowing asynchronously dividing cells to incorporate radioactive thymidine into their chromosomes for about one tenth of their generation time and then preparing autoradiographs to determine how many cells were labeled. Labeled cells would have been in the process of replicating their chromosomes when they were exposed to radioactive thymidine. It was found that when the generation time was short (i.e., on glucose or succinate medium), virtually all the cells were labeled; but when the generation time was long (i.e., on proline or acetate medium), only about half the cells were labeled. This indicated that when limited by a long generation time, bacterial cells do not replicate their chromosomes throughout their entire cell cycle. It was also found that as the generation time increases, the time required for chromosome replication also increases. These experiments indicate that the energy available to a cell controls its overall metabolism, including its rate of chromosome replication.

DNA synthesis and protein synthesis

The discovery that chromosome replication is regulated by cell growth raises the question of whether DNA synthesis is in some way specifically protein dependent.

This problem has been approached experimentally as follows. Cultures of *E. coli* 15T$^-$ were placed in a medium deficient in their required amino acids but containing ^{3}H-thymine. At various times after incubation, samples of cells were taken from the culture and tested for radioactivity. The cells were found to contain an increasing amount of radioactivity for a period of time that roughly equaled the bacteria's generation time. Thereafter, no additional amount of radioactivity was found in the cells. The conclusion was drawn that even if bacteria are starved of required amino acids, a limited amount of DNA synthesis can occur. A further hypothesis was that continuous chromosome replication was protein dependent. The task remained to gather experimental evidence that would elucidate this last point.

To study the protein-dependency of DNA synthesis, an experimental procedure was established that would permit protein synthesis while inhibiting chromosome replication. This was achieved by placing the bacteria in a medium that lacked thymine. In such a medium, DNA transcription and mRNA translation take place, but no DNA replication can occur. Two experiments were conducted. In one, cells were starved of their required amino acids for 80 minutes and then placed in a medium containing these amino acids and ^{3}H-thymine. In the other experiment, cells were starved of thymine for 80 minutes and then placed in a medium containing ^{3}H-thymine. Samples of the cultures were taken at various time intervals thereafter and tested for radioactivity. The results showed that almost three times as much radioactivity is incorporated after thymine starvation as after amino acid starvation. The increased incorporation of ^{3}H-thymine by previously thymine-starved cells indicated that these cells were undergoing more chromosome replication than were the cells starved of required amino acids. These findings were taken to indicate that thymine-starved cells continue to produce proteins required for the initiation of DNA synthesis and that these proteins begin to function once thymine becomes available. However, cells starved of required amino acids could not produce these proteins and thus were not able to undergo rapid chromosome replication when the required amino acids became available to them.

The nature and function of the protein necessary for the initiation of chromosome replication is at present speculative. On the basis of experiments that are not germane to our present discussion, geneticists have thought that the protein may function as an "attachment site" of the newly formed chromosome to the cell membrane. The lack of such attachment sites would, under this hypothesis, prevent the formation of new chromosomes. Cell growth would then control DNA synthesis by limiting the area available for such attachment sites. From our discussion we can clearly see that *E. coli* chromosome replication is severely regulated by cell metabolism and that continuous DNA synthesis can be associated specifically with the cell's ability to carry on protein synthesis. We are not in a position to indicate whether the above pattern also governs chromosome replication in organisms other than *E. coli.*

REGULATION OF DNA TRANSCRIPTION

Many of the developmental changes that occur in the life cycle of an organism are effected through the regulation of DNA transcription. To a large extent, differentiation is determined by the initiation of certain transcriptions at particular times in development, with the corresponding termination of other transcriptions whose proteins are no longer necessary. We shall consider some examples of developmental changes that are associated with changes in transcription.

Enzymes of E. coli

We have considered the regulation of enzyme production in *E. coli* in Chapter 9. Whether this type of gene regulation is to be considered as a true developmental phenomenon may be questioned. It does, however, represent an excellent illustration of how changes in protein production can be effected through control of DNA transcription.

In *E. coli,* two kinds of control over enzyme synthesis have been discovered. One type of gene regulation involves the so-called *inducible enzymes,* and the other, the so-called *repressible enzymes.* Inducible enzymes are produced only when the substrate on which they will act is present. The example we have used previously is β-galactosidase, which catalyzes the hydrolysis of lactose into galactose and glucose. The enzyme is produced only when lactose is present in the medium. In the absence of lactose, enzyme synthesis virtually stops. There are a large number of such inducible enzymes. They are normally involved in the splitting of some compound that is used as a source of energy or as a source of molecular fragments for some synthetic process. Repressible enzymes are produced only when the end product they form is absent. As an example, the enzymes involved in histidine synthesis are produced only when the amount of free histidine in the cell is drastically reduced. In the presence of free histidine, enzyme production stops. Repressible enzymes, of which there are many, are normally involved in synthesizing some end product, very often an amino acid.

The phenomena of enzyme induction and repression were discovered independently of one another, but investigators quickly recognized that the two might be considered to be different aspects of the same phenomenon. As we have discussed in Chapter 9, the model of gene regulation devised by Jacob and Monod assumes that there are two general types of genetic elements: *controlling genes* and *structural genes*. The controlling genes exist as two types: *regulators* (e.g., *i* and *p*) and *operators* (e.g., *o*). Regulators may or may not be located close to the genes they regulate. Operators are always located close to, and, in fact, may be part of, one of the structural genes they control. The operator and its structural genes constitute an *operon*. The *i* regulator gene acts through the medium of a repressor substance. In an inducible system, the repressor, in the absence of an *inducer* (e.g., lactose), associates with the operator gene, thus preventing DNA transcription. If an inducer is present, it combines with the repressor, rendering the repressor inactive. This frees the operator and permits the structural genes to be transcribed. In a repressible system, the repressor, in the absence of a *corepressor* (e.g., histidine), does not combine with the operator gene, and DNA transcription takes place. If a corepressor is present, the repressor substance combines with it and the repressor-corepressor complex blocks the operator. Under these conditions, no DNA transcription occurs. The operon model for induction and repression is shown in Fig. 11-1. It would not be surprising to find that the various operons form integrated systems,

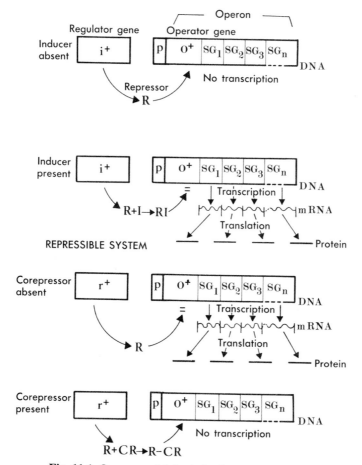
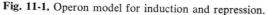

Fig. 11-1. Operon model for induction and repression.

with the end products of some operons acting as inducers of others. We do not know at present how widespread the operon type of gene regulation is among organisms. Some systems of gene regulation in multicellular organisms appear to be consistent with an operon model.

RNA polymerase: core enzyme and sigma factor

In bacteria, and presumably in all other organisms, the various types of RNA (e.g., rRNA, tRNA, mRNA) are all synthesized as a result of the action of a single enzyme, DNA-directed RNA polymerase. The total dependence on this enzyme for all the above types of RNA found in a cell even extends to the transcription of the genomes of those viruses that are present in the cell in DNA form. Although it has been known for some time that there are "controller" genes which determine the parts of the genome that will be transcribed, it has recently been demonstrated that RNA polymerase itself has a regulatory function.

R. R. Burgess and co-workers reported in 1969 that they had been able to separate from the RNA polymerase of *E. coli* a protein component with which the enzyme is usually associated. Without this protein component, which they called *sigma* (σ), the remaining portion (i.e., *core enzyme*) of the RNA polymerase was able to transcribe various types of DNA (e.g., calf thymus DNA) but was not able to transcribe the DNA of phage T4. When the sigma factor was reassociated with the core enzyme of RNA polymerase, the complete enzyme was able to transcribe many types of DNA including that of T4. Other experiments indicated that sigma functions in the initiation phase of transcription and that this protein serves to attach its RNA polymerase molecule to a promoter locus. Further investigations demonstrated that during phage T4 infection, the host RNA polymerase sigma factor is replaced by a virus-induced sigma factor. The replacement of host sigma factor by that of the phage results in the preferential transcription of the virus genome rather than the genome of the host. These experiments showed that the RNA polymerase sigma factor does have an important function in regulating transcription of the DNA of both the bacterium and its invading viruses.

A report on the role that RNA polymerase core enzyme may play in the regulation of DNA transcription during one of the life-cycle changes of an organism was published in 1970 by R. Losick and colleagues. They studied the bacterium *Bacillus subtilis*, which can exist either in a vegetative or in a spore state. Tremendous alterations in the cell's biochemistry are associated with this change of states, as vegetative-type genes and sporulation-type genes are alternately switched on and off. It was found that with the change of states, there is a corresponding modification in the activities of the cell's RNA polymerase. This is seen in the inability of RNA polymerase from sporulating cells to transcribe the DNA of phage ϕe, whereas the RNA polymerase from vegetative cells can do so. A hypothesis of a possible connection of sporulation control with viral DNA-transcription control received support from the discovery that *B. subtilis* mutants which have lost the ability to sporulate continue to transcribe phage ϕe DNA indefinitely. It was also discovered that the difference in activities between vegetative and sporulation RNA polymerase was a function of the core enzyme of the polymerase molecule and not of its sigma factor. This was clearly seen when the transfer of the sigma factor from a vegetative to a sporulation-type RNA polymerase molecule did *not* result in the transcription of phage ϕe DNA.

From the above findings there appear to be at least two mechanisms involving RNA polymerase for the regulation of DNA transcription. One mechanism depends on the dissociable protein factor sigma, while the second mechanism is a function of the core enzyme of the polymerase molecule. Although the existence of sigma factors and core enzyme differences has not yet been demonstrated for eucaryotic cells, it is quite

Table 11-2. Differential puffing in region 14 of chromosome 3 of *Chironomus tentans*

	Locus					
	1	**2**	**3**	**4**	**5**	**6**
Salivary gland	–	–	+	–	–	+
Malpighian tubule	–	–	+	+	+	+
Rectum	+	+	–	+	+	+
Midgut	–	–	–	–	–	–

+ = puffing of various degrees; – = no puffing. From Beermann, W. 1956. Nuclear differentiation and functional morphology of chromosomes. Cold Spring Harbor Symp. Quant. Biol. **21**:217-232.

conceivable that they may well be involved in the coordinated switching on and off of genes during development.

Puffs in chromosomes

In some tissues (salivary glands, malpighian tubules, rectum, midgut, etc.) of the bodies of dipteran flies, the homologous chromosomes are permanently synapsed. The cells of these tissues do not divide but only enlarge while their chromosomes are duplicated regularly. This process of chromosome duplication without cell division is called *endomitosis,* and the chromosomes are called *polytene* (i.e., many-stranded) *chromosomes.* Studies indicate that these giant chromosomes arise by successive doublings of the original chromosomes. However, they are not constant in width. Especially striking is the occurrence of enlarged, nonbanded areas called *puffs.* (See Fig. 11-3.) An examination of a particular region of a polytene chromosome in different tissues of the body shows that the pattern of puffs varies with the tissue. An example of the results of such a study in the dipteran fly *Chironomus tentans* is shown in Table 11-2. It was concluded that the puffs must be regions of the chromosomes that are involved in DNA transcription. In addition, geneticists hypothesized that the difference in puff pattern reflected a difference in protein synthesis in the various tissues.

A clear-cut demonstration that puffs are associated with protein production was provided by W. Beermann in 1961. He studied two species of flies, *Chironomus tentans* and *C. pallidivittatus.* In *C. pallidivittatus* there is a small specialized sector of the larval salivary gland, consisting of four cells, that produces a granular secretion. In *C. tentans* the granular component is not present in the secretion of these four special cells. An examination of the polytene chromosomes in the salivary gland cells shows that they contain huge, ball-like puffs, which are called Balbiani rings. In the fourth chromosome of all the salivary gland cells in both species there are three Balbiani rings. In *C. pallidivittatus,* the tip of the fourth chromosome in the special granule-secreting cells always shows an additional Balbiani ring. In *C. tentans,* this additional puff is absent. There is therefore the concomitant occurrence of a specific salivary puff and the synthesis of a specific protein. The relationship of secretion to Balbiani ring formation was further demonstrated by mating members of the two species to one another, thereby producing hybrids. These hybrids produce only about half the number of granules in their salivary secretion as do the members of *C. pallidivittatus.* It was found that in the special salivary gland cells, the fourth chromosome derived from *C. pallidivittatus* forms a Balbiani ring, while the homologous chromosome from *C. tentans* does not. Diagrams that illustrate these findings are shown in Fig. 11-2.

Although the observations just cited demonstrate that the formation of puffs and protein synthesis are concomitant events, they do not tell us whether the puffed regions of the chromosomes are involved in DNA transcription. Autoradiographic studies dealing with this question were reported in 1959 by C. Pelling on *Chironomus* and by Rudkin and Woods on *Drosophila*. They injected ³H-thymidine into fly larvae. These larvae were dissected shortly thereafter, and their salivary glands were prepared for autoradiography. The thymidine apparently was distributed evenly throughout the polytene chromosomes. When ³H-uridine was used in this experimental procedure, the uridine was found only in the puffs, Balbiani rings, and nucleoli. It was further found that injecting actinomycin D first and then ³H-uridine led to a drastic inhibition of uridine incorporation and a shrinkage of all puffs. These results clearly demonstrated that the puffs of dipteran flies are regions of DNA transcription.

If the regulation of transcription is a mechanism of gene control of development, we should find some instances in which chromosomal puffs vary in the same tissue from one stage of a life cycle to another. This would apply only to tissues that produce different proteins at different times of the life cycle. This is exactly what has been found in many cases, where puffs appear and disappear in regular fashion during development. An example of such a sequence in *Drosophila hydei* salivary glands is shown in Fig. 11-3.

Extensive investigations have been made of puffs during the developmental stages of many dipteran flies. It has been observed that some puffs do not change at all during development, others are present only during molting periods, while still other puffs are restricted to the pupal molt (i.e., metamorphosis). A special interest has developed in the relationship of puff formation to molting, since ecdysis represents the transition from one developmental stage to another. Molting in insects is induced by the hormone ecdysone, which is secreted by the prothoracic glands. In a series of experiments reported by U. Clever in 1964, tests were made of the effects of ecdysone injection on puff formation of the salivary gland chromosomes of *C. tentans*. After injection of ecdysone into the last larval stage, two puffs (I-18-C and IV-2-B) develop within an hour. Other puffs

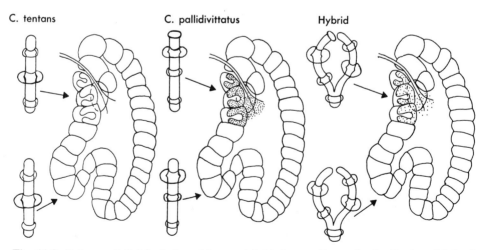

Fig. 11-2. Patterns of Balbiani rings (shown at left) from salivary gland cells (on right) of *Chironomus tentans, C. pallidivittatus,* and their hybrid. (From Beermann, W. 1963. Amer. Zool. **3**:23-32.)

appear in a sequential fashion within 10, 24, and 48 hours, as shown in Fig. 11-4, *A*. The order of appearance of the puffs is the same in normal larvae as it is in these experimentally treated larvae. Experiments in which actinomycin D and ecdysone were incorporated into the salivary glands simultaneously showed no puff formation for 15 to 20 hours after ecdysone injection. This was the period of time during which mRNA synthesis was

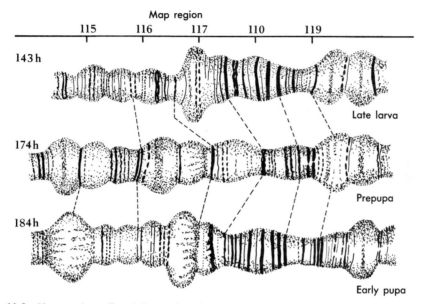

Fig. 11-3. Changes in puffs of *Drosophila hydei* salivary glands during development. (From Berendes, H. D. 1965. Chromosoma **17**:35-77.)

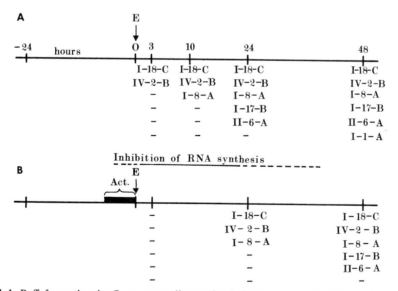

Fig. 11-4. Puff formation in *C. tentans* salivary gland chromosomes. **A,** After ecdysone injection. **B,** After actinomycin D and ecdysone injections (**E** = ecdysone; **Act.** = actinomycin D). (From Clever, U. 1964. Science **146**:794-795. Copyright 1964 by the American Association for the Advancement of Science.)

inhibited by the antibiotic (Fig. 11-4, *B*). These results make it apparent that ecdysone stimulates puff formation through the induction of DNA transcription.

The experimental findings obtained when ecdysone was used alone raised the possibility that there may be other sources of stimulation of puff formation. Some of the puffs did not appear until 24 to 48 hours after ecdysone injection (Fig. 11-4, *A*). These later appearing puffs may not have been responding to ecdysone but to some other stimulus. On the possibility that protein synthesis might be required for this later puff formation, experiments utilizing puromycin were conducted. In one experiment, ecdysone was injected into larvae 1 hour after the injection of puromycin. When they were examined 24 hours later, most of these larvae showed only those puffs that would have normally appeared earlier than 15 hours after the injection of ecdysone. The puffs that would have normally appeared 24 hours after the injection of ecdysone were missing. It was concluded that induction of those puffs which appear later in the last larval stage requires previous protein synthesis. The nature of the protein produced is at present unknown.

Our consideration of puff formation as an example of regulation of DNA transcription in development has revealed a complex situation. We have found that polytene chromosomes undergo differentiation during development, through a sequential formation of puffs. These puffs have been demonstrated to be sites of DNA transcription and have been found to vary from tissue to tissue. Some of the puffs are induced by the hormone ecdysone, while other puffs require protein synthesis for their formation.

Hemoglobins in man

Human hemoglobins provide an unusually clear-cut example of developmental changes caused by the transcription of one gene in early life, followed by the transcription of other genes later on. Beginning at a relatively early stage of embryogenesis, the genes that determine the production of the alpha and gamma chains of hemoglobin are most active, and the formation of fetal, F, hemoglobin results. The production of alpha chains continues at a high level throughout life. However, toward the end of fetal life, there are both a reduced synthesis of gamma chains and an initiation of transcription by the genes determining the production of beta and delta chains. The beta chains, when linked to alpha chains, form hemoglobin A, while the combination of delta and alpha chains yields hemoglobin A_2. By the end of the first year of life, gamma chain production has ceased completely, beta chain synthesis has reached a very high level (98% of the nonalpha chains), and delta chain production is maintained at a steady low level (2% of the nonalpha chains).

The termination of gamma chain production and the substitution of beta and delta chain syntheses can be considered a classic example of developmental changes through the regulation of transcription. It would be logical to assume that an operator type gene is involved in this regulation of transcription. However, nothing is known of the inducer or corepressor that controls the postulated operator gene.

Dosage compensation

One of the problems raised in connection with the regulation of DNA transcription pertains to the genes located in the sex chromosomes. In most organisms the males and females have either different numbers or different types of sex chromosomes. Yet, it has been found that the amounts of gene end products produced by sex-linked genes are the same in both sexes. The equality of amounts of gene products produced by genes in hemizygous and homozygous individuals was termed *dosage compensation* by H. J. Muller in 1932.

Dosage compensation in mammals. An explanation of the mechanism involved in dosage compensation in mammals was advanced by M. Lyon in 1961. She hypothesized that in each cell of the female mammalian embryo, one of the two X chromosomes present becomes inactivated.

Which of the chromosomes becomes inactivated was conjectured to be random; however, the descendants of a particular cell would always have the same X chromosome inactivated. Under this hypothesis, the adult mammalian female is a mosaic for clones of cells, each with one or the other X inactivated.

Support for the inactive-X hypothesis has been quite substantial. It has been observed that female cats and mice that are heterozygotes for X-linked coat color genes always exhibit a mosaic phenotype of the two colors involved. It has also been found that women who are heterozygous for the X-linked gene responsible for the formation of G-6-PD have two lines of cells, one possessing the enzyme and the other lacking it. Investigations of women heterozygous for the X-linked muscular dystrophy gene have revealed that their muscles are composed of two populations of muscle fibers in approximately equal numbers, one group normal and the other dystrophic.

We need to consider the concept of an inactive X chromosome in more detail: (1) it is desirable to know if there is any evidence, morphological as well as physiological, that one of the X chromosomes is inactive; (2) it will also be important for the acceptance of this hypothesis if, experimentally, we can demonstrate that the inactivation of one of the X chromosomes is at random. Morphological indication of the existence of an inactive X chromosome was provided by the discovery of a deeply staining body, called Barr body or sex chromatin body, in the nuclei of cells of female mammals (Chapter 5). This body does not appear in the zygote but does form in cells during early embryogenesis (e.g., twelfth day of human development). The Barr body was observed to be a highly coiled chromosome. A demonstration that the Barr body is inactive in transcription was provided by D. E. Comings in 1966, using human female fibroblasts. The cells were cultured in ³H-uridine and subsequently autoradiographed. The amount of uridine that had been incorporated into the Barr body was measured and compared to the amount of uridine incorporated into a comparably sized portion of another chromosome chosen at random. From these data, the maximum rate of RNA synthesis by the DNA of the sex chromatin body was estimated to be approximately 18% of the rate of RNA synthesis by a comparable amount of DNA in some other chromosome. Thus, we have both morphological (Barr body) and physiological (uridine incorporation) evidence that one of the X chromosomes of a mammalian female is inactive in transcription.

The evidence necessary to demonstrate that there is a random inactivation of one of the X chromosomes was provided by Mukherjee and Sinha in 1964. They studied chromosome duplication in cultured leukocytes from a female mule. In the mule, the paternal donkey and maternal horse X chromosomes can be clearly distinguished morphologically. Autoradiographic studies utilizing ³H-thymidine revealed that, out of 33 metaphases studied, 17 late-replicating chromosomes were identified as the X chromosomes from the male donkey parent, while in 16 cases they came from the female horse parent. The experiment quite clearly shows, on the chromosome level, that which had been postulated from earlier studies on the cellular level. All the foregoing evidence supports the hypothesis that dosage compensation in mammals is achieved through the random inactivation of one of the X chromosomes in an early stage of embryogenesis. The mechanism by which this is effected is unknown.

Dosage compensation in Drosophila. Although dosage compensation exists in *Drosophila,* its mechanism must be radically different from the X inactivation of mammals. Cells of female *Drosophila* do not show any Barr bodies in their nuclei, and X-linked genes in *Drosophila* do not show mosaic expression in the heterozygote. Nevertheless, in most X-linked mutants, the hemizygous male phenotypically resembles the homozygous female. An examination of the salivary gland chromosomes of

Table 11-3. Effect of *tra* and *dsx* on G-6-PD activity

Genotype	Phenotype	G-6-PD activity
D/tra; attached-X/Y	Female	1.2
tra/tra; attached-X/Y	Pseudomale	2.4
D/dsx; attached-X/Y	Female	2.0
dsx/dsx; attached-X/Y	Intersex	3.4

From Komma, D. J. 1966. Effect of sex transformation genes on glucose-6-phosphate dehydrogenase activity in *Drosophila melanogaster*. Genetics **54**:497-503.

the male shows that its single X chromosome is almost as wide as are the paired X chromosomes in the female. An analysis of the DNA content of the sex chromosomes by UV absorption, however, shows that the X chromosome of the male contains the same amount of DNA as a single X chromosome of the female. If the width of a chromosome is indicative of its level of DNA transcription, as seen in puffing of chromosomes, this should mean that the single X chromosome of the male is as active as are the two X chromosomes of the female.

An experiment to provide this information was reported by Mukherjee and Beermann in 1965. They dissected the salivary glands from larvae of wild type *D. melanogaster* and incubated the glands in ³H-uridine. The glands were then squashed, prepared for autoradiography, and examined for uridine incorporation into the salivary gland chromosomes. The results showed that at the level of chromosomal RNA synthesis, as measured by uridine incorporation, the single X chromosome in males works with an efficiency close to that of the two X chromosomes in the female. This, of course, does not indicate whether dosage compensation involves an enhancing effect in the male or a depressing effect in the female. To settle this last point, a triploid stock of flies was treated as just described. In triploid nuclei, two of the X chromosomes are paired, but the third remains unpaired. Here it was found that the unpaired X chromosome of the triploid female showed far less RNA syn-

thesis than the single X of the male. These experiments indicate that dosage compensation in *Drosophila* is achieved through enhanced DNA transcription of the single X chromosome of the male. The mechanism by which this is achieved is largely unknown.

An experiment designed to test the effects of sex-transforming genes on dosage compensation in *Drosophila* was reported by D. J. Komma in 1966. In this experiment, the X-linked gene that produces G-6-PD was studied. Some stocks of flies were established that contained various combinations of the recessive gene *transformer, tra,* which in homozygous condition transforms an XX female into a pseudomale. Other stocks were established that contained various combinations of the recessive gene *doublesex, dsx,* which in homozygous condition causes both XX and XY embryos to develop into intersexes (Chapter 5). Both *tra* and *dsx* are on the third chromosome, but they are not alleles. Some of the stocks also contained the third-chromosome dominant gene *Dichaete (D),* which is a wing mutation that is lethal in homozygous condition. *Dichaete* is used in the present experiment to identify those flies that are heterozygous for either *tra* or *dsx.* All the stocks had their females as "attached-X females," which means that their two X chromosomes are attached to the same centromere. It will be recalled that in *Drosophila,* XXY individuals are fertile females, and in Komma's experiment these individuals are designated as "attached-X/Y." The unit of measurement in this experiment was the level of activity of the enzyme G-6-PD. The critical data are shown in Table 11-3. They indicate that both pseudomales and intersexes show increased G-6-PD activity as compared to their normal sisters. This finding suggests a relationship between G-6-PD activity and sexual differentiation and implies that dosage compensation for this enzyme is dependent on sexual differentiation. In this same investigation another enzyme, 6-PGD, which is also produced by an X-linked gene, was studied but was not found to show a relationship between enzyme activity and sexual differentiation.

It is obvious that the mechanism of dosage compensation in *Drosophila* requires a good deal more study before any general statement can be made about it.

REGULATION OF mRNA TRANSLATION

Experimental evidence for the role of regulation of mRNA translation in developmental processes is extremely scarce. It is quite clear that the availability of required amino acids, tRNAs, or ribosomes could affect protein synthesis. What is lacking are examples of such processes acting in the various stages of a life cycle. Our discussion, earlier in this chapter, of chromosome replication in an *E. coli* auxotroph illustrated how the need for an amino acid in protein synthesis can limit DNA synthesis. However, no indication exists of such a situation in nature, as part of the normal life cycle of any organism.

Hemoglobin synthesis in rabbit reticulocytes

An example of what might be the operation of a regulatory mechanism at the ribosome level was provided by H. M. Dintzis in 1961. He incubated normal rabbit reticulocytes for a short time with ^{14}C–amino acids. The cells were then lysed and the alpha and beta chains of the cells' hemoglobin were separated and analyzed for radioactivity. The radioactivity of the beta chains was found to be from three to five times higher than that of the alpha chains. This result would be expected if there were more completed or nearly completed alpha chains than beta chains on the polysomes at the beginning of the incubation period. Under these conditions, one would also expect to find labeled beta chains associated with unlabeled alpha chains after a short pulse with ^{14}C–amino acids, and this was indeed the case. These observations suggest that the rate-limiting step in the assembly of the hemoglobin molecule in normal rabbit reticulocytes is that rate of synthesis and thus the availability of beta chains. Here we have an example of how regulation at the ribosomal level could operate to control protein synthesis. However, the point must be made that the above phenomenon would also occur if there were differences in the rates of transcription of the alpha and beta hemoglobin genes.

Selective translation in virus-infected cells

Earlier in this chapter we discussed the selective transcription that occurs in *E. coli* cells which have been infected with phage T4. We now want to examine the evidence for the occurrence of selective translation in T4-infected cells. A series of investigations on this aspect of translation was reported by Hsu and Weiss in 1969. They performed in vitro experiments with ribosomes from normal and T4-infected cells. In their studies, they compared the translational efficiencies of the ribosomes from these two sources with different RNA templates.

The results showed that when poly-U is used as a template, virtually the same amount of polyphenylalanine is formed by each type of ribosome. When T4 mRNA was the template, similar amounts of polypeptides were formed with both types of ribosomes. However, when mRNA from *E. coli* or RNA from phage MS2 (an RNA virus) was used, the amount of polypeptides synthesized on ribosomes from T4-infected cells was much less than that synthesized on ribosomes from normal cells. This indicated that the ribosomes of infected cells had been changed as a result of T4 infection.

As with all studies of this type, it was of interest to learn whether the observed changes in ribosome activity were associated with the synthesis of a protein. To determine this, chloramphenicol was added to the bacterial culture medium just prior to T4 infection. It was found that the ribosomes isolated from these cells after T4 infection

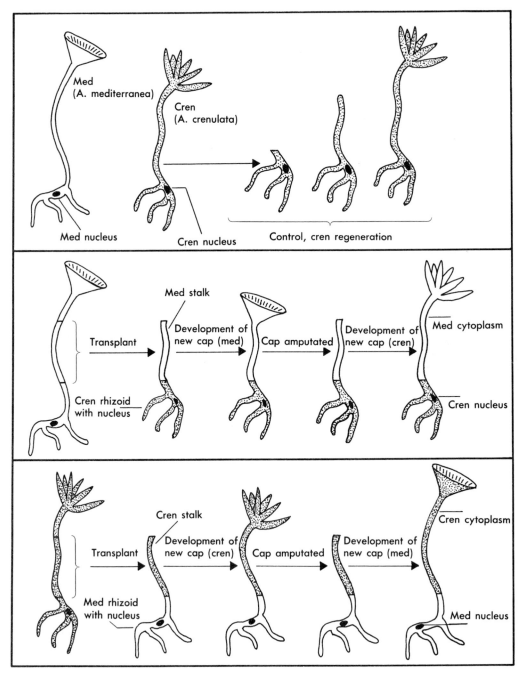

Fig. 11-5. Nucleus-cytoplasm interaction in *Acetabularia*.

exhibited the same translational characteristics as ribosomes from uninfected cells. The conclusion drawn was that protein synthesis was necessary in order to change the translational characteristics of normal ribosomes. This conclusion was supported by the discovery that ribosomes from infected cells contain a heat-labile factor that is not present in normal ribosomes. This heat-labile factor could be combined with normal cell ribosomes, which then act the same as the ribosomes from T4-infected *E. coli* cells.

These studies have shown that phage T4 infection induces a change in host ribosomes which results in a selective translation of T4 mRNA. The discovery of this phenomenon lends strong support to the hypothesis that regulation of mRNA translation may be an important factor in normal cell development.

Phosphatase synthesis in an enucleate cell

An experiment demonstrating regulation of protein synthesis in an enucleate cell was reported by Spencer and Harris in 1964. They worked with a unicellular green alga, *Acetabularia crenulata*. The cell body of this organism consists of a basal rhizoid in which the nucleus is located, a narrow stalk some two inches long, and a terminal cap (Fig. 11-5). It has been found that if one removes the cap, and after a few days removes the rhizoid, there will be a regeneration of the cap by the enucleated cell. It takes some six weeks before the new cap begins to form. These investigators studied the synthesis of three phosphatases having pH optima of 5.0, 8.5, and 12.0, respectively. The synthesis of these enzymes was studied during normal growth of the cells (i.e., from zygote to maturity) and during the period of regeneration that follows removal of the cap. In addition, a comparison of enzyme production was made between regenerating cells that contained a nucleus and similar cells from which the nucleus had been removed. In all three cases, synthesis of the pH 12 enzyme is delayed until the cap is about to be formed; then it proceeds at a faster rate

than the formation of the other two phosphatases. The amounts of pH 12 enzyme synthesized are the same in all three instances. It seems clear from these results that the pH 12 enzyme is specifically associated with cap formation. In seeking to explain these results, one must realize that for the enucleated cells, DNA transcription has to end with the removal of the nucleus. To be sure, long-lived mRNA can still be available. However, it is clear that mRNA translation for the pH 12 enzyme must, by some as yet unknown mechanism, be inhibited until the beginning of cap formation. One must then conclude that the cytoplasm of *A. crenulata* contains a mechanism of long-term duration, for the regulation of protein synthesis. Here, too, there is no certainty that such a mechanism is involved in the normal development of this organism.

NUCLEUS-CYTOPLASM INTERACTION

The interaction of nucleus and cytoplasm in development has long been a subject of interest to geneticists. Involved in this type of investigation have been questions about the degree of control the nucleus continues to have over a differentiated cytoplasm and whether a specialized cytoplasm can in turn control the nucleus. Research in this area of developmental genetics has included many varied organisms and techniques. We shall examine a few examples that illustrate some basic principles.

Cap formation in Acetabularia

Acetabularia has been an important experimental organism in studies on nucleus-cytoplasm interactions. Much of this type of research has involved the cap of the alga, which has a distinctive morphology that differs from species to species. One species, *A. mediterranea,* has a cap that is disk-shaped, while another species, *A. crenulata,* has a cap that is branched (Fig. 11-5). As discussed earlier, by amputating the cap one can obtain a nucleated cell and by amputating both cap and rhizoid one can obtain an enucleated cell. Studies on the regenerative capacities of these fragments were reported by J. Hämmerling in 1953. He found that an enucleated cell whose cap had been removed was capable of regenerating a new cap. The newly formed cap could then be removed and the process could be repeated for a limited period of time. On the other hand, nucleated fragments continued to regenerate caps indefinitely, after amputations. The regenerated caps, in both instances, were always of the type characteristic of the species. It seems apparent that the enucleated fragment contains, for a limited time, the genetic information necessary for cap formation but that, eventually nuclear activity is required for continued cap production.

In another experiment designed to test the role of the nucleus in cell differentiation, a stalk of *A. mediterranea* was grafted to the nucleated base of *A. crenulata,* and reciprocal transfer of a stalk was made from *A. crenulata* to the nucleated base of *A. mediterranea.* In both cases caps were regenerated that reflected the species of the stalk. On repeated amputation of the newly formed caps, others were formed that reflected the species of the nucleus. The experimental findings are shown in Fig. 11-5. These results confirmed and extended those described previously. They show that the cytoplasm does contain enough genetic information to continue its previous type of activity regardless of the presence of a foreign nucleus. They show also that the control the nucleus has over cell differentiation can take effect only when preexisting genetic information has been exhausted and needs replacement.

Nuclear transplantation in Amphibia

A limitation of the experiments performed on *Acetabularia* is that one is restricted to working with a single nucleus and for a rather short period of time. A better test of nucleus-cytoplasm interaction would be to test identical nuclei after various stages of increased cytoplasmic differentiation. A technique suitable for such an experiment was reported by Briggs and King in 1952 In this procedure, a section of an amphibian embryo (e.g., blastula, gastrula, or tail bud) is taken and placed in a solution that causes the cells to separate from one another. Since all the cells of the embryo are derived by successive mitoses from a zygote, all the nuclei carry identical genetic information, mutations excepted. As recipient cells, unfertilized eggs are chosen, their haploid nuclei are removed, and an embryonic nucleus plus a small amount of cytoplasm are injected into each enucleated egg cell. These eggs now have identical diploid nuclei in their cytoplasms and, having been previously activated by the prick of a needle, will begin to divide. One can then note the frequencies of developing embryos that are normal and those that exhibit abnormal development and death.

In most of the experiments, nuclei were taken from the endodermal (yolk) area of *Rana pipiens* embryos. The blastula nuclei gave completely normal embryos, thus demonstrating that blastula nuclei are not irreversibly differentiated. However, when endoderm nuclei from the gastrula and later stages were studied, a progressive increase in the frequency of developmental abnormalities and deaths was observed as nuclei from more advanced stages were used (e.g., early gastrula, 23%; late gastrula, 80%; tail bud, 96%). These results indicate that the nuclei of older embryos become irreversibly differentiated and can no longer transcribe all the genetic information needed for normal development. The types of developmental abnormalities were analyzed. It was found that although endodermal and mesodermal structures were normal, ectodermal structures were poorly developed. The nervous system and sense organs showed many defects, and in some cases the epidermis was so incomplete that it scarcely covered the embryo. These findings indicate that the endodermal nuclei have become differentiated to produce only endoderm-forming proteins and the genes producing ectoderm-forming proteins have been completely repressed. The mechanism by which this repression has occurred is as yet unknown. However, what is clear is that a differentiated cytoplasm can produce an irreversible situation in its differentiated nucleus.

Experiments similar to those performed on *Rana pipiens* have been carried out using the African clawed frog, *Xenopus laevis*. It was found that an irreversible condition did *not* develop in the nuclei of differentiated cells. Nuclei from cells that had undergone considerable visible differentiation could be injected into enucleated eggs, and normal embryos, or even adults, were obtained. These observations are extremely important in indicating that nuclear differentiation need not be irreversible in all cases. These results demonstrate the need for further study of the phenomenon of nuclear differentiation, especially of those factors that affect its reversibility.

TISSUE DIFFERENTIATION

Tissue differentiation in multicellular plants and animals is one of the more spectacular aspects of development. It involves the formation of distinctly different types of cells. The specialized cells not only are morphologically different but also show chemical, immunological, and behavioral differences as well. One of the questions raised for differentiated cells is whether they can revert to a generalized state and become specialized in some other direction. The ultimate in such a process would be the ability of a differentiated cell to produce an entire organism. Cells that could do this would be said to be *totipotent,* that is, they would have retained their ability to utilize their entire genome. We shall examine some of the findings that have been made of totipotency in plants and animals.

Totipotency in plants

An excellent example of totipotency in plants has been demonstrated in the case of phloem cells of the carrot. Single phloem cells are obtained from the roots of the plant. These are placed in a liquid medium in which they divide and grow, forming cell clusters. Some of these cell aggregates form organized areas, or nodules, from which roots are produced. When a rooted nodule

is transferred to agar, it ultimately forms a shoot. The plantlet thus formed is in all respects a normal carrot plant that will mature and produce a storage organ, flowers, and seeds. The cycle may then be repeated, with similar results obtained each time. A diagram of the vegetative growth cycle is shown in Fig. 11-6. It will be noticed that the cultured cells are the sole link between one vegetative growth cycle and the next, thereby demonstrating the totipotency of cultured carrot phloem cells. No assurance can be given, of course, that all differentiated plant cells will demonstrate this versatility.

Totipotency in animals

No example of totipotency in animals has been demonstrated. It would appear that differentiated cells lose their totipotency on becoming specialized and thus pass a point of no return. However, a number of cases of change of specialized characteristics of one cell type and transformation into other cell types have been found. One example of this occurs when a limb of a salamander is amputated. The muscle cells of the stump change in characteristics and form mononucleated cells that contribute to the blastema from which a new limb regenerates.

Another example of transformation of a specialized cell occurs in the production of a cancer cell. A detailed discussion of cancer is much beyond the scope of this book. However, it is of interest to note that two main changes occur in a cancer cell. One change involves rapid growth and division. This implies a change in the regulatory mechanism that normally restricts cell multiplication to the needs of the organism. The second change involves the invasive characteristic of cancer cells. Cancer cells are not confined to the original tissue in which they arose but spread to other tissues where they proliferate.

Tumor induction by viruses

As was mentioned in Chapter 1, tumors in both plants and animals have been associated with various types of viruses. All

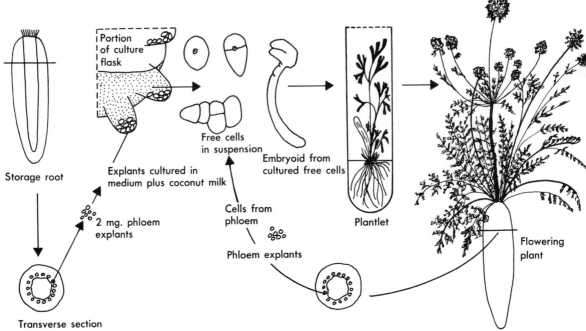

Fig. 11-6. Vegetative growth cycle of the carrot plant. (From Steward, F. C. 1964. Science **143:** 20-27. Copyright 1964 by the American Association for the Advancement of Science.)

those viruses thought to cause plant tumors have thus far been found to have their nucleic acid in the form of double-strand RNA. The viruses involved in animal tumors are found to be either of double-stranded DNA or single-stranded RNA composition. Theoretically, one should anticipate finding examples of both single- and double-stranded viruses of either RNA or DNA composition that can produce tumors in one or more species. Much more is known about the viruses that cause tumors in animals than those that affect plants, and for this reason, we shall confine our discussion to the former group.

Examples of double-stranded DNA tumor viruses are (1) polyoma virus, whose host cells include those of the mouse, hamster, and rat; (2) adenovirus 12, which attacks the cells of the hamster, rat, rabbit, and man; and (3) SV 40, which has been found in the cells of the mouse, hamster, rat, rabbit, monkey, and man. A viral infection of a host cell by one of these viruses can have either of two results. In one case, called *productive infection,* the virus multiplies rapidly within the cell. The cell eventually dies, and the virus particles are released. In the second case, called *transformation,* the virus enters the cell but cannot be identified or isolated thereafter. Presumably, like an episome, it enters the cell's genome and causes the transformation of the cell by altering the transcription pattern of the cell's genes, while in addition transcribing its own genes. Ample evidence exists that at least part of the viral DNA persists in the tumor cells and is multiplied and distributed to all daughter cells. One example of such evidence is the appearance of a new antigen at the surface of a transformed cell. The relationship of this antigen to the virus can be demonstrated by experiments involving the virus SV 40. SV 40–induced tumor cells can ordinarily be implanted into hamsters, where they grow, divide, and form large tumors. In one experiment, hamsters were first inoculated with the SV 40 virus. Attempts were made later to implant SV 40 tumor cells into these animals, but without success. Evidently inoculation with the virus caused the production of antibodies by the animal. These antibodies, in turn, attacked any graft of cells containing the same antigen. We may conclude from these results that a virus-induced tumor cell represents a cell that has been transformed through the addition of genetic material from the virus. We do not know how many human cancers, if any, are virus induced. Nor do we know whether the above procedure could be used to develop an immunity in man to any demonstrated cases of virus induced cancers.

Examples of single-stranded RNA tumor viruses are (1) Rous sarcoma virus (RSV), which invades the cells of the chicken, mouse, hamster, rat, and monkey; (2) Rauscher leukemia virus (RLV), which causes leukemia in mice; and (3) mouse mammary tumor virus (MMTV), which infects mouse mammary gland cells. When, as an example, a host cell is invaded by a Rous sarcoma virus, the cell survives but becomes transformed into a cancer cell. One also finds that new single-stranded RNA tumor viruses are produced and leave the transformed cell without killing it. The unique feature of this host cell–tumor virus system is that the RNA virus makes a DNA copy of itself that becomes incorporated into one of the host cell's chromosomes and is replicated along with the host DNA. It was demonstrated that the virus replicates through a DNA intermediate by placing actinomycin D into a tissue culture medium containing infected cells. This resulted in the termination of RNA virus production. Also terminated was the production of all the host cell's RNA molecules that are synthesized on a DNA template. When the experiment was performed on cells that were infected with other types of single-stranded RNA viruses (e.g., RNA poliovirus), new RNA viruses continued to be produced, although the host cell's formation of RNA molecules ceased. It is evident from this and other experiments that actinomycin D does not affect the synthesis of RNA that is made on an RNA template. These findings mean that a polymerase enzyme must be present that will use RNA as a template to synthesize DNA. Subsequently RNA tumor virus particles were found to contain a DNA polymerase that uses the RNA strand of the virus as a template.

Questions immediately arose as to the applicability of this information to human cancers. In a series of exciting experiments reported by S. Spiegelman and co-workers in 1972, the enzyme RNA-directed DNA polymerase (reverse transcriptase) was used to prepare quantities of DNA that were derived from, and hence homologous to, MMTV (mouse mammary tumor virus) and RLV (Rauscher leukemia virus). These two types of DNA were then used in hybridization experiments with RNA derived from human breast cancer cells, white blood cells of leukemic patients, normal breast cells, fibrocystic (benign tumor) breast cells, normal white blood cells, normal spleen cells, normal liver cells, etc. It was found that of the 29 human breast cancers tested, 67% contained RNA that hybridized with DNA complementary to the RNA of MMTV. None of the 30 control preparations from normal or benign breast tissues contained detectable quantities of this type of RNA. The breast cancer RNA samples that hybridized with the DNA derived from MMTV did not hybridize with the DNA complementary to RLV. In studies involving human leukemia, the white blood cells of 89% of the 27 leukemic patients studied contained RNA that was homologous to that of RLV but not to that of MMTV. With regard to the normal tissues studied, none of the 34 different types of normal tissues used in these experiments contained RNA that could hybridize with the DNA derived from either MMTV or RLV. The conclusions drawn by the authors are understandably, and correctly, conservative. They point out that their results do *not* constitute a proof of a viral etiology of human cancers. The experiments do show that human cancer cells contain RNA that is specifically homologous to that found in viruses that cause cancers in other animals. Further research in this field is certainly indicated.

Lactic acid dehydrogenase

As stated above, tissue differentiation includes chemical as well as morphological differences. Studies of the enzyme lactic acid dehydrogenase (LDH) provide an excellent example of the chemical differentiation of tissues. This enzyme protein consists of four polypeptide chains. Each chain can be one of two types, A or B. From this arrangement, five different proteins may be formed: A_4, A_3B_1, A_2B_2, A_1B_3 and B_4. These proteins are referred to as LDH-5, 4, 3, 2, and 1, respectively. Different molecular forms of an enzyme are called *isozymes*. Each of these isozymes has its own rate of electrophoretic migration, LDH-1 being fastest, followed by the 2, 3, 4, and 5 forms. Each isozyme has its characteristic physical and enzymatic properties, and which ones are present in a given tissue is a characteristic of that tissue at a particular stage of development. For example, all human embryonic tissues have, predominantly, A_4 (LDH-5). Adult skeletal muscle retains its predominance of A_4, but adult heart muscle has predominantly B_4 (LDH-1). Examples of electrophoretic patterns of LDH isozymes from adult mouse tissues are shown in Fig. 11-7, and the changes that occur during development in tongue tissue are shown in

Fig. 11-8. The A and B polypeptide chains appear to be under separate genetic control, since separate mutations involving each chain have been found. Which isozyme is formed appears to depend on the relative activity of the two genes. The factors that determine the rates of DNA transcription of these two genes are thus far unknown. LDH isozymes represent a fascinating illustration of chemical differentiation in tissues, based on the molecular forms of an enzyme. A number of such cases of tissue specialization are known, and undoubtedly more remain to be discovered.

ORGAN SYSTEM DIFFERENTIATION

Our discussions of genes and development have thus far been mainly restricted to the mechanisms by which the syntheses of gene products are regulated and how this affects chromosome, cell, and tissue differentiation. When we study organ system differentiation, however, other factors have to be considered. There are complex interactions of gene products, cells, and tissues that are essential for the development of an organ system. At each step, the environment continues to play its important regulatory function, much as it did in going from DNA to gene product. One very successful line of research of organ system differentiation has been the study of *lethal genes*. The term *lethal* is applied to those changes in

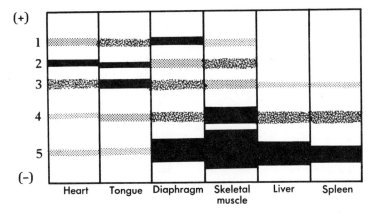

Fig. 11-7. Diagram of electrophoretic patterns of LDH isozymes from adult mouse tissues. (From Markert, C. L., and Ursprung, H. 1962. Develop. Biol. **5:**363-381.)

the genome of an organism which produce effects severe enough to cause death. Here, too, different environments may modify the action of a "lethal gene" so as to render the organism viable. Lethal genes are important tools in developmental genetics in that they interfere with normal development and, in doing so, they permit the analysis of normal developmental patterns. We shall consider a few examples of lethal genes and see that the lethal effects of these genes trace back to widely divergent causal factors and operate through equally divergent pathways.

Danforth's Short-tail (Sd) in mice

The gene *Sd*, in heterozygous condition, causes a shortening or absence of the animal's tail. In homozygous condition, the animals die at birth or just before because of an almost total lack of kidneys. Dominant genes such as *Sd* that have lethal effects when homozygous are discovered as a result

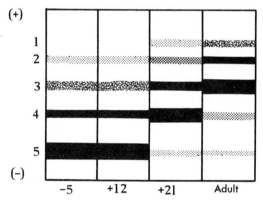

Fig. 11-8. Diagram of changing electrophoretic patterns of LDH isozymes during the development of mouse tongue tissue. (From Markert, C. L., and Ursprung, H. 1962. Develop. Biol. **5:** 363-381.)

Sd / + by Sd / +

Sd/Sd	**Sd / +**	**+ /Sd**	**+ / +**
Dies	Short-tail		Normal

Fig. 11-9. Diagram of a cross involving a homozygous lethal gene.

of the inability of the geneticist to establish a stock homozygous for the gene. In Fig. 11-9 are diagrammed the results of a cross between two mice, heterozygous for *Sd*.

Note that the cross of animals heterozygous for this dominant gene does not yield the expected 3:1 ratio of viable dominant to recessive phenotypes but rather a 2:1 ratio. This is usually the first indication that lethal homozygotes are being formed. In the case of mice, examination of developing or newborn litters will reveal the dead individuals. These are the dominant homozygotes that constitute the missing class of offspring.

Dissection of the *Sd/Sd* offspring reveals a variety of abnormalities of the kidneys and ureters. One feature, however, is constant. Whenever there is a kidney, a ureter is found attached to it. Although one may find ureters in various stages of development without kidneys, the reverse is not true. A possible explanation would be to assume that the ureter, which arises as a bud from the mesonephros, acts as a kind of *organizer* in inducing the formation of the metanephric part of the mammalian kidney. The *Sd/Sd* genotype, under this hypothesis, causes an interference with the normal elongation and branching of the mesonephric bud. The absence of the kidney results from the failure of the mesonephric organizing tissue to make adequate contact with the tissue it normally induces to form the capsules and secretory tubules of the kidney. The correctness of this hypothesis was tested in tissue cultures. A report on such an experiment was published by Glucksohn-Waelsch and Rota in 1963.

Reciprocal combinations of normal and "lethal" ureter with normal and "lethal" kidney mesenchyme were made in tissue cultures. All four combinations showed a differentiation of kidney tubules. This demonstrated that both the lethal ureter and lethal kidney tissues were capable of their respective normal functions: induction and differentiation. The *Sd/Sd* genotype, through some as yet unknown mechanism, retards the growth of the ureters so that they do not reach the kidney mesenchyme and, hence, do not induce kidney formation. Here we have an example of how the study of the effects of a lethal gene has produced an understanding of the normal developmental requirements for the formation of a critical organ of the body. It is hoped that, eventually, the primary action of the *Sd* gene will be discovered and the pathways by which its protein products cause the retarded growth of the ureters will be understood.

Creeper in fowl

In the chicken, a dominant gene, *Creeper (Cp),* in heterozygous condition causes a shortening of the bird's legs so that the animal appears to be creeping when it walks. A cross of two Creepers yields viable offspring in the ratio of 2 Creepers: 1 normal. In addition, 25% of the eggs fail to hatch. *Creeper* thus appears to follow the typical pattern of a dominant gene which in homozygous condition is lethal. An examination of the *Cp/Cp* embryos disclosed that some 90% died within three days of incubation while the remaining 10% survived until about nineteen days, which is about the time of hatching from the egg. The effects of this gene have been studied in detail since 1930 by W. Landauer and co-workers.

An examination of the heterozygote *Cp/+* shows that the only effect of the gene is a shortening of the tibia of the leg (68% of normal). The effects of *Cp/+* cannot be shown before the seventh day of incubation, when limb retardation begins to manifest itself. It is not possible to compare the heterozygote with the early-dying homozygote, since the latter dies before limb cartilages are formed. In the early-dying *Cp/Cp,* death is caused by an oxygen deficiency due to the failure of underdeveloped yolk vessels to make proper connections with the yolk circulation. It has been observed that before the animal dies, the development of the head stops prematurely and the rudiments of the eye remain noticeably small. In the late-dying *Cp/Cp,* death is caused by a severe anemia because no blood-forming cells are produced in the animal's underdeveloped bones. Other characteristics of the late-dying embryo include a drastic shortening of the tibia (21% of normal), phocomelia (stumplike limbs), severely shortened skull, and microphthalmia (very small eyes).

The existence of two critical developmental phases (at three days and nineteen days) for Creeper homozygotes led investigators to study the percent of mortality in *normal* fertile eggs from different breeds, on each day of embryonic development. It was found that here also there were two sensitive phases, one between the second and the fourth days of incubation and the other between the eighteenth and twentieth days. It is of interest to note that the lethal factor *Creeper* becomes effective at the same stages in development that are vulnerable for normal embryos. This would imply that these two stages of embryogenesis represent critical periods during which any great interference with the normal developmental pattern leads to the death of the organism. It is probably correct to assume that such sensitive stages of development represent periods of radical changes in the physiology of the organism. The ability of the organism to pass through such sensitive stages in effect tests the adequacy of the organism's developmental processes for the formation of a viable individual.

In all studies of lethal genes, it is informative to see whether one can overcome the expected lethal effects of a genotype. This can be attempted by altering the environment or by changing the genotypic milieu of the gene. The ability of environment to affect the developmental process has been tested by incubating the eggs at two different temperatures. At an incubation temperature of 38° C., 6.5% of the Creeper homozygotes reached the late stage, whereas at 35° C., 17.2% did so. The reduced temperature during the first three days of embryogenesis permitted the development of a larger number of viable embryos than was possible at the higher temperature. It is not known what embryological processes were changed or in what manner. However, the reduced temperature did not result in Creeper homozygotes' developing past the nineteenth day of incubation. The role of the genotypic milieu on the effects of the Cp gene was tested by crossing Creepers to a genetically unrelated breed of chickens (Bantam fowl) and obtaining homozygotes from the F_2. In the original Creeper line, 6.5% of the Creeper homozygotes reached the late stage, but Creeper homozygotes derived from a cross with Bantam fowl showed 18% to 41% reaching the late stage. We should not be surprised that the effects of the Cp gene are determined, at least to some extent, by the other genes of the genome, since we are not considering the action of the gene at the DNA transcription level but rather the effects of the gene at some organ system level. In this experiment, as was true of the temperature experiment, none of the Creeper homozygotes developed past the nineteenth day of incubation.

Our ability to alter the expression of a gene either through changes in the environment or through manipulation of the rest of the genome leads to the possibility that an individual may possess a particular genotype without expressing it. Geneticists, in considering this possibility, have found it necessary to characterize each gene as to the percent of individuals that exhibit the trait for which they have the gene. This characteristic of the gene is called its *penetrance*. It is measured from zero to 100% and is taken as an indication of the ability of the gene to penetrate the environmental and genetic factors that might act to suppress its expression. The concept of penetrance is extremely useful when we are faced with dominant genes that are not expressed in all their carriers. Many human genetic traits are believed to be caused by dominant traits with incomplete penetrance. With regard to Creeper chickens, we would say that the *Creeper* gene shows 100% penetrance.

In addition to penetrance, another concept has become important to the developmental geneticist: even when a gene has 100% penetrance, it does not affect the particular trait in every individual to the same degree. This, in turn, may be due to differences in the internal environments of

the various individuals or to differences in the rest of their genomes. The degree to which a gene affects a trait, compared to the maximum possible alteration of the trait, is called the *expressivity* of a gene. Expressivity is difficult to quantify in some cases. In the case of *Creeper,* we would say that there are two levels of expressivity of the lethal homozygotes: 90% at the early stage and 10% at the late stage. Although it is useful to establish as two separate concepts the phenomena of penetrance and expressivity, they may very well have a common determining basis in embryogenesis.

One question that always arises in a study of a lethal gene is whether the death of the organism is due to the effects of the gene on the individual cells of the organs of the body or whether the lethal effects are due to a lack of integration of potentially normal functioning structures (as was true of Sd/Sd embryos). One experimental approach to this problem is to transplant rudimentary structures from affected to normal individuals and observe whether the rudiments develop into normal or abnormal structures. The results of such experiments on Creepers were reported by V. Hamburger in 1941. He found that normal leg rudiments transplanted to normal hosts result in legs barely shorter than normal legs. Leg rudiments transplanted from $Cp/+$ individuals to normal hosts show only $Cp/+$ development. Leg rudiments from Cp/Cp individuals (i.e., 24 to 33 hours of incubation) transplanted to normal hosts develop to the later stages of embryogenesis and show the extremely short tibia and phocomelia of the late-dying embryos. This means that the limb rudiments of the embryos dying at the early stage do contain the potentiality of developing to the later stage of differentiation. However, the lack of normal development of $Cp/+$ and Cp/Cp leg rudiments indicates that the genes have caused certain changes in the cells of the limbs that cannot be altered by the normal environment of the host.

Transplantation experiments were also conducted with eye rudiments. Normal eye rudiments transplanted to normal hosts develop normally. Eye rudiments from Cp/Cp individuals transplanted to normal hosts develop normally. Eye rudiments from $+/+$ individuals transplanted to a Cp/Cp host that reached the late-dying stage developed into the typical Cp/Cp eye (microphthalmia). According to these findings, the abnormalities arising in the eyes of inviable embryos are due to external causes. The Cp/Cp eye becomes abnormal only if its rudiment is part of an inviable organism. In order to check on whether the poor development of the Cp/Cp eye, when in its own body, was due to a faulty blood supply, an eye rudiment of a $+/+$ individual was transplanted into the lateral coelom of a $+/+$ embryo. The coelom does not contain as rich a blood supply as the head. The eye developed with all the characteristics of the Cp/Cp eye, implying that a faulty blood supply might be the cause of the microphthalmia of Creeper homozygotes.

One of the experimental approaches used by some developmental geneticists in analyzing the actions of lethal genes is to alter the environment of a normal developing organism in such a way that the resultant phenotype of the individual resembles a known mutant. Such an individual is called a *phenocopy,* since its phenotype mimics that of a known mutant. The importance of this approach to developmental genetics hinges on the assumption that the environmental change produces its effects through the same pathway as the mutant. If the phenocopy and the mutation lead to identical forms by affecting different phases of development, the phenocopy experiment has no direct relation to the effect of the mutation.

At the beginning of our discussion of the *Creeper* gene, it was pointed out that the yolk vessels of the Cp/Cp embryos are underdeveloped, with the result that the body of the embryo fails to be connected in a normal way to the yolk circulation. Abnormal conditions of pressure in the embryo lead to short circuits' being formed between the aortae and the cardinal veins. As a result of these short circuits, the blood returns to the heart before it has passed over the sur-

face of the yolk. Thus the embryo does not obtain the proper supply of oxygen and nutrients. The lack of these materials causes the eventual death of the embryo. In 1941 J. M. Cairns succeeded in reproducing all the abnormalities of the early-dying Creeper embryos in normal embryos merely by severing the large vessels of the yolk. Although the phenocopy thus produced corresponds in all details to the Creeper mutant, the operation is by no means equivalent to the effects of the gene. The mutation in the Creeper embryos must become active at an early stage by inhibiting the original formation of the vessels, whereas the phenocopy operation is performed at a later stage and merely produces what are the secondary consequences of the effects of the *Creeper* gene. This experiment, which may well be representative of many other phenocopy experiments, does not help us to understand the primary action of the *Creeper* gene.

Our consideration of the Creeper syndrome has revealed a gene with many effects (pleiotropism). Some of these result in direct changes of the cells of a particular organ (e.g., leg bones), while other effects of the gene result in alterations of the internal environment to such an extent that potentially normal tissues are unable to develop in their characteristic fashion (e.g., eye). This example of the effects of genes on development indicates the complexities of embryogenesis and the various experimental approaches used to gain an understanding of organ system differentiation.

T locus in mice

The last example of genetic control of organ system differentiation that we shall consider will be the T locus in mice. Most of the investigations of this locus have been carried out by L. C. Dunn and co-workers, starting in 1936. The T locus alleles are quite varied in their types and effects. They include three dominant genes, all of which are homozygous lethals, and an ever-increasing number of newly discovered recessive genes which, as homozygotes, may be either lethal or viable. As is true of all well-studied genetic systems, the T locus in mice is extremely complex. We shall only touch on some of the developmental aspects of the effects of this locus.

With allowance made for occasional exceptions, the interactions of the dominant, recessive, and wild type alleles at this locus can be summarized as follows:

Genotype	Phenotype
+/+	Normal tail
T/+	Short tail
T/T	Embryonic lethal
+/t	Normal tail
T/t	Tailless
t/t	Embryonic lethal

In the above listing, T may be any one of the three dominant alleles, and t may represent any one of the many recessive lethal t alleles. In this discussion we shall not consider the recessive viable t alleles. When heterozygotes involving the wild type and a mutant allele are mated to one another (e.g., $T/+ \times T/+$ or $+/t \times +/t$), one fourth of the offspring die as embryos. The dissection and study of these embryos at the various stages of gestation has revealed the abnormalities that caused the death of the embryo. It is found that each homozygous-lethal allele has its own syndrome of abnormalities and a characteristic period when death is most likely to occur. These are diagrammed in Fig. 11-10.

We shall first consider the development of embryos heterozygous for T (either $T/+$ or T/t). The most obvious phenotypic effects of these genotypes involve the tails of the animals. In these heterozygotes the tail develops normally until about the eleventh day of gestation. At this time a constriction develops around the middle of the tail in $T/+$ embryos, and at base of the tail in T/t embryos. Distal to the constriction, all

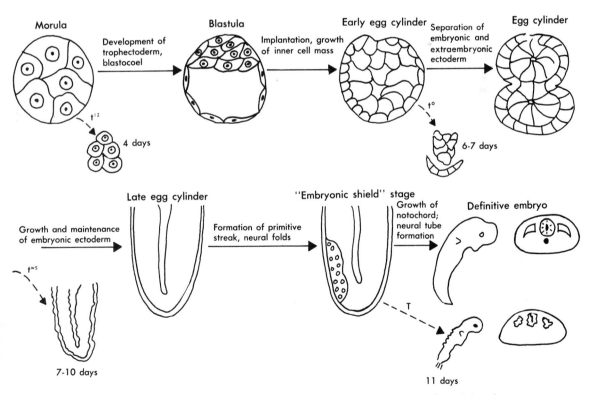

Fig. 11-10. Diagrams of developmental stages of the mouse, showing points of interference by T locus alleles. (From Bennett, D. 1964. Science **144**:263-267. Copyright 1964 by the American Association for the Advancement of Science.)

the tissues atrophy and are resorbed. There are other developmental abnormalities present in the lumbar and sacral regions of these embryos. These abnormalities involve both the notochord and the neural tube. The notochord is very irregular in size and often contains a lumen, while the neural tube shows abnormalities of size and shape. The neural tube abnormalities appear to depend on abnormalities of the notochord in its vicinity, since neural abnormalities, unlike notochordal abnormalities, tend not to occur alone.

Studying the T/T homozygotes, one finds a severe accentuation of the above effects. The 9- to 10-day-old embryo lacks the whole posterior body structure behind the anterior limb buds. The neural tube is greatly kinked and folded, and, histologically, no trace of notochordal cells can be found. These embryos have virtually no allantoic connections to the placenta. It is presumably the failure of this system which leads to their death at about 11 days of gestation, which is the time when the umbilical circulation begins to function in normal embryos.

The abnormalities involving the notochord make it appear that T affects the chordamesoderm, which is the primary inductive system of the embryo. Experimental confirmation of this hypothesis was provided by D. Bennett in 1958. Geneticists had previously shown that somites (mesodermal segments) from normal mouse embryos would form cartilage in tissue culture only under the influence of an inducer such as the ventral spinal cord from an embryo of the same developmental stage. In the present case, T/T somites were found to be unable to form cartilage in the presence of normal ventral spinal cord. However, ventral spinal cord from T/T embryos was quite capable of inducing cartilage formation in somites from normal mouse embryos. These results support the supposition that T has its effects primarily on the notochord-mesoderm system.

In contrast to the association of T with chordamesoderm, the recessive alleles of this locus seem to have their effects primarily in tissues of ectodermal origin. In addition, most of the recessive alleles cause death earlier in the gestation period. The earliest-acting lethal allele is t^{12}. Embryos homozygous for this allele progress normally to mid-morula but then degenerate without further development. Normal embryos at this stage show many cytoplasmic changes in their cells, which are apparently related to their transition into blastocysts. The t^{12} homozygotes do not appear to be able to carry out the new functions associated with this transition. Experiments were reported by B. Mintz in 1964 on the study of developing mouse embryos in a chamber designed for their cultivation. Through the use of ^{3}H-uridine and ^{3}H-leucine in the culture medium, she found that both RNA and protein synthesis was much reduced in the cells of t^{12} homozygotes as compared to normal cells. However, we do not as yet know what is the initial effect of the t^{12} gene.

We shall next consider the t^0 allele. Embryos homozygous for t^0 become implanted in the uterus and appear to develop normally until the stage when, in normal embryos, the inner mass of the former blastocyst becomes differentiated into an extraembryonic ectoderm and an embryonic ectoderm. In t^0/t^0 embryos the differentiation of the two types of ectoderm does not take place. Instead, the embryos deteriorate and are resorbed. Although the critical stage of development for these embryos is the formation of the definitive ectoderm, we do not know what the protein product of the gene may be or how it affects subsequent development.

The last allele that we shall consider is t^{w5}. Embryos homozygous for this allele pass successfully through the stage of separation of embryonic and extraembryonic ectoderm and through the beginning of the formation of the egg cylinder. At this point the embryonic ectoderm portion of the egg cylinder deteriorates and the embryo dies.

As stated earlier, the recessive alleles of the T locus all appear to have early effects and appear to be involved mainly with ectodermal abnormalities. We do not know the gene products of any of these alleles nor even if the effects of these alleles are related to one another. It is known that compounds of the alleles (e.g., t^0/t^{12}, t^0/t^{w5}, t^{12}/t^{w5}) are viable. This complementation would imply that these alleles have different functions. However, this complementation is not always perfect, and such animals are often poorly viable. Although the study of the T locus has provided a very important understanding of the pathways of embryonic differentiation and how they can lead to abnormalities, it has not yet provided us with information on the primary products of the various T alleles nor on the biochemical pathways involved in the production of the different abnormalities.

SUMMARY

In this chapter we have reviewed the roles of genes in development. Most of our discussions have been concerned with the problem of "regulation," since this represents the mechanism by which most developmental phenomena are integrated into a functional pattern. We have considered regulation of DNA replication and DNA transcription and have found that in both processes protein synthesis may be involved. The regulation of DNA transcription raised the question of dosage compensation, and we reviewed two different solutions to this problem as found in mammals and in *Drosophila*. This was followed by a discussion of the regulation of mRNA translation and its possible role in development.

After considering the problem of the regulation of DNA and RNA functions, we turned to some examples of nucleus-cytoplasm interactions as found in *Acetabularia* and Amphibia. We then reviewed some aspects of tissue differentiation, including the phenomenon of totipotency in plants and animals. The chapter was concluded with a discussion of organ system differentiation, using lethal genes as a tool for studying organogenesis. As examples, we reviewed the following traits: Danforth's short-tail in mice, Creeper in fowl, and T locus in mice. In these studies, where the embryological

effects can only be studied at a level far removed from the primary product of the gene, our task of pinpointing cause and effect becomes extremely difficult. We found ourselves struggling with the end points of complex biochemical pathways without having a clear picture of the pathway itself or, especially, of its beginning.

Questions and problems

1. With respect to Lark's experiments on the control of DNA replication:
   a. What relationship exists between a cell's generation time and its rate of chromosome replication?
   b. What specific factor is responsible for the different rates of replication on the various growth media?
   c. What relationship exists between a cell's ability for chromosome replication and its synthesis of protein?
   d. Why is the above relationship necessary?
2. Compare and contrast inducible and repressible enzyme systems in terms of the Jacob and Monod operon model.
3. Define the following:
   a. Polytene chromosome
   b. Chromosome puff
   c. Endomitosis
   d. Balbiani ring
   e. Ecdysone
   f. Puromycin
4. Draw a hypothesis correlating the phenomenon of ecdysone-induced puffs and the Jacob and Monod model of gene regulation.
5. Assess the evidence supporting the inactive–X chromosome theory of human dosage compensation.
6. Contrast the mechanism of dosage compensation in mammals with that in *Drosophila*.
7. With respect to Hämmerling's experiments dealing with nucleus-cytoplasm interactions:
   a. What is responsible for the enucleated fragment's ability to form caps?
   b. An enucleated *A. crenulata* stalk is grafted onto a nucleated *A. mediterranea* base. What will the newly formed cap look like?
   c. With reference to b, what will the newly formed caps look like after repeated amputation?
   d. What information regarding the genetic control of development can be gained from Hämmerling's experiments?
8. State in your own words, on the basis of information presented in this chapter, what is meant by the term "nuclear differentiation," as used by Briggs and King.
9. Define the following:
   a. Transformation
   b. SV 40
   c. Productive infection

d. Isozymes   f. Organizer
e. Lethal genes

10. From information presented in this chapter and from your knowledge of episomes and lysogeny, offer a reasonable hypothesis as to how viruses may alter the host's genetic material and cause the formation of neoplasms.

11. Differentiation is a broad term. Discuss, with examples:
    a. Nuclear differentiation
    b. Chemical differentiation
    c. Cellular differentiation
    d. Organ differentiation

12. How do the concepts of penetrance and expressivity alter expected phenotypes?

13. Why are genes with 100% penetrance and uniform expressivity very valuable in the study of a gene's properties?

14. By what mechanisms might mRNA translation be controlled? How could such control serve to regulate developmental processes?

15. Amplify, illustrate, and discuss the following statement: All the gene mutations of the *T* locus in mice appear to interfere with links in a chain of processes leading to, and deriving from, the chordamesoderm or ectoderm.

16. How can studies of lethal genes be used to elucidate normal embryonic developmental processes? Discuss this statement with reference to the Creeper mutation in fowl, and the *Sd*, short-tail, mutation in mice.

17. If all cells, theoretically, have the same genotype, why do different cells differentiate at different times?

18. Some genes act early in development, while others act later in life. From information presented in this chapter, discuss the comparative importance of these two classes of genes.

19. Embryos homozygous for the dominant *Creeper* gene usually die at about 3 days of incubation, at which time they are greatly retarded in development as compared with normal embryos of the same age. When strains of Creeper fowls, related only very distantly, are crossed, the embryos homozygous for the dominant gene may live beyond 72 hours. How can this be explained?

20. Compare totipotency as found in plant cells with that found in animal cells.

References

Ames, B. N., and Hartman, P. E. 1963. The histidine operon. Cold Spring Harbor Symp. Quant. Biol. **28**:349-354. (A discussion of the histidine gene cluster of *Salmonella typhimurium* in terms of the operon model.)

Axel, R., Schlom, J., and Spiegelman, S. 1972. Presence in human breast cancer of RNA homologous to mouse mammary tumour virus RNA. Nature **235**:32-36. (Report on the hybridization of nucleic acids from humans and mice.)

Baglioni, C. 1961. The genetic control of fetal and adult hemoglobins. Nature **189**:465. (A discussion of the "switch" from fetal to adult hemoglobin.)

Bennett, D. 1964. Embryological effects of lethal alleles in the t-region. Science **144**:263-267. (An excellent review of the embryological effects of alleles at the T-locus in mice.)

Berendes, H. D. 1965. Salivary gland function and chromosomal puffing patterns in *Drosophila hydei*. Chromosoma **17**:35-77. (Detailed consideration of the techniques used and the results obtained in this type of study.)

Burgess, R. R., Travers, A. A., Dunn, J. J., and Bautz, E. K. F. 1969. Factor stimulating transcription by RNA polymerase. Nature **221**:43-46. (Original report of the discovery of the sigma factor.)

Gall, J. C. 1963. Chromosomes and cytodifferentiation, p. 119-143. *In* Locke, Michael (ed.). Cytodifferentiation and macromolecular synthesis. Academic Press, Inc., New York. (A description and study of the lampbrush chromosomes of amphibian oocytes, which parallel the giant chromosomes of dipterans, described by Beermann and Clever.)

Grüneberg, H. 1960. Developmental genetics of the mouse. J. Cell. Comp. Physiol. **56**(supp. 1):49-60. (A description of the genetic abnormalities occuring in the mouse.)

Hämmerling, J. 1963. The role of the nucleus in differentiation, especially in *Acetabularia*. Symp. Soc. Exp. Biol. **17**:127-137. (A description of the experiments done with *Acetabularia*, involving reciprocal grafting of enucleated and nucleated sections derived from different species.)

Hsu, W.-T., and Weiss, S. B. 1969. Selective translation of T4 template RNA by ribosomes from T4-infected *Escherichia coli*. Proc. Nat. Acad. Sci. U. S. A. **64**:345-351. (Report of selective translation in virus-infected cells.)

Jacob, F., Brenner, S., and Cuzin, F. 1963. On the regulation of DNA replication in bacteria. Cold Spring Harbor Symp. Quant. Biol. **28**:329-348. (A brilliant report of experiments designed to determine the sequence and initiation site of replication of the *E. coli* chromosome.)

King, T., and Briggs, R. 1956. Serial transplantation of embryonic nuclei. Cold Spring Harbor Symp. Quant. Biol. **21**:271-290. (A description of the experiments designed to determine the degree of totipotency of nuclei of various "ages.")

Komma, D. J. 1966. Effect of sex transformation genes on glucose-6-phosphate dehydrogenase activity in *Drosophila melanogaster*. Genetics **54**:497-503. (Experiments designed to test the effects of sexual differentiation on dosage compensation.)

Kufe, D., Hehlmann, R., and Spiegelman, S. 1972. Human sarcomas contain RNA related to the RNA of a mouse leukemia virus. Science **175**: 182-185. (Hybridization experiment involving human and mouse nucleic acid.)

Lark, K. G. 1966. Regulation of chromosome replication and segregation in bacteria. Bacteriol. Rev. **30**:3-22. (A description of experiments using various growth media to determine the relationships between macromolecular syntheses and chromosome replication.)

Losick, R., Shorenstein, R. G., and Sonenshein, A. L. 1970. Structural alteration of DNA polymerase during sporulation. Nature **227**:910-913. (Description of the role of RNA polymerase in selective transcription.)

Maaløe, O. 1966. The control of macromolecular synthesis. W. A. Benjamin, Inc., New York. (A study of the relationships between DNA, RNA, and protein synthesis.)

Markert, C. L. 1963. The epigenetic control of specific protein synthesis in the differentiation of cells, p. 65-84. *In* Locke, Michael (ed.). Cytodifferentiation and macromolecular synthesis. Academic Press, Inc., New York. (A description of the chemical differentiation occurring during embryogenesis in terms of the enzyme lactate dehydrogenase.)

Steward, F. C., Mapes, M., and Mears, K. 1958. The growth and organized development of cultured cells. II. Organization in cultures grown from freely suspended cells. Amer. J. Bot. **45**: 705-708. (A description of the growth and development of single cultured cells into whole plants.)

Temin, H. M. 1971. Mechanism of cell transformation by RNA tumor viruses. Ann. Rev. Microbiol. **25**:609-648. (Detailed review of the viruses, their life cycles, and the cell transformation process.)

12 Behavior genetics

Behavior genetics is a study of the control that heredity has over an organism's actions. The importance of this area of investigation is evident when one considers that it is through behavior that the individual is able to obtain food, escape enemies, find a mate, and reproduce. Analysis of the genetic control of a behavior is complicated by the fact that the primary action of the gene can affect (1) the sensory organs, thus changing information input; (2) an intermediary system (nervous, endocrine), thus altering coordination and perceptual capacities; or (3) the effector organ, thus affecting the response. In most cases we know neither the initial gene product nor the pathway from receptor to effector. The problem of working from phenotype to gene action, through a complex pathway, is no different for behavior genetics than it is for developmental genetics. This is more clearly recognized when one considers that behavior is in reality an extension of development. The ability to act must await the development of the structures involved in the action. In addition, the kind of action of which the organism is capable is determined by the versatility of the structure itself.

Historically, an unfortunate and sterile argument arose over the relative roles of heredity and environment in the determination of an organism's behavior. It was argued that every action of an animal could be classified either as "instinctive" or as "learned." Into instinctive behavior went all those actions for which the animal's history prior to receiving a stimulus is relatively unimportant. The common reflexes would belong to this category. Into learned behavior went all those actions whose form is dependent on the animal's earlier experiences. Problem-solving situations would belong to this category. It was considered that instinctive behavior is a product of the genes while learning is a product of the environment. The futility of the instinct/learning dichotomy becomes apparent when we consider that gene action is itself controlled by the organism's environment. It follows that the organism is in reality never free of its environment and that the organism's behavior, like the rest of its phenotype, is a reflection of its gene-guided development in its particular environment.

Three major lines of research have been pursued in studies on behavior genetics. One approach involves a study of the effects of single genes or chromosomes on behavior. This type of investigation is limited both by the fact that many behavioral traits are polygenic in character and by the fact that most genes are pleiotropic. This latter aspect of the problem necessitates making the distinction between changes of behavior that are caused by structural changes and those that are purely behavioral in character. In many cases geneticists have not been able thus far to be certain about this important distinction. A second approach has been to study differences in behavior between different inbred lines or strains of organisms. Since these lines are produced by many generations of brother-sister matings, it is assumed that each strain is homozygous for most of its genes. The various strains will, by chance, have become homozygous for different alleles at many loci. This approach, however, cannot tell you either which genes or how many genes are involved in the behavioral differences between strains. Such information can sometimes come from data on the F_1, F_2, and backcrosses between the strains. Here too, however, polygenic characters become extremely difficult to analyze. The third approach has been to attempt to modify a behavioral trait, usually in opposite directions, by selection in an initially variable population. Such experiments have usually been successful after a number of generations of selection, indicating the polygenic nature of the traits involved. However, unless some detailed genetic analysis of the selected lines is made at the end of the selection process, the genetic control over the behavior cannot be specified. We shall now consider some examples of the three approaches to behavior genetics.

STUDIES ON SINGLE-GENE EFFECTS
Waltzing in mice

Perhaps the oldest recorded behavioral mutant is that called "waltzer." It was described in China in 80 B.C. in the *Annals* of the Han Dynasty as a mouse ". . . found dancing with its tail in its mouth." We do not know whether the hereditary nature of this behavior was understood at that time. Today we know that a recessive gene is involved and that animals homozygous for this gene run in circles for long periods of time. When at rest, they exhibit a head-shaking behavior; in addition, they are deaf. Investigations by M. S. Deol in 1956 and by others have revealed abnormalities of the inner ear, which include degeneration of the semicircular canals that results in the circling and head-shaking phenomena and degeneration of the chochlea that results in deafness. Waltzer is a relatively straightforward example of a structural anomaly producing a particular behavioral trait. However, there is as yet no explanation of the primary gene action nor of the developmental pathway involved in the production of the phenotype. It is of interest that, in addition to the classic waltzer mutant described above, other nonallelic mutants have been discovered—some of which show the circling and head-shaking characteristics but can hear and others that, although deaf, do not circle or shake their heads. In these last two types of mutants, it was found that degeneration occurred only in the structure controlling the respective behavior. The existence of a group of genetically independent factors that can individually cause a similar behavioral phenotype implies that the development of the inner ear is a complex process that is controlled by a large number of genes.

Nest-cleaning in honeybees

The nest of the honeybee consists of a tremendous number of wax cells arranged in combs. Eggs are laid, one to a cell, and the bee undergoes its complete metamorphosis inside the cell. Since the honeybee uses the same cells over and over again for the rearing of broods, the cleaning of these cells after each cycle of brood is an important activity of adult worker bees. If a larva or pupa dies in its cell, however, an unusual cleaning problem is presented, since the cell must first be uncapped and the remains must then be removed. One of the causes of death of a larva or pupa is an infection, called *American foulbrood* (AFB), which is caused by the bacterium *Bacillus larvae*. It has been known for some time that some colonies of honeybees are resistant to AFB but others are not. More recently, it was found that the workers of the resistant colonies remove AFB-killed larvae and pupae from the cells very quickly, whereas the workers from the sensitive colonies allow the dead individuals to remain in their cells indefinitely. These dead larvae and pupae become a reservoir of the pathogen and lead to new infection.

W. C. Rothenbuhler in 1958 began a long-range study of the nest-cleaning behavior in honeybees. He concentrated on two lines on honeybees that differed greatly in the response of their colonies to brood

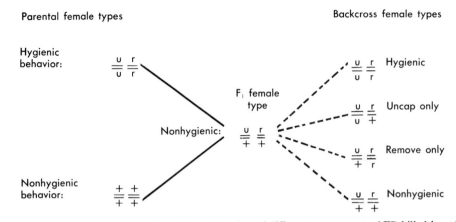

Fig. 12-1. Genetic hypothesis offered in explanation of different responses to AFB-killed brood. (From Rothenbuhler, W. C. 1964. Amer. Zool. 4:111-123.)

killed by AFB. A *Brown* line uncapped cells and removed dead larvae quickly, while a *Van Scoy* line did not uncap cells and left dead larvae in the brood nest indefinitely. These two lines were crossed. It will be remembered that in the bees, as in many other Hymenoptera, drones (males) develop from unfertilized eggs and are consequently haploid in orgin and fixed with respect to the transmission of genetic factors. Queens (females) arise from fertilized eggs, which makes them diploid in origin and similar to any other diploid in the transmission of genetic factors. The worker bee is a reproductively undeveloped female, which also arises from a fertilized egg.

The F_1 generation workers from a cross of the two lines allowed AFB-killed larvae to remain in the brood nest as had the workers of the Van Scoy line. A technique of mating single F_1 drones to queens of the original lines was employed to secure colonies, each of which contained genetically identical worker bees. Matings of F_1 drones to queens from the Brown line produced four types of colonies, with respect to their workers. All the worker bees of any one backcross colony belonged to one of the following four behavioral types: (1) uncappers of cells and removers of dead brood, (2) uncappers only, (3) removers only after human uncapping, (4) neither uncappers nor removers. From these results, a two-locus

hypothesis was developed. It states that uncapping of a cell containing dead brood is dependent on homozygosity for a recessive gene *(u)*, and removing of the dead brood is dependent upon homozygosity for a different recessive gene *(r)*. A diagram of the kinds of females obtained from the crosses described above is shown in Fig. 12-1. Unlike the situation described for the behavioral mutant waltzer, there is no known structural or physiological difference between the worker bees that perform any of the tasks involved in the removal of a dead brood and those that do not. However, here as elsewhere in genetic studies, it is impossible to prove a negative statement. One must acknowledge that as our means of detecting structural and physiological differences improve, we shall most probably discover some alteration in an intermediary system of one type of bee as compared to the other. With reference to our continuing search for gene differences on the molecular level, here, too, we have thus far discovered no differences between the two types of bees.

Courtship behavior in Drosophila

Courtship in *Drosophila melanogaster* provides us with another example of single-gene effects on behavior. When a male is put with a virgin female, he will very soon approach her and tap her with his foreleg. If she stands still, he begins to court. At the beginning of courtship, the male stands facing the female, usually in the position from which he has approached. This part of the courtship pattern is called *orientation*. While facing the female, the male extends one wing until it is at right angles to his body. Then the wing is vibrated rapidly up and down for a few seconds before being returned to rest. This action may be repeated at short intervals, and is called *vibration*. Eventually, the male moves into a position behind the female and licks her genitalia with his proboscis. This portion of the courtship pattern is called *licking*. After some licking, the male attempts to copulate. This involves curling his abdomen under his body and thrusting the tip of his abdomen toward the vaginal plates of the female, meanwhile grasping her abdomen with his forelegs as he mounts her. He will not succeed in copulating with the female unless she spreads her wings, permitting him to mount, and also spreads her vaginal plates, allowing him to insert his phallus.

In 1956 M. Bastock reported on her studies of the effect of the recessive X-linked mutant "yellow" on the courtship pattern of this species. One of the effects of this gene is to give the body of the fly a distinct yellow rather than the normal light brown color. Single-pair matings of wild and yellow males with wild and yellow females were observed for one hour. The percent of successful matings observed are shown in Table 12-1. The wild males were clearly more successful than the yellow males in matings involving both kinds of females. Analysis of the orientation, vibration, and licking patterns of both types of males revealed that they differed in the vibration portion of the courtship pattern. Yellow males were found to vibrate their wings for shorter periods of time and at longer time intervals than the wild males. This lessened activity on the part of the yellow males appears to increase the chance of the female's responding to some other environmental stimulation and moving away from the male.

One of the problems always raised in studies on male courtship is the role of the

Table 12-1. Percentage of mating success in one hour in *Drosophila melanogaster*

Males	Females	
	Wild	Yellow
Wild	75	81
Yellow	47	59

From Bastock, M. 1956. A gene mutation which changes a behavior pattern. Evolution **10**:421-439.

female. Female *Drosophila* can repel a courting male by one or more of the following actions: (1) kicking the male, (2) twisting her abdomen so that the male cannot copulate, (3) flicking her wings so that the male cannot mount, (4) jumping or flying away. An analysis of the female courtship behavior did not reveal any difference in female repelling movements while being courted by either type of male.

Another question raised is whether the yellow males are just too weak to court the females as actively as the wild males. This would occur if one of the effects of the gene was to produce a general muscular weakness in yellow males. In order to examine this possibility, yellow and wild males were paired with females of a different species, *D. simulans,* which do not permit *D. melanogaster* males to copulate with them under single-pair conditions. Both yellow and wild males were found to court the *D. simulans* females with equal fervor.

The experiments just described led to the hypothesis that the yellow males lacked sexual drive toward both yellow and wild *D. melanogaster* females. This does not tell us why these males lack sexual drive. The hypothesis simply indicates that one must look for the answer in the male rather than in the female and in some part of the male other than its musculature. In the case of the gene *yellow,* we are dealing with a pleiotropic gene. One of its effects is on body color, and another, seemingly unrelated, effect is on male courtship behavior. We are much in the dark about the primary enzymatic effect of the gene not only as it relates to the courtship pattern but even as it relates to body color.

Phenylpyruvic idiocy in man

Man has a dietary requirement for the amino acid phenylalanine, which is used in many metabolic pathways including those leading to the production of melanin and thyroxine. As was pointed out in Chapter 10, a number of hereditary defects can occur in the biochemical pathways controlling the metabolism of phenylalanine and tyrosine. For behavior genetics, a most striking effect occurs when an individual does not have the enzyme necessary for the conversion of phenylalanine to tyrosine. Under these conditions, the phenylalanine accumulates in the blood, and some of it is converted to phenylpyruvic acid. Some of both these compounds accumulates in the cerebrospinal fluid, while the rest is excreted in the urine. People with this enzyme deficiency are feebleminded and have other phenotypic characteristics such as light pigmentation. We thus see that the gene is pleiotropic. Affected individuals are called *phenylpyruvics* or *phenylketonurics* and the disease is called *phenylketonuria*. The gene involved is an autosomal recessive, and the enzyme missing in the liver of homozygotes, where most of the phenylalanine is normally metabolized, is phenylalanine hydroxylase. Children who are homozygous for this gene can be spared the deleterious effects of the gene if the amount of phenylalanine in the diet is severely restricted during the early years of life. Sufficient phenylalanine must be included in the diet for protein synthesis, but the amount must be insufficient for any appreciable quantity to be accumulated and converted into phenylpyruvic acid. After about the age of 6 years, such children can be given a normal diet, and neither the increased level of phenylalanine in the body nor the production of phenylpyruvic acid will affect their mental capacities.

Since the mental defect can be circum-vented by dietary restriction of phenylalanine intake, it is probable that the pathological effects are due to the accumulation of phenylalanine and phenylpyruvic acid rather than to the enzyme deficiency. Furthermore, since the accumulation of these compounds in the cerebral fluid will not impair the mental capacities of older children, one may assume that there is a particular period in the life of the child during which the brain tissue is sensitive, in some as yet unknown fashion, to one or both of these chemical compounds. We have therefore in phenylketonuria a genetic disease whose behavioral effect can be overcome by the environmental factor, diet. In cases where the behavioral effect is manifested, it is caused by an accumulation of an essential amino acid and the end product of a biochemical reaction involving this amino acid. Furthermore, the mental deficiency occurs only if the biochemical block is not obviated during a particular period in the development of the brain. However complex the above example may appear, it represents one of the simplest and best-understood situations in behavior genetics.

Chemotaxis in Escherichia coli

Although we normally think of behavior genetics in terms of large organisms, and more specifically large animals, no such restriction is actually involved in this area of study. In fact, a large and fascinating body of information has been building up on the genetic control of the movements of microorganisms in response to changes in their environment. A major portion of our knowledge in this field has been due to the efforts of J. Adler and co-workers, who have been reporting on their experiments since 1966.

It has been known for some time that motile bacteria such as *E. coli* are attracted to a variety of chemicals. This can be demonstrated in the following manner. A drop of a bacterial suspension is placed on a slide. A chemical solution (e.g., glucose) is put into a capillary tube that is then placed on the slide with its open end lying in the

Table 12-2. Results of competition between attractants

Capillary tube	Percentage of bacteria	Bacterial suspension	Percentage of bacteria
Fucose and galactose	0	Galactose	100
Galactose and fucose	0	Fucose	100
Galactose and glucose	0	Glucose	100
Glucose and galactose	40	Galactose	60

middle of the drop containing the bacteria. After incubation for 60 minutes at 30° C., the capillary tube is removed from the bacterial suspension. A series of dilutions is made of the contents of the capillary tube. Each of these dilutions is then placed separately into a tube containing melted nutrient agar, and each mixture is poured into a different Petri dish. The resultant colonies in each dish are counted, and after taking into account the magnitude of the individual dilutions, one obtains an estimate of the number of bacteria in the capillary tube. In this procedure, each bacterial colony is assumed to be derived from a single bacterium.

An examination of the number of bacteria in the capillary tube shows that when only water is present in the tube, one gets an average of about 3000 bacteria. These have undoubtedly been swept in by water currents which are produced when the tube is placed in the bacterial suspension. When, however, a solution of glucose or some other attractant compound is used, an average of about 150,000 bacteria is obtained. No difficulty is encountered in distinguishing chemicals that are attractants for bacteria from those that are not.

One of the questions raised in this type of study is whether the organism is attracted only to those chemicals that it can utilize in its metabolism. The importance of this question for behavioral studies lies in the distinction between whether the organism's movements are a response to the attractant itself or are they a response to the organism's own internal metabolism. In this

latter instance, the bacteria would be moving into the capillary tube because they would be responding to a continuing increase in their metabolic levels as they move toward and into the tube. A number of experiments have been conducted to answer the question. In one experiment, glycerol was placed in the capillary tube. Glycerol is readily metabolized by *E. coli*. In fact, the bacteria can grow quite well in a medium containing glycerol as the sole source of carbon and energy. However, investigators found that glycerol fails completely to attract the bacteria. A contrasting situation was observed when fucose was used. This compound is a galactose analogue that cannot be used as a source of carbon and energy by bacteria. When fucose was placed in the capillary tube, *E. coli* entered the tube in vast numbers. Both of these experiments indicate clearly that *chemotaxis* is not a response of the organism to its own metabolism but rather a response to the attractant itself.

Our discussion thus far would indicate that the chemotactic response of an organism is a result of some event that has occurred at the surface of the cell. This raises the possibility that the cell membrane may contain specific chemoreceptors for the chemicals involved. Competition experiments designed to test this possibility were set up using different attractants. In these experiments one attractant is placed in the capillary tube, while another is added to both the capillary tube and the drop of bacterial suspension. The results are shown in Table 12-2.

These data indicate that galactose and fucose use the same receptor sites and, as a result, chemotaxis to one completely inhibits movement toward the other. In this situation, all the chemoreceptor sites of the cell are occupied by the attractant in the bacterial suspension, and as a result, there is no opportunity for the bacteria to be attracted to the capillary tube. However, a different situation prevails when galactose and glucose are used. Although glucose completely inhibits taxis toward galactose, the reverse is not the case. This suggests that all galactose receptors are also glucose receptors but that there are some glucose receptors that will not be responsive to galactose.

Mutants of *E. coli* have been discovered that are not responsive to certain compounds toward which the wild type bacteria will move. A number of these mutants are concerned with the sugar galactose. Involved in galactose metabolism are a number of structural genes, one of which produces a "permease" type enzyme (methyl-β-galactoside permease). This type of enzyme functions to transport galactose from the cell surface into the cell. One of the components of the galactose transport system is a protein called the *galactose-binding protein*. Molecules of this protein appear to be part of each of the cell's galactose chemoreceptors and serve to bind galactose molecules to the cell surface.

From experiments whose details we shall not discuss, mutants have been discovered which (1) fail to show taxis toward galactose, (2) are defective in galactose transport within the cell, and (3) lack the galactose-binding protein. Other mutants have been found that exhibit normal taxis toward galactose but are blocked in the transport of galactose into the cell. However, these last-mentioned mutants contain normal levels of the galactose-binding protein. These mutants are probably deficient in one or more components that, in addition to the galactose-binding protein, are required, for transport but are not required for chemotaxis. In addition to the above types of mutants, one other type has been isolated that fails to show taxis toward galactose but has a full level of galactose-binding protein and is normal in the transport of galactose. This last type of mutant would appear to be deficient in one or more additional components that are required for chemotaxis but not for galactose transport. The information just presented clearly indicates that the central molecule in this system is the galactose-binding protein, which is involved in both chemotaxis toward, and cellular transport of, galactose. How widespread may be this dependence of chemotaxis on a binding-type protein is at present unknown. Other points that also remain unknown are the actual mechanism of chemoreception and the way in which chemoreception leads to changes in the beating of the bacterium's flagella, thus resulting in a change in the cell's movement and producing the behavior we characterize as chemotaxis.

Chromosome complement and human aggression

In the field of human behavior genetics, a new area of research has developed, dealing with the possible effects of different karyotypes on aggression. It will be recalled that in chapter 6 various types of human sex chromosome aneuploidy were discussed. These included males with an XXY karyotype and those with an XYY karyotype. The incidence of each of these two types of aneuploidy in the male population has been estimated to be about 0.14%. In 1966 M. D. Casey and co-workers reported on a study they had conducted of two English prison mental hospitals that contained patients who were mentally subnormal (i.e., average measured I.Q. was 77) and, in addition, required special security on account of persistent violent or aggressive behavior. Among the 942 male patients, a buccal smear cytological examination revealed 21 individuals (2.2%) who were sex-chromatin positive (i.e., showed Barr bodies). A chromosomal examination demonstrated that 12 were XXY, 7 were XXYY, and 2 were XXY/XY mosaics. These results indicated that the incidence of Klinefelter's type males in the prison-hospital population was almost twenty times greater than in the general population.

In another study in 1967, W. H. Price and P. B. Whatmore reported that among 315 men in a Scottish prison mental hospital for those with dangerous, violent, or criminal propensities, a chromosomal examination revealed 11 individuals who exhibited some form of sex chromosome aneuploidy. These included 9 who were XYY, 1 who was XXYY, and 1 who was an XXY/XY mosaic. The incidence of XYY patients in this prison-hospital population was 2.9%, again a twentyfold increase over that in the general population. A comparison of XYY patients with XY patients who were in the same maximum-security prison mental hospital showed that the XYY males were involved in crimes against property to a greater extent (88%) than were XY males (63%). It was also found that the disturbed behavior of XYY patients revealed itself in an earlier average age of first conviction (13 years) than was the case for XY patients (18 years). Lastly, it was discovered that the incidence of convicted criminals among the siblings of XYY males was dramatically less (3%) than amongst the siblings of XY males (19%). The discovery of a unique crime pattern among XYY males may prove to be just as important a finding as that of an increased incidence of this particular chromosomal type in prison mental hospitals.

Similar studies have more recently been carried out in the penal institutions of Australia, France, and the United States, with comparable findings. One known physical characteristic of both XXY and XYY males is their unusual height. They tend to be at least 6 feet (182 cm.) in height in the United States, and their counterparts in the British Isles are somewhat shorter (i.e., 5 feet 10 inches, or 178 cm.). This difference in height is also found in the general populations of the two countries and probably reflects differences in nutrition. One hypothesis was advanced that linked the aggres-

sion of these aneuploid individuals to their abnormal tallness, which was thought to make them subject to social stresses that could result in criminal acts. However, a study of XY males in penal institutions revealed that their height distribution was no different from that of the general population to which they belonged.

A large and comprehensive chromosomal study was made of the inmates from all Scottish prisons, reform schools, and juvenile delinquency centers. Its findings were reported in 1971 by Jacobs and colleagues. They found that the incidence of sex chromosome aneuploidy in the general prison population was the same as that found in the nonprison population. However, the incidence of XXY and XYY persons who are mentally ill and who become institutionalized for criminal behavior is apparently twenty times greater than the incidence of these individuals in the general population.

Attorneys for XXY and XYY individuals accused of crimes have been quick to point out the association of abnormal chromosome complement and aggressive behavior. In such cases, the defense has rested heavily on arguments of legal nonresponsibility because the defendant possessed a particular karyotype. Thus far, there has been no definitive resolution of the legal complications raised by the discovery that certain chromosome complements are associated with some forms of human aggression.

STUDIES ON INBRED LINES
Alcohol preference in mice

Early studies on the role of heredity in alcohol preference measured preference in a two-choice situation (water vs. alcohol solution). However, this experimental procedure did not yield a preference score that permitted statistical evaluation. In 1964 J. L. Fuller proposed an approach that utilized a six-choice situation. A set of six concentrations of alcohol was provided, such that the strength of the solution increased by a factor of two at each step (from 0.5 up to 16 vol.%). A preference score was devised that measured essentially the concentration of alcohol below which the animal obtained 50% of its fluid intake. Four inbred strains and the six possible hybrids between them were tested in this manner. The results obtained are shown in Fig. 12-2. Each hybrid is represented twice, on the curves of each parent, while the pure strains are shown only once. The use, in this experiment, of a variety of unrelated mouse strains permits the geneticist to make a broader generalization about the species as a whole than is possible with only a pair of strains. The four strains used showed great variation in their preference scores. This would indicate that the strains are genetically quite different from one another. In considering the hybrids, it was found that F_1 offspring with a C57Bl/6J parent had high alcohol preference scores while those with a DBA/2J parent, except for the cross including C57Bl/6J, had low scores. The offspring of A/J or C3HeB/J mice always resembled the other parent. The conclusion to be drawn from this experiment is that high preference for

alcohol may be dominant, recessive, or neither, depending on the particular mating. This great variability in the genetic contributions of the four strains to F_1 preference scores suggests, although it does not prove, that the physiological factors underlying the observed preferences may vary from strain to strain. The genetic bases for the preference differences among the strains are completely unknown.

One of the characteristic effects of alcohol is a depression of the central nervous system which eventually leads to "sleep." This anesthetic effect of alcohol is known, in man, to depend on such variables as dose, food intake, and habituation. It may also depend on physiological differences among individuals in brain sensitivity or on metabolic variations in the absorption, distribution, and detoxification of the alcohol. A study was reported by Kakihana and co-workers in 1966 on the relationship of genotype, ethanol tolerance, and sleeping time in mice. Male mice from inbred strains C57Bl/Crgl (black) and BALB/cCrgl (albino) were given sleep-inducing intraperitoneal injections of alcohol. As soon as the injection was completed, the animal was hung from a wire mesh and the time until he fell to a foam rubber pad was measured. This constituted the animal's "fall time." Next, the mouse was placed on his back in a trough. The time until the animal righted himself was recorded and considered his "sleep time." In addition to the above observations, analyses were made of the blood-alcohol and brain-alcohol of the animals when they had awakened.

The fall times of the two types of mice were essentially the same. This was interpreted as meaning that the rate of alcohol absorption, distribution, and detoxification was the same in the two strains under the conditions of this experiment. On the other hand, the median sleeping time of the black mice, 38 minutes, was significantly less than the median sleeping time of the albino mice, 138 minutes. Furthermore, it was found that, on awakening, the black mice had significantly higher blood-alcohol and brain-alcohol levels than the albino mice. This was

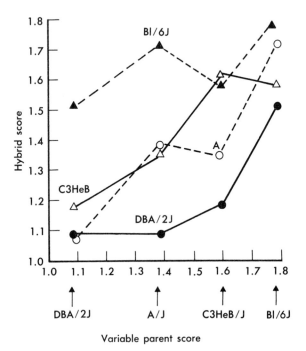

Fig. 12-2. Alcohol scores of constant-parent groups as a function of scores of the variable parents. (From Fuller, J. L. 1964. Amer. Zool. **4:**101-109.)

taken to mean that the observed differences in sleeping time were due to an increased brain tissue tolerance of the black mice for alcohol. It is of interest to note that in an earlier study the black mice had exhibited a greater preference for alcohol than had the albino mice. This means that the strain showing the greater preference for alcohol in a choice experiment also showed the greater resistance to the effect of injected alcohol in a sleeping time experiment. Nothing is known of the physiological basis for the differences in preference and tolerance of alcohol by the mice.

Open-field behavior in mice

When mice are placed in a new environment, they normally begin to explore their surroundings. It also has been observed that the mice characteristically begin to drop fecal boluses. The amount of exploration conducted by the mice in a given time interval can be taken as a measure of their level of activity, while the number of fecal boluses dropped in the same time interval can be considered as a measure of the animals' emotionality.

In a study of these aspects of behavior, mice are placed in a flat enclosed space free of barricades or other types of obstructions —hence, the term *open-field*. The floor of the field is divided, by lines, into squares. The number of lines crossed and the number of fecal boluses dropped in a given amount of time are then recorded. Such an experiment was reported by DeFries and associates in 1966. Two inbred strains were used, BALB/c (albino) and C57B1/6 (black). The albino strain showed a pattern of relatively low activity and high defecation rate in the open field when compared to the black strain. To permit statistical evaluation of the results, the raw scores were transformed into square roots. The mean transformed activity scores of the albino and black strains were 4.46 and 16.06, respectively, while those for defecation rate were 2.88 and 1.07.

Animals from the two inbred strains were crossed to one another to produce an F_1

generation. The F_1 mice were then used to produce F_2, F_3, and F_4 generations by random matings. Inbred C57B1/6 mice are non-agouti black *(aaBBCC)*, whereas BALB/c mice, although albino, are homozygous for the agouti gene *(AAbbcc)*. In the F_2 and subsequent generations, five coat colors occurred: black-agouti, brown-agouti, black, brown, and albino. Although open-field activity and defecation were known to be influenced by genes at many loci, it was of interest to compare the scores when the data were grouped according to coat color. The activity and defecation scores among the four pigmented classes in the F_2, F_3, and F_4 generations did not differ (approximately 12.53 and 1.80, respectively, in every generation); but pigmented animals were consistently more active and defecated less than the albinos (approximately 10.28 and 2.50, respectively, in every generation). The differences observed were statistically significant. From these results, it was concluded that the observed differences in open-field behavior between the two strains were due, at least in part, to the effect of the single gene at the *c* locus. Analysis of variance led to the conclusion that the *c* locus may account for 12% of the strain differences in both traits. It was further hypothesized that this single-gene effect is mediated through the visual system of the animal, and that a lack of eye pigment made the albino mice more sensitive to light than the pigmented mice. Since the experiment had been conducted under a bright, white light, the behavioral results were considered to reflect differences in fear reactions of the two types of mice to the light. A second experiment was performed to test this hypothesis.

Animals from the F_4 control population were randomly mated to produce an F_5 generation. The F_5 mice were tested either under a white light as were the earlier generations or under a red light, which provides little or no visual stimulation for mice. The mean transformed activity scores of albino and pigmented animals under white light were 8.80 and 12.91, respectively, whereas the activity scores for albino and pigmented animals under red light were 13.28 and 14.10. The mean transformed defecation scores of albino and pigmented animals under white light were 2.10 and 1.95, respectively, while the defecation scores of albino and pigmented animals under red light were 1.73 and 1.76. The differences in scores between albino and pigmented mice are significant only under a white light test situation. These results were taken to demonstrate that the single-gene effect of the *c* locus operates through the visual system of the mouse and that albinos are more photophobic than pigmented animals.

In this experiment, we have seen an example of behavioral differences, between inbred strains, whose genetic basis has been partially analyzed. The gene for pigmentation was found to be responsible for a significant amount of the difference in behavior between the two strains. It was further demonstrated that the effect of the gene was mediated through the visual system of the animal. For the present, this represents as much as we can say about the genetics of open-field behavior in these mice.

Fighting behavior in mice

Young male mice normally begin to fight a stranger when they are between 32 and 36 days of age. In contrast to this, male mice reared together in the same litter will, normally, not begin to fight one another until about 55 days of age. The pattern of activity that constitutes a typical fight includes (1) *sniffing,* especially of the opponent's genital region; (2) *hair-fluffing,* in which the hair stands on end over most of the body; (3) *tail-rattling,* in which the tail is switched from side to side; (4) *mincing,* in which the mouse circles around or near the opponent; (5) *attack,* in which there is biting of the tail, back, and sides of the opponent's body; (6) *submission* by the defeated male, which can include squealing, running away, or freezing (the defeated mouse rears up on its hind feet, holding out its front paws toward the agressor and remains motionless until attacked, when it may squeal and run away).

A large number of studies have been made on the effects of genotype on fighting behavior in mice. One experimental approach has been to take males from two different inbred strains and fight them once a week in round-robin fashion, so that every male of one strain meets each male of the opposite strain but once. Contests are classified as "no-fight" (neither male attacks the other during a period of 30 minutes), "draw" (either or both males attack but no submission behavior is elicited), or "victory" (one male attacks and elicits a submission reaction from the other). In addition to classifying each contest as indicated above, the number of attacks initiated by each male is recorded, as is the time interval between placing the animals into the cage and the first attack (attack latency). If a victory is achieved, the animals are separated immediately. The males are isolated, each in its own cage, except during the actual contests. One can utilize the above experimental procedure not only to test for the genetic control of fighting behavior but also to test for the interactions of environment and heredity in this behavior.

A study on interstrain fighting in male mice was reported by Levine and co-workers in 1965. The males used came from two inbred strains: an albino strain (ST/bJ) and a pigmented, black-agouti, strain (CBA/J). In the first experiment eleven males from each strain were isolated at weaning (30 days of age) and remained isolated until 90 days of age, at which time the fights were begun. The results obtained are shown in Table 12-3, column A. It is quite clear that, under these experimental conditions, the black-agouti males were the better fighters. This superiority is evident in their percentage of victories, in the percentage of times they were first to attack, and in their shorter average attack latency.

A second experiment was performed to test the effects of early social experience, with males of the same strain, on subsequent fighting behavior. In this study, after weaning, the males were housed with others of their own strain until one week before the

Table 12-3. Interstrain fighting in male mice

	A* Males isolated at weaning	B* Males of same strain kept together	C† Males of opposing strains kept together
Victories (percent)			
ST	11	0	13
CBA	85	72	40
Draws (percent)	3	6	2
No-fights (percent)	1	22	45
Times first to attack (percent)			
ST	17	3	28
CBA	83	97	72
Average attack latency (minutes)			
ST	10	‡	10
CBA	3	5	8

*From Levine, L., Diakow, C. A., and Barsel, G. E. 1965. Interstrain fighting in male mice. Anim. Behav. **13**:52-58.

†From Levine, L. (unpublished data).

‡Too few cases for reliable estimate.

start of the contests, at which time the males were isolated. The results obtained are shown in Table 12-3, column B. Here, too, the black-agouti males were the superior fighters as seen in the percentage of victories and in the percentage of times they were first to attack. In fact, their fighting ability as measured by these two criteria was essentially the same as in the first experiment. The only significant effect of this type of early social experience on the pigmented males' fighting behavior was an increase of their average attack latency. However, an examination of the effects of prefight social experience with males of their own strain on the albino males' fighting behavior revealed that their fighting ability was virtually abolished. The albino males were never victorious and initiated very few of the attacks. It seems evident that the increase in the percentage of no-fights in this experiment is due to the decreased aggressiveness of the albino males. The second experiment indicated that the different hereditary backgrounds of the

two types of mice produced different behavioral responses to a subsequent fighting situation, as a result of early social experience with males of their own strain. This illustrates the need to consider genetic differences in assessing the effect of an environmental factor on behavior.

A third experiment was conducted to test the effects of early social experience, with males of the opposing strain, on subsequent fighting behavior. In this study, after weaning, the males were housed with others of the opposing strain until one week before the start of the contests, at which time they were placed in individual cages. The results of the third experiment are shown in Table 12-3, column C. Once more the black-agouti males were the better fighters, as reflected both in the percentage of victories and in the percentage of times they initiated attacks. However, in both of these aspects of fighting behavior, their performance was significantly poorer than in the first two experiments. In addition, their average attack latency was significantly longer in this experiment than in the two previous experiments and was essentially the same as the attack latency of the albino males. The albino males' fighting behavior appeared to be unaffected by early social experience with males of the opposing strain when compared to that of albino males isolated since weaning. In the third experiment, the extremely high percentage of no-fights is clearly a reflection of the decreased aggressiveness of the pigmented males. The conclusion drawn from this experiment was that the different genetic constitutions of the two types of mice produced different behavioral responses to a subsequent fighting situation, as a result of early social experience with males of the opposing strain.

The conclusions drawn from the second and third experiments are deceptively similar. Although it is true that in both cases early social experience reduced subsequent aggressive behavior, it must be noted that each type of prefight association affected the two types of mice differently and to a different degree. A comparison with males isolated since weaning shows that early association with males of their own strain virtually eliminated subsequent aggressive behavior of albino mice but did not affect black-agouti mice to any significant degree. By contrast, early association with males of the opposing strain did not affect the aggressive behavior of albino mice at all but severely reduced that of the pigmented mice. Here we have a differential effect of environmental factors on different inbred strains. The results indicate that caution must be exercised in attempting to set down broad principles concerning the interactions of heredity and environment in aggressive behavior. This caution must also be exercised in attempting to extrapolate the findings from one species (e.g., mouse) to another species (e.g., man).

STUDIES INVOLVING SELECTION
Selection for maze-learning in rats

Rats, like mice, will begin to explore any new environment into which they may move or be placed. In their explorations, they may have to deal with the problem of getting around or over physical barriers. The ability of the animals to solve such problems and the ability to remember the solutions have

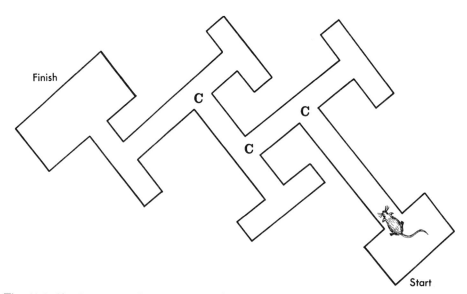

Fig. 12-3. Simple T maze of two units and three choice points, **C.** The animal cannot see the end of any blind alley from a choice point. Correct turns here occur in a simple *left-right-left-right* order.

been the focal points of much research in animal behavior. An apparatus that permits the study of the problem-solving ability of an animal is the maze. In its simplest form, the maze consists of a succession of T-shaped compartments that lead into each other. Such a maze is shown in Fig. 12-3. When a rat is placed in a maze for the first time, it shows a great deal of exploratory behavior. The rat carefully inspects all the passages, including the blind alleys. Eventually it comes to the end of the maze where there is a piece of food for it to eat. The next time the rat is put into the maze, it will normally go into fewer blind alleys than before, and with each repetition it tends to take a more and more direct path. If the rat has been deprived of food for a number of hours before each trial, one finds that the number of errors per trial decreases rapidly and the rat may eventually achieve errorless runs.

Using maze-learning ability in rats as a criterion, R. C. Tryon reported in 1940 the results of an extensive selection experiment. He had bred rats selectively for performance on 19 trials in a specialized automatic maze. The brightest rats (smallest number of errors) were mated with the brightest and the dullest with the dullest. The results are shown in Fig. 12-4. It is clear that two different lines were established, one superior to the other in terms of errors made in learning the maze. Although it is not shown in this illustration, by generation 8 there was practically no overlap between the distributions of these two groups. By that time the dullest bright rats were about equal to the brightest dull rats. According to these results, there seemed no doubt that maze-learning ability in rats is genetically determined. The gradual separation of the two selected lines implied that the behavioral trait was controlled by a multiple-factor genetic system. A cross of bright rats with dull rats yielded an F_1 whose maze-running characteristics were intermediate to the parent strains.

After the two selected lines had become well separated, both types of rats were tested in a variety of ways. The results of these tests were reported by L. V. Searle in 1949. He found that bright rats learned better than dull rats in hunger-motivation problems. However, in escape-from-water tests, the dulls were superior to the brights. From this, he concluded that the selection for brightness in rats had not produced a generally superior strain of rats but rather a strain that was superior only in certain learn-

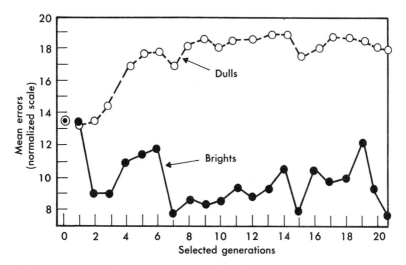

Fig. 12-4. Results of R. C. Tryon's selective breeding for maze-brightness and maze-dullness. (From McClearn, G. 1962. Chap. 4. *In* Postman, L. (ed.). Psychology in the making. Alfred A. Knopf, Inc., New York.)

ing situations. Tests were also conducted to discover differences between the selected lines in emotionality. Bright rats were found to be more emotional in open spaces, while the dull rats were more emotional in the maze. The selection for one characteristic had resulted in strains that differed in their emotional reactions to different situations. It was obvious from these findings that in the selective breeding process characteristics other than those deliberately sought had become incorporated into the developing lines.

Studies have also been made of the bright and dull rats with respect to their brain levels of cholinesterase. This enzyme determines the rate of breakdown of acetylcholine, a chemical that must be present at a synapse in order to get neural transmission. The results of such investigations by Krech and collaborators were first reported in 1954. It was found that the bright rats had a higher level of cholinesterase activity in their cerebral cortex than the dull rats. The investigators proposed that the greater cholinesterase activity in the bright rats reflected a greater production of acetylcholine in these animals and that the increased acetylcholine at the synapses led to a greater efficiency of neural transmission.

To further investigate the relationship between enzyme and behavior, the strains were crossed, and an F_1 and F_2 were obtained. When the F_2 was tested for maze performance and for brain cholinesterase level, animals with the greater enzyme activity tended to make the greater number of errors. This was interpreted to mean that the genes for acetylcholine production and those for cholinesterase production were not linked. This conclusion was supported by the results of a different experiment that involved selective breeding for cholinesterase activity without regard to any behavioral characteristics. It was found that the animals selected for high cholinesterase activity performed more poorly in the maze than did those selected for low cholinesterase activity. The results imply that raising the level of cholinesterase activity while leaving the level of acetylcholine unaltered will cause a too rapid

breakdown of the acetylcholine for efficient synaptic transmission. These findings would also indicate that the behavioral selection by Tryon involved a selection for greater production of both compounds. To date we have no information on the number or interrelationships of the genes involved in the various behavioral traits that characterize these selected lines of rats.

Selection for geotaxis in Drosophila

Geotaxis, the response to gravity, is a basic behavioral characteristic of all organisms. Study of the genetic control of geotaxis in *Drosophila* has been greatly facilitated by the construction of a multiple-unit classification maze by J. Hirsch. A modified form of the maze, as described in 1959, is shown in Fig. 12-5. It is made of thick Plexiglas into which alleys are cut. The alleys form a linked series of T units fanning out from the initial T. Mazes may be constructed to include any number of units. For studying geotaxis, the maze is placed in a vertical position opposite a fluorescent lamp, which serves to attract the flies. Flies are placed in a tube that is attached to the initial T. Attracted by the light, they begin to move through the maze. Those flies that choose to go up at each trial position will finally enter the uppermost collecting tube, and conversely those that choose to go down at each trial position will finally enter the lowermost collecting tube. Flies that make both types of choices at the various trial positions will enter intermediate collecting tubes. Each tube represents a final composite score that indicates the number of up and down choices made by the individuals in that collecting tube. Males and females are run through the maze separately. In the selection for negative geotaxis, females and males from the upper tubes are mated to provide the next generation of flies. Conversely, in the selection for positive geotaxis, females and males from the lower tubes are mated to provide the next generation of flies.

A report on the genetic control of geotaxis in *D. melanogaster* was published by Erlenmeyer-Kimling and co-workers in

1962. A heterogeneous population was used as a foundation stock, and selection for both positive and negative geotaxis was conducted for some 65 generations. The results are shown in Fig. 12-6. It was apparent that the response to selection had been prolonged, asymmetrical, and irregular. The extent of the response to selection, together with the prolonged and gradual nature of the strain divergence, strongly suggested that a polygenic system was involved.

As can be seen in Fig. 12-6, the two selected lines had diverged quite widely by generation 25. Periodically thereafter, as reported by Hirsch and Erlenmeyer-Kimling in 1962, samples of flies from the unselected foundation population and from each of the selected lines were crossed to a laboratory tester stock that contained dominant genes on its chromosomes X, II, and III and also had inversions in these same chromosomes. As will be recalled from Chapter 7, inversions have the effect of suppressing crossing-over between homologous chromosomes and thus preserving the integrity of both homologues. The F_1 flies obtained from each of the crosses were backcrossed to flies from the unselected and selected populations. The backcross offspring fell into eight classes, which represent all combinations of the three chromosome pairs that can result from the backcross. The breeding

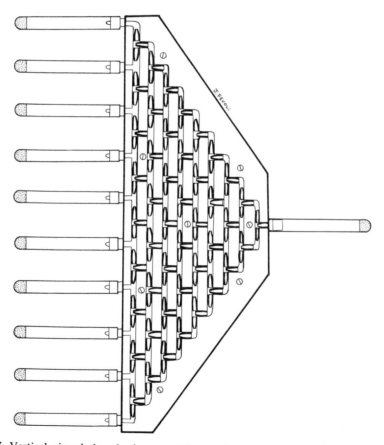

Fig. 12-5. Vertical nine-choice plastic maze. Flies are introduced into the vial at right and are collected from the vials at left. Small traplike funnels, each having its larger opening continuous with the alley surfaces and a small one debouching in midair, discourage backward movement in the maze.

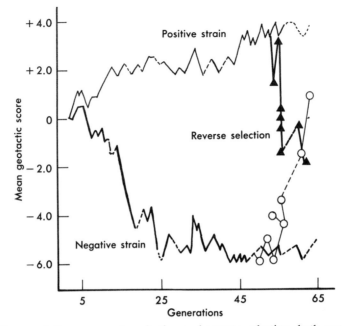

Fig. 12-6. Course of the response to selection and reverse selection, both sexes combined. (From Erlenmeyer-Kimling, L., Hirsch, J., and Weiss, J. M. 1962. J. Comp. Physiol. Psychol. 55:722-731.)

scheme for the chromosomal assay is shown in Fig. 12-7. Chromosome IV was disregarded because of its small size.

The females obtained from the above backcross were run through the maze and scored as to tube and genotypic class. The effect of a particular choromosome on geotaxis can be estimated by comparing the geotactic scores of all genotypic classes that are structurally homozygous with the scores of those which are structurally heterozygous for a given chromosome pair. Such an analysis revealed that, in the unselected population, chromosomes X and II contribute to positive geotaxis and chromosome III to negative geotaxis. In response to selection for positive geotaxis, while there are no important changes in chromosomes X and II, the effect of chromosome III on geotaxis changes from negative to slightly positive. In response to selection for negative geotaxis, all three chromosomes show marked changes: the positive effect of chromosomes X and II is greatly diminished, while the negative effect of chromosome III is much

enhanced. The discovery that all three major chromosomes are involved in geotaxis lends support to the hypothesis that this behavior is polygenically controlled. However, the number of genes involved and their interrelationships are unknown. In addition, we have no information on any structural or chemical differences that may exist between the flies of the two selected lines.

Selection for mating speed in Drosophila

The courtship pattern of *Drosophila* was discussed earlier in this chapter. Other studies of the phenomenon have included selection for increased and decreased mating speed. A report of such an investigation was made in 1961 by A. Manning; it describes an experiment in which lines of *D. melanogaster* were selected for fast and slow mating speeds over some 25 generations. In each generation, fifty pairs of virgin flies were placed in an empty vial. The first ten pairs of flies to mate were removed and became the parents of the next generation of a fast (F) line, while the last ten pairs to mate

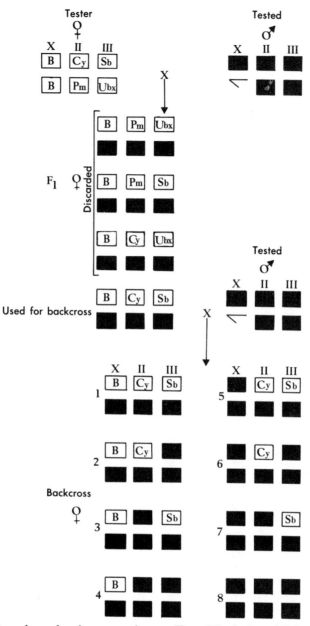

Fig. 12-7. Breeding scheme for chromosomal assay. (From Hirsch, J., and Erlenmeyer-Kimling, L. 1962. J. Comp. Physiol. Psychol. **55**:732-739.)

became the parents of the next generation of a slow (S) line. Two fast lines (FA and FB) and two slow ones (SA and SB) were established. The response to selection was almost immediate, and from F_3 onward the S and F lines were quite separate. The sepa-

ration between the selected lines widened for some 7 or 8 generations, but thereafter there was no persistent change in average mating speeds. After F_8, the mean mating speed of a fifty-pair population was about 80 minutes in the slow lines, 10 minutes in

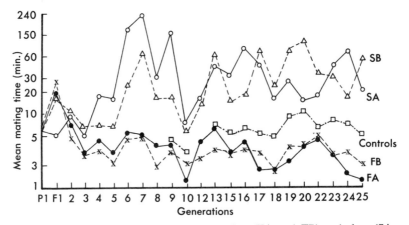

Fig. 12-8. Course of the response to selection for fast (**FA** and **FB**) and slow (**SA** and **SB**) mating speeds. Mean mating time is plotted on a logarithmic scale. (From Manning, A. 1961. Anim. Behav. **9**:82-92.)

the control population, and only 3 minutes in the fast lines. Data are shown in Fig. 12-8.

There were great fluctuations in mating speed between generations. These fluctuations usually affected all the selected lines in parallel fashion and were considered to reflect some common environmental factor, the nature of which was unknown. As in the previously discussed selection experiments, the pattern of divergence of the selected lines implies that a polygenic system controlled the behavior. The data in Fig. 12-8 also indicate that selection had a greater effect toward slowness than rapidity. Rapid mating is probably a feature that contributes to the "biological fitness" of a population; one would expect that even in the foundation population mating speed would, by natural selection, be held fairly close to the possible maximum. Most experimental populations show a similar asymmetrical response to selection.

To determine whether the behavior of both sexes had been affected by the selection process, the mating speeds of males from an F line with females from an S line and vice versa were studied. In both cases the mating speeds were found to be intermediate. This was taken as evidence that both sexes had been affected by the selection process.

It was also of interest to establish whether the courtship patterns of males from the various lines differed. The licking portion of the courtship pattern was studied as a measure of courtship intensity, since frequency of licking had been found to be positively correlated with the length of bouts of wing vibration. Females from an unselected stock population were paired singly with males from the various lines for about 200 seconds, and the numbers of licking movements were recorded. The results are summarized in Table 12-4 and are expressed as the average frequency of licking movements in 100 seconds of courtship. The data show clearly that the S-line males have a lower-intensity courtship than either controls or F-line males. Bear in mind that even a small difference in the licking frequency represents a considerable difference in the amount of wing vibration. It is quite probable that the deficiency of the S-line males in this respect is sufficient to account for their poor mating performance.

Differences in female sexual behavior are more difficult to measure. The general finding was that females from the F lines tend to stand still when courted by a male whereas those from the S lines tend to flick their wings, kick the males, jump, and fly away when courted. Females from the S lines require more courtship before

Table 12-4. Mean frequency of licking movements

	SA	SB	Controls	FA	FB
F_9	9.1	10.5	14.3	13.6	13.6
F_{27}	6.4	—	11.5	13.6	—

From Manning, A. 1961. The effects of artificial selection for mating speed in *Drosophila melanogaster*. Anim. Behav. **9**:82-92.

accepting males than do females from the F lines.

No investigations were made of the physiological bases of the behavioral differences in the various lines. However, some studies were conducted that reflected the general metabolic levels of the different types of flies. It was found that the development time of all the selected lines remained the same as the controls' throughout the experiment. In addition, the egg production of females from the various lines was the same. Flies from the F and S lines were also found to perform equally well with the controls in forced-flight tests. These tests represent a good measure of muscular efficiency and endurance. From the above findings, it was concluded that there was no generalized alteration of the insect's total metabolism in the selected lines. Although experimental evidence was lacking, the investigators hypothesized that the genes accumulated in the selected lines were affecting neural thresholds in mechanisms concerned with the control of sexual behavior. No information is available on the number of genes that might be involved.

ROLE OF INTERMEDIATE SYSTEMS IN BEHAVIOR

At the beginning of this chapter we pointed out that the regulation of a behavior can be exercised through any of a number of different pathways. For each of the behavior patterns discussed above we have reviewed the information available on its genetic control. We now want to consider some studies that have dealt specifically with the role of intermediate systems in behavior.

Role of nervous system in wasp behavior

A study that investigated the role of the nervous system in some of the behavioral characteristics of a wasp, *Habrobracon juglandis,* was reported by P. W. Whiting in 1932. In this species, as in many Hymenoptera, the males are fatherless and haploid, while the females are biparental and diploid. Occasionally (1:1000 to 1:10,000, depending on the strain studied), binucleate eggs are formed in which one nucleus is fertilized and the other is not. Female parts develop from the fertilized nucleus, and male parts from the unfertilized nucleus. Such individuals, whose bodies are part male and part female, are called *gynandromorphs* (sex mosaics). The most useful sex mosaics are those which arise from the cross of a female homozygous for a recessive gene and a male carrying the dominant allele. Female parts of such gynandromorphs show the dominant trait of paternal origin, whereas male parts show the recessive trait of maternal origin. The distribution of male and female parts in the mosaics depends on chance distribution of nuclei in early embryology. Both fertilized and unfertilized nuclei undergo several cleavages before becoming fixed in a definite region of the embryo. Gynandromorphs may be male anteriorly and female posteriorly, or the reverse. In other cases the dividing plane may cut the animal into sexually different right and left halves. In general, a plane may be at any angle to the axes of the body, and the sexually diverse parts may be equal or unequal in amount. A further finding was that a plane of division may separate male from female parts very definitely or there may be considerable invasion of one region by tissue of the opposite sex.

As can be visualized from this brief description of gynandromorph formation, sex mosaics furnish excellent material for determining the region of the body that controls the reactions characteristic of each sex. For example, if the head is male and the abdomen is female, it is possible to determine whether brain or gonads govern behavior, provided some reliable behavioral criteria are available. Consistent differences in normal behavior between the sexes are known. One of these involves the reactions of male and female wasps to the caterpillar (larva) of the flour moth *Ephestia*. Male wasps are entirely indifferent to these larvae and will simply move away from them if accidental contact occurs. In contrast to this, a female wasp will approach and sting a caterpillar with her ovipositor, thereby paralyzing it. She will then sting the paralyzed larva a number of times, apply her mouth to one of the punctures and suck the fluids out of the caterpillar. Another difference in behavior between male and female wasps is that involved in courtship and mating. A male wasp normally courts a female by flipping his wings in her presence and then mounting her while beating his wings and antennae rhythmically. The female wasp will normally engage neither in courting another wasp nor in attempting to mount another wasp.

Gynandromorphs having male heads and female abdomens were found to avoid contact with caterpillars. This type of sex mosaic, however, courted and mounted normal females. In addition, these gynandromorphs would not lay any eggs despite the fact that their abdomens were enlarged with them. Gynandromorphs with female heads and male abdomens would inspect larvae with their antennae and thrust out their abdomens as if to sting. The abdomens might also be held under the caterpillars for long periods as if eggs were being laid. This type of sex mosaic would neither court nor mount normal females. A most interesting situation develops if the brain itself is a sex mosaic. Under these conditions, there are various types of disturbances in the characteristic behaviors of the two sexes. The most unusual of the behavioral disturbances has been called "wires crossed." Gynandromorphs showing this type of behavior court moth larvae and attempt to sting wasp females.

It is quite clear that, in the case of the wasp, behavior is determined by the genotype of its brain cells, which act as the intermediate system. Only when gonadal tissue and brain tissue have the same genotype will normal behavior result.

Role of endocrine system in rodent behavior

In the search for behaviors associated with the endocrine system, the most promising stage of development to study is the onset of sexual maturity. At that time sex hormones are produced in large quantities, and new behaviors appear. One of these is fighting, and another is mating. We shall examine some of the experiments that indicate the role of the endocrine system as the intermediate agent of the genes.

Fighting behavior in male mice. We have discussed the characteristics of mouse fighting earlier in this chapter. We shall now consider the role of the male sex hormone in fighting behavior. A report of such a study was made by E. Beeman in 1947. Male mice (C57 black and Bagg albino) that had not been trained to fight were castrated, the source of most of the testosterone of the body thereby being removed. Twenty-five or more days later attempts were made to train the mice to fight others of the same strain, using the round-robin technique. The castrated males never fought. Pellets of testosterone propionate were then implanted subcutaneously in each animal to provide the male hormone. The mice soon developed the usual amount of aggressiveness. The pellets were then removed, and although most of the males stopped fighting immediately, a few continued to fight. This implied that although the male sex hormone was necessary for the initiation of the fighting behavior, it was not always necessary for continuation of a previously developed fighting pattern.

Another experiment was performed to check the above conclusion. Males were first trained to fight, and the dominant animals were then castrated and tested for fighting immediately afterward. The animals continued to fight without interruption. These results seem to confirm the conclusion drawn from the first experiment. A question that still remains is whether the fighting behavior, once established, can be maintained in the complete absence of testosterone or whether it needs the small amount of the male sex hormone that is produced by the adrenal glands. Experiments designed to test this last possibility have not yet been reported. Should such experiments demonstrate that trained fighters continue to fight in the complete absence of male sex hormone, they would indicate that the initial fights had produced permanent changes elsewhere in the body. The system most likely involved would be the *nervous system,* since it is the mechanism through which the organism receives the stimuli and coordinates the reactions that are reflected in fighting behavior.

Evidence for the role of the nervous system in fighting comes from experiments in which very fine platinum electrodes are inserted in various parts of an animal's brain. A weak electric current is then run through each electrode in turn to determine the behavior controlled by that portion of the brain. The observations made thus far indicate that many areas of the brain affect fighting behavior and that no one area predominates in controlling aggression.

Emerging from this review of fighting in male mice is the general picture of the endocrine system as the primary controlling agent over the development of fighting behavior. However, as fighting becomes an established behavior, concomitant changes take place elsewhere in the animal's body (most probably in the nervous system) that permit the behavior to continue in the absence of most, and possibly all, of the male sex hormone. We can see the importance of studying the earliest manifestations of a behavior in order to ascertain the primary effects of the genes controlling the behavior.

Mating behavior in female guinea pigs. The sexual behavior of the female guinea pig is better known than that of any other rodent. One of the first signs of the animal's readiness to mate is increased restlessness and greater activity. This is followed by true sexual activities. At first the female acts very much like a male, mounting other females and males if they are present. This type of behavior appears to stimulate any male present, who in turn tries to mount the female and is eventually accepted by her. The female who accepts a male assumes a posture *(lordosis)* that will permit copulation. The female is receptive to males for about 8 hours and then abruptly ceases her mating activities. This behavior pattern has been correlated with the amount of a female sex hormone, *estrogen,* produced by the animal. If the ovaries are removed, thereby eliminating the production of female sex hormone, all mating activity ceases. If estrogen is injected into a nonreceptive female, sexual behavior will follow. It is assumed that the action of the hormone consists in lowering the threshold of stimulation of the nervous system of the female and that once mating begins, there is an intense reciprocal stimulation between male and female. How sex hormones affect the nervous system is not yet known. However, the fact is known that, unlike fighting, the initial mating does not result in any permanent alteration of the nervous system. Estrogen must be produced anew for each mating cycle.

In 1957 Goy and Young reported on the patterns of sexual behavior of female guinea pigs from inbred strains 2 and 13, and from a genetically heterogeneous group referred to as strain T. Previous experiments had shown that females whose ovaries had been

Table 12-5. Responses of spayed female guinea pigs injected with different amounts of estrogen (α-estradiol benzoate) but with the same amount (0.2 I.U.) of progesterone

I.U. of estrogen	Strain	Percentage in estrus	Heat latency (hours)	Heat duration (hours)	Lordosis duration (seconds)	Number of mounts (average)
25	2	41	6.0	6.0	10.3	0
	T	15	6.5	5.0	10.5	15.5
	13	0	—	—	—	—
50	2	100	4.5	6.2	13.9	2.5
	T	85	6.4	3.3	10.4	19.4
	13	53	6.1	4.2	23.9	46.2
100	2	98	3.4	7.2	15.6	0.9
	T	97	6.0	4.3	11.0	8.7
	13	90	5.8	5.0	25.2	37.7
400	2	100	3.0	8.4	14.2	0.3
	T	92	4.7	6.0	10.9	14.4
	13	100	4.4	6.1	23.0	55.9
800	2	100	2.6	9.6	13.9	0.5
	T	92	5.1	6.3	9.6	21.1
	13	100	4.6	6.8	21.6	39.2

From Goy, R. W., and Young, W. C. 1957. Strain differences in the behavioural responses of female guinea pigs to alpha-oestradial benzoate and progesterone. Behaviour **10**:340-354.

removed would exhibit normal mating behavior if injected with suitable quantities of an estrogen and progesterone. Therefore, in this experiment, only spayed and subsequently injected females were used. Five measures of mating behavior were used: (1) percentage of females brought into heat by the hormonal treatment *(percentage in estrus);* (2) time interval, in hours, between the injection of progesterone and the elicitation of the first lordosis *(heat latency);* (3) number of hours lordosis can be elicited *(heat duration);* (4) length of time, in seconds, of maximum lordosis *(lordosis duration);* and (5) amount of malelike mounting behavior per estrus period *(number of mounts).* The data obtained are shown in Table 12-5.

A most striking strain difference occurs in the percentage in estrus induced by the various levels of the estrogen. For strain 2

females, 100% were in heat when only 50 I.U. of estrogen were injected. The strain T females reached the same level when 100 I.U. were given, but strain 13 females did not show 100% response until 400 I.U. had been administered. An examination of heat latency shows that it tends to decrease, in all strains, with increased levels of estrogen. Heat duration, on the other hand, tends to increase with increased levels of estrogen. The other two measures of mating behavior, lordosis duration and number of mounts, appear to be relatively unaffected by estrogen level.

Our discussion of mating behavior in female guinea pigs leads to the conclusion that the endocrine system is the intermediate agent through which the female genotype produces this behavior. We have further found that although some components of the mating pattern are estrogen-dose dependent, other components are largely dose independent. The observation that inbred strains vary in their response to different doses of estrogen is extremely important. It demonstrates that the control the female sex hormone has over mating behavior is itself limited by the ability of the organism to respond to a given dose of estrogen. This capacity to respond to different drug levels is also genotype controlled and may be under the influence of different genes from those that determine hormone production. To date, no experiments designed to investigate this point have been reported.

ROLE OF NUCLEIC ACIDS AND PROTEINS IN LEARNING AND MEMORY

Early views on learning and recall mechanisms were based on electrophysiological data and relied heavily on analogies with electronic computers. The importance of the electrical activity of neurons in both learning and the initiation of long-term memory has much support from experimentation. However, it has not seemed plausible to expect that the many different electrical activities of neurons could be stored for the long periods of time involved in long-term memory. A more recent theory on the mechanism of memory storage, advanced by H. Hydén in 1960, suggests that the nucleic acids and proteins of the neurons contain the learned information. The hypothesis is based on the fact that growth of both the nerve cell body and its processes is one of the long-term responses a neuron makes to repeated stimulation. Since cell growth involves the synthesis of nucleic acids and proteins, it was felt that different types of stimulation would induce the formation of specific nucleic acids and proteins. Thus cell growth, the response to learning, would provide for the storage of the learned information through the specificity of the molecules produced in the growth process.

Our discussion on the role of nucleic acids in learning and memory will center on messenger-RNA. This would appear logical, since mRNA is the mechanism by which

proteins are specified. However, our knowledge of the independent actions of ribosomes that sometimes occur in the translation of mRNA should lead us to anticipate that other factors also may affect learning and memory. We shall now review some of the experimental evidence linking learning and memory to nucleic acids and proteins.

Learning and messenger-RNA

One of the experiments relating learning to changes in messenger-RNA was reported by Hydén and Egyházi in 1964. In this experiment, Sprague-Dawley rats were required to reach into a cylinder to obtain food pellets. Each animal was permitted to show, by 25 reaches, which paw it preferred to use. The criterion of handedness was the use of the same paw in at least 23 of the 25 reaches. Thereafter, each right-handed rat was forced to use its left paw in order to reach the food. This requirement to use the left paw was accomplished by placing a wall parallel to, and jutting out in front of, the left side of the food cylinder, thereby preventing the use of the right paw. During the 4-day period of the experiment, each rat on the average performed 400 to 500 reaches. Investigators had previously reported that a shift in handedness is achieved by as few as 200 forced reaches and that the animal continues to use the "new" paw even when tested nine months after the transfer. After the 4-day experimental period, the rats were sacrificed and nerve cell bodies were removed from right and left sides of the area of the cerebral cortex that controls handedness. These neurons were analyzed for RNA content. Neurons from the right side of the cortex were called *learning,* and those from the left side were called *controls.* If the transfer to left-handedness had caused any change in the RNA content of cortex cells, it would be in those cells of the right side of the brain. The data obtained, shown in Table 12-6, indicated that after the experiment the right side of the cortex had a significantly larger amount of RNA per cortical neuron than the left side. It was still necessary to demonstrate that the difference

Table 12-6. RNA content of cortical neurons from left and right sides

	Controls (left side)	Learning (right side)	p
RNA ($\mu\mu$g)	22 ± 2.3	27 ± 2.5	0.02

From Hydén, H., and Egyházi, E. 1964. Changes in RNA content and base composition in cortical neurons of rats in a learning experiment involving transfer of handedness. Proc. Nat. Acad. Sci. U. S. A. **52:**1030-1035.

Table 12-7. RNA base composition of cortical neurons from left and right sides

	Controls (left side)	Learning (right side)	p
$\dfrac{G + C}{A + U}$	1.72	1.51	0.02

From Hydén, H., and Egyházi, E. 1964. Changes in RNA content and base composition in cortical neurons of rats in a learning experiment involving transfer of handedness. Proc. Nat. Acad. Sci. U. S. A. **52:**1030-1035.

in RNA content represented an increase that had occurred in the cells of the right side of the cortex during the transfer of handedness. To demonstrate this, right-handed rats that had not been subjected to the transfer experiment were sacrificed, and their cortical neurons analyzed for RNA content. No significant difference between right and left sides of the cortex was observed. These findings supported the hypothesis that the learning process resulted in an increased production of RNA in those nerve cells involved in the learning. However, up to this point, nothing was known of the type of RNA produced. If it were ribosomal RNA, the implication would be that this kind of learning involves merely an increase in the metabolic level of the neuron. If the RNA produced proved to be mRNA, it would imply that this type of learning involves the production of one or more specific proteins and that the basis for the change in handedness is to be found in the protein content of the neuron. To settle this important point, the base composition of the RNA from cortical neurons of left and right sides was analyzed. The data (Table 12-7) show a significant difference in the G + C to A + U ratio of the RNA from the two sides of the cortex. An analysis of the base composition of RNA from the cortical neurons of right-handed rats not subjected

to the transfer experiment showed no significant difference between right and left sides of the cortex. These findings demonstrated that the new RNA formed in this learning process was of the mRNA type. All of the foregoing lent support to the theory that mRNA or, more likely, protein provides the chemical substrate for storage of information in the central nervous system.

More recently (1970), a report was published by V. E. Shashoua on experiments involving learning in goldfish. A significant difference was found in the uridine to cytidine ratio of the RNA formed in the brain during the acquisition of new swimming skills, as compared to the RNA formed under nonlearning conditions. These results confirm and extend the information gained from the experiments on change of handedness in rats. Quite clearly they serve to reinforce the "mRNA-protein" theory of learning and memory.

Effect of actinomycin D on brain RNA synthesis and memory

With the advent of the mRNA-protein theory of learning and memory, experiments were performed to test whether learning and memory can occur in animals in whom brain mRNA synthesis is inhibited. One such study was reported by Barondes and Jarvik in 1964. In this experiment, CF-1

Table 12-8. Influence of shock on subsequent activity in mice treated with mannitol or actinomycin

	Mannitol no shock	Mannitol shock	Actinomycin no shock	Actinomycin shock
Crossings per minute (average)	10.7	5.4	11.0	4.3
p	< 0.02		< 0.001	

From Barondes, S. H., and Jarvik, M. E. 1964. The influence of actinomycin-D on brain RNA synthesis and on memory. J. Neurochem. **11**:187-195.

(Carworth Farms) female mice received intracerebral injections of antinomycin D of sufficient strength to result in 83% inhibition of brain RNA synthesis. A control group received intracerebral injections of D-mannitol. Four hours later all animals were placed, individually, in a training cage. The floor of the cage was covered with four rectangular metal plates. Whenever an animal stepped so that its feet rested on any two of the plates, it completed a circuit. The completion of the circuit activated a relay that resulted in the movement's being recorded. In addition, if the experimenter so desired, the completion of the circuit could also cause the mouse to receive an electric shock. The animals were allowed to explore the training cage for 1 minute. During the following minute, half the animals in each group were shocked each time they stepped on a pair of plates, but the other animals received no shocks as they moved about the cage. The shocked animals soon learned that they could avoid further shocks by remaining in one spot.

The type of situation described is called a *passive avoidance conditioning* situation. It requires that the animal do nothing in order to avoid an unpleasant experience. At the end of the minute, each mouse was returned to its home cage. One hour later, each animal was placed in the training cage for 1 minute, and counts were taken of the number of crossings between plates. If the previously shocked mice remembered the ear-

lier experience, they should move around less than the nonshocked animals. If the actinomycin-treated and shocked animals had their memory-producing mechanism impaired by the actinomycin, they should move about as much as the nonshocked mice.

According to the data obtained (Table 12-8), it is evident that despite the large inhibition (83%) of brain RNA synthesis, the actinomycin-treated mice learned and remembered as well as the controls. Two possible explanations of the data can be considered. Perhaps the residual RNA synthesis may be sufficient to mediate memory storage. Another possibility is that memory storage of the type studied here is not dependent on the synthesis of a new or unique mRNA.

The results of this experiment do not necessarily contradict the findings of that on change of handedness in rats. The present experiment tested the mouse's memory only an hour after the training period, whereas the change of handedness covered a 4-day training period. We can speculate from this that memory storage may be a time-dependent process, and we may have to distinguish between so-called *short-term memory* and *long-term memory*. If there is evidence for the existence of these two phases of memory storage, we can further anticipate that although short-term memory must precede and develop into long-term memory, the two processes need not depend on the same physical-chemical mechanism. We shall now consider some experiments that examine the time-dependent aspect of memory storage and its relation to protein synthesis.

Effect of puromycin on brain protein synthesis and memory

In studies of the effect of puromycin on brain protein synthesis and memory, the learning and retention of a *discrimination-avoidance response* are tested. In this type of experiment, mice are trained in a Y maze that has a grid floor through which an electrical shock can be applied. The mouse is placed in the stem of the Y. If the animal fails to move out of the stem within 15 seconds *(error of avoidance)*, it is shocked. The shock causes the mouse to leave the stem of the Y immediately. If the animal fails to enter the "correct" arm of the Y *(error of discrimination)*, it receives shock until it moves into the correct arm. Training is continued in one session until the mouse has achieved nine correct responses out of ten attempts (the *criterion*). The same procedure (i.e., shock given for error in performance) is used to test the mouse's memory of the training experience *(retention testing)*. Memory is evaluated in terms of the difference in numbers of trials and errors during training and testing for retention *(savings percentage)*.

An experiment of the above type was reported by Barondes and Cohen in 1966. Male Swiss albino mice had puromycin injected into the temporal region of the cerebral cortex. Such injections diffuse into the hippocampus and the posterior third of the cortex, including the entorhinal cortex. Five hours after the injection, a time at which protein synthesis had already been inhibited more than 80% for several hours, the mice were trained to choose the left limb of a Y maze to escape or avoid shock. A control group of mice that had received temporal injections of NaCl were similarly trained. Both groups required essentially the same number of trials to reach criterion. When puromycin-treated animals were evaluated for memory 15 minutes after they had finished learning the task, they had a savings

percentage of 82%, which was indistinguishable from that of the controls. However, at 45 minutes and at 90 minutes after learning, the savings percentage (35%) of the experimental group had decreased significantly, but that of the controls remained the same. At 180 minutes after learning, the savings percentage of the puromycin-treated mice was less than 7%, and again the controls remained at their earlier high level.

The difference between experimental and control groups indicates that the impairment of memory is not due to some nonspecific effect of the temporal injections. To determine whether the results were due to some nonspecific toxic effect of intracerebral puromycin, other mice were injected with the antibiotic in the frontal region of the cerebral cortex. Injections here diffuse into the anterior third of the cortex. It was found that puromycin injections in the frontal region of the cerebral cortex do not interefere with memory storage.

Thus the memory-inhibiting effect of puromycin appears to be specific for the temporal region of the cortex. Another question raised by the experiment was whether the experimental mice had permanently lost their ability to retain a learned response. Studies showed that animals trained 8 hours after temporal puromycin injections learned normally and demonstrated the normal savings percentage when tested 15 minutes after training. Thus at a time when the experimental animals had forgotten what they had learned 3 hours earlier, they still retained their capacities for learning and for retention of a new experience.

• • •

These experiments are consistent with the hypothesis of the existence of short-term and long-term phases of memory storage. Short-term memory, probably lasting no more than an hour, appears to be independent of protein synthesis in the temporal lobe of the cerebral cortex. Long-term memory probably begins to develop within minutes of the learning experience and is dependent on protein synthesis in the temporal region of the brain. The possible nature of short-term memory requires some further comment. Short-term memory is believed to have its basis in the electrical activity of those nerve cells which participate in a learning process. In memory storage, the electrical activity of the nerve cell must induce, in some unknown fashion, the formation of mRNA, which then results in the production of protein. It is in the protein that long-term memory is thought to be stored. How this protein produces its memory effect is at present unknown.

A word of caution must be expressed about the above conclusions on the role of nucleic acids and proteins in learning and memory. This field of study is young, very exciting, and highly speculative. As research in this area continues, we must anticipate the emergence of new discoveries that may require drastic changes in our present ideas. Such a changing situation is to be welcomed as evidence of the development of a new field of knowledge.

SUMMARY

In this chapter we have considered the area of genetics involved with the control of an organism's actions. This has included (1) the effects of single genes as seen in waltzing mice, nest-cleaning in honeybees, courtship behavior in *Drosophila,* and phenylpyruvic idiocy in man; (2) the behavioral differences to be found between inbred strains as seen in alcohol preference, open-field behavior, and fighting behavior in mice; (3) the modification of a behavioral trait through selection as seen in selection for maze-learning in rats, for geotaxis in *Drosophila,* and for mating speed in *Drosophila.* In too few cases has it been possible to link the behavior of the organism to an enzyme or other specific gene product, although it has been possible to link the behavior of that organism to a gene-controlled change in its development or reaction to previous experience. However, in experiments on learning and memory, it has been possible to correlate the behavior with messenger-RNA and protein synthesis. Here also the specificity of the gene product is at present unknown.

Questions and problems

1. Citing specific examples, discuss how single genes can affect behavior by:
   a. Acting on sensory organs
   b. Acting on steps in intermediary metabolism
2. On the basis of information presented in this chapter, assess the effect of heredity vs. environment upon behavior.
3. What steps would you take to determine whether maze-learning ability in rats was genetically determined? Outline an experimental method and describe the results you might expect.
4. Discuss the experiments conducted by Hirsch and Erlenmeyer-Kimling (1962), which were designed to test the possibility of a polygenic system's involvement in the geotactic behavior of *Drosophila*.
5. How have gynandromorphs contributed information regarding the site of control of sexual behavior? Discuss this point in reference to what has been learned about wasp behavior.
6. Discuss the role of sex hormones on behavior, with specific reference to fighting behavior in male mice and to mating behavior in female guinea pigs.
7. Design an experiment that could conceivably determine whether the fighting behavior in mice, once established, can be maintained in the complete absence of testosterone or whether it needs the small amount of male sex hormone that is produced by the adrenal glands.
8. What experimental evidence exists for the mRNA-protein theory of learning and memory?
9. On the basis of experiments reported in this chapter, what can you say about the effects of early social experience on aggression?
10. Fig. 12-6 shows that lines of *Drosophila* originally selected for extremes in geotaxis were subjected to reverse selection and responded to it. What does this indicate about the genetic constitution of the selected lines with respect to the alleles that determine geotaxis?
11. Recent work in the field of schizophrenia research has concentrated on searching for a chemical factor, possibly appearing in the blood of schizophrenics but not of normal individuals, that might account for the altered behavior in afflicted individuals. In light of information presented in this chapter, assess the validity of this type of research.
12. From the discussion of chemotaxis in *E. coli*, draw a diagram picturing the relationship of

the galactose-binding protein to both chemotaxis and galactose transport.

13. Discuss how an organism's behavioral traits can be modified through selection, citing specific examples found in this chapter.

14. What are the advantages and the limitations of studying behavior traits through the use of inbred lines?

15. Describe an experiment you would conduct to determine whether phototactic responses in *Drosophila* were genetically controlled. Outline your procedure, list the materials you would need, and speculate as to the results you might obtain.

16. Discuss the term "pleiotropic gene." How do pleiotropic genes "complicate" the study and analysis of behavior traits?

17. How are the fields of developmental and behavior genetics interrelated? Give examples.

18. Define the following terms as they apply to experiments concerned with memory and protein synthesis:

   a. Error of avoidance
   b. Discrimination-avoidance response
   c. Error of discrimination
   d. Criterion
   e. Retention testing
   f. Savings percentage

19. What genetic hypothesis has been offered in explanation of different responses to American foulbrood-killed larvae observed in honeybees?

20. What conclusions have been drawn from the study of alcohol preference in mice as to the genetic basis of this preference in mice?

References

Caspari, E. W. (ed.). 1964. Behavior genetics. Amer. Zool. 4:97-174. (Papers delivered at a symposium on behavior genetics during the AAAS meetings in Cleveland, Ohio, 1963. Organisms discussed include mice, honeybees, *Drosophila,* and dogs.)

Fuller, J. L., and Thompson, W. R. 1960. Behavior genetics. John Wiley & Sons, Inc., New York. (Excellent review of the subject and a classic in this field.)

Goy, R. W., and Young, W. C. 1957. Strain differences in the behavioural responses of female guinea pigs to alpha-oestradiol benzoate and progesterone. Behaviour 10:340-354. (Basic paper on the sexual behavior of the female guinea pig.)

Hafez, E. S. E. (ed.). 1962. The behavior of domestic animals. Ballière, Tindall, & Cox, Ltd., London. (Publication containing a good deal of information on the behavior genetics of sheep, goats, swine, horses, cats, chickens, turkeys, and ducks.)

Hazelbauer, G. L., and Adler, J. 1971. Role of the galactose binding protein in chemotaxis of *Escherichia coli* toward galactose. Nature New Biology 230:101-104. (A report of the experiments demonstrating the central role of the galactose-binding protein in chemotaxis and cellular transport.)

Hydén, H., and Egyházi, E. 1964. Changes in RNA content and base composition in cortical neurons of rats in a learning experiment involving transfer of handedness. Proc. Nat. Acad. Sci. U. S. A. 52:1030-1035. (Paper showing the relationship of learning to nucleic acid composition of nervous tissue.)

Jacobs, P. A., Price, W. H., Richmond, S., and Ratcliff, R. A. W. 1971. Chromosome surveys in penal institutions and approved schools. J. Med. Genet. 8:49-58. (A report of the most comprehensive chromosomal study made to date of a prison population.)

Jakway, J. S. 1959. The inheritance of patterns of mating behaviour in the male guinea pig. Anim. Behav. 7:150-162. (Basic paper on the sexual behavior of the male guinea pig.)

Manning, A. 1965. Drosophila and the evolution of behavior, p. 125-169. *In* Carthy, J. D., and Duddington, C. L. (eds.). Viewpoints in biology. Butterworth & Co., London. (Excellent discussion of the evolutionary process as it relates to behavior, using experimental evidence from *Drosophila* research as illustrative material.)

McGill, T. E. (ed.). 1965. Readings in animal behavior. Holt, Rinehart & Winston, Inc., New York. (Collection of reprints, including papers on behavior genetics, on the development of behavior, and on social behavior, ethology, and evolution.)

Scott, J. P. 1958. Animal behavior. University of Chicago Press, Chicago. (Book written for the general reader and containing a chapter on heredity and behavior.)

Scott, J. P. 1966. Agonistic behavior of mice and rats: A review. Amer. Zool. 6:683-701. (Excellent review of the recent literature on the genetic and environmental bases of aggression in mice and rats.)

Scott, J. P., and Fuller, J. L. 1965. Genetics and the social behavior of the dog. University of Chicago Press, Chicago. (Excellent treatise on the genetic basis of dog behavior.)

Shashoua, V. E. 1970. RNA metabolism in goldfish brain during acquisition of new behavioral patterns. Proc. Nat. Acad. Sci. U. S. A. 65: 160-167. (A report of experiments that support the mRNA-protein theory of learning and memory.)

Whiting, P. W. 1932. Reproductive reactions of sex mosaics of a parasitic wasp, *Habrobracon juglandis*. J. Comp. Psychol. 14:345-363. (Classic paper demonstrating the role of the nervous system in wasp behavior.)

13 Genes in populations

Up to this point, we have considered the gene solely in terms of the individual. Our discussions of genotype, phenotype, gene functions, protein synthesis, metabolism, development, and behavior have all been concerned with effects on the organism. We now have to study one more aspect of the gene: its distribution in populations. This will require a change in our points of reference. We shall no longer be restricted to a single genotype per individual; we shall need to consider many genotypes per population. We shall no longer be restricted to a single life cycle; we shall need to consider the fates of genotypes from generation to generation. Obviously, a population is made up of individuals, and what happens to the population is a reflection of what happens to its individuals. However, we shall find instances in which certain genotypes are eliminated from a population without any apparent threat to the survival of the population. Thus, one may lose various types of individuals but retain the population.

A discussion of genes in populations, especially over a number of generations, will bring us to the subject of evolution. We usually think of evolution in terms of the structural and physiological changes that accompany the origin of new forms of life from older, preexisting forms. However, these changes must be recognized as due, in the last analysis, to changes in the types and frequencies of genes in the population. It is with the genetic changes involved in evolution that we shall be concerned. The study of the mechanisms by which such changes are effected constitutes the field of *population genetics*. Since the study of evo-lution is essentially a study of diversity, we can consider population genetics as having three facets of research: (1) investigations on the *origin of genetic diversity* (mutation in the broadest sense, including any heritable change); (2) investigations on the *structuring of genetic diversity* (selection, migration, genetic drift); and (3) investigations on the *segregation of genetic diversity* (race and species formation). In the present chapter, we shall consider the origin and structuring of genetic diversity. In Chapter 14, we shall study the segregation of genetic diversity.

HARDY-WEINBERG EQUILIBRIUM

Before discussing mutation, selection, migration, and genetic drift, it is important for us to consider the "rule" that governs the expected distribution of genotypes in a given population at a given time. The basic algebraic formula that describes the expected frequencies of various genotypes in a population is known as the *Hardy-Weinberg equilibrium*. We shall now examine this most important theorem of population genetics in some detail.

Let us assume that there are only two possible alleles, *A* and *a,* at a particular locus on an autosome. Let us, in addition, take "p" as the frequency of the *A* allele in the population and "q" as the frequency of *a.* Under these conditions, $p + q = 1$, since this is the totality of these genes in the population. If p is the frequency of the *A* allele in the population, then the chance that an individual chosen at random will have the *AA* genotype is p^2. The chance that the individual will have the *aa* genotype is q^2, and the chance that he will be *Aa* is 2pq.

Table 13-1. Punnett square showing results of random combinations of sperm and eggs

Eggs	Sperm	
	p (A)	q (a)
p (A)	p^2 (AA)	pq (Aa)
q (a)	pq (Aa)	q^2 (aa)

Summary: p^2 (AA) + 2pq (Aa) + q^2 (aa) = 1

The "2" in 2pq comes from the fact that there are two ways of forming the heterozygote, since each allele can be contributed to the zygote either through the egg or through the sperm. We can represent the formation of the different zygotes by a Punnett square (Table 13-1) if we consider not only the genes but also their frequencies in the population. Since this is an autosomal trait, the frequency distribution of the two alleles will be the same in the gametes of both sexes. It must be stressed that Table 13-1 does *not* represent a single mating but rather the outcome of all possible matings involving these alleles.

Since p^2, q^2, and 2pq each represents the chance that a randomly chosen individual will have that particular genetic constitution, the terms also express the proportion of these different genetic constitutions in the population. For the condition in which a given locus has but two alleles, these three genotypes represent the totality of genotypes in the population for this locus. From this has been formulated the *Hardy-Weinberg equilibrium:*

$$p^2 + 2pq + q^2 = 1$$

As shown in Table 13-1, this expresses the results of random combinations of sperm and eggs carrying either of these alleles. The above equation represents the binomial expansion of $(p + q)^2$, and the genotypes then have a binomial distribution, which is the result expected if chance alone determines the distribution.

The Hardy-Weinberg equilibrium can be used to determine gene frequencies in a population. As an example, let us consider the antigenic property of red blood cells produced by the *MN* locus. Two alleles are involved, *M* and *N,* which exhibit codominance. Hence, we have three possible genotypes: *MM, MN, NN.* Let us assume that the blood-typing of a population with anti-M and anti-N serum has yielded the following data:

MM	MN	NN	Total
76	92	32	200
p^2	2pq	q^2	1

It will be simplest to work from the *NN* genotype. Thirty-two out of 200 individuals represent 16% (0.16) of the population. The *NN* genotype also represents q^2. If we take the square root of 0.16, we shall then have the value of q, which will be 0.4. Once we have the value of q, we can calculate the value of p from p + q = 1, which answer is 0.6. Due to the existence of codominance between the *M* and *N* alleles, the accuracy of these estimates of allelic frequencies can be checked directly from the data obtained. There were (76 × 2) + 92 = 244*M* and (32 × 2) + 92 = 156*N* alleles, in a total of (200 × 2) = 400 of these genes in the population. From these numbers, we can calculate that the frequency of *M* is $\frac{244}{400}$ = 0.61, and that of *N* is $\frac{156}{400}$ = 0.39. These represent very good agreements with those estimates based on the square root of the *NN* genotype frequency.

If we take p = 0.6 and q = 0.4 as our values of allelic frequency, it is simple to calculate the expected proportions of the three genotypes and their expected distribution among the 200 tested individuals:

	Expected proportion	Expected distribution	Observed distribution
p^2 *(MM)* =	0.36	72	76
2pq *(MN)* =	0.48	96	92
q^2 *(NN)* =	0.16	32	32

The expected distribution here is very

close to the observed distribution. A test by the chi-square method (Chapter 3), using one degree of freedom, will indicate whether the differences involved are significant. If the differences between the expected and observed distributions are significant, it will indicate that these genes are not neutral but rather are contributing, in some fashion, to the chances for survival of the various genotypes.

We want to consider another aspect of the Hardy-Weinberg equilibrium. If we start with the proportions p² *(MM)* = 0.36, 2pq *(MN)* = 0.48, and q² *(NN)* = 0.16, we can determine what will happen to the gene frequencies in the next generation. The frequencies of the genes *M* and *N* among the gametes will be as follows:

$M = p^2$ (from *MM*) + pq (from *MN*) = p
$= 0.36 + 0.24 = 0.6$
$N = q^2$ (from *NN*) + pq (from *MN*) = q
$= 0.16 + 0.24 = 0.4$

The proportions of the gametes formed with *M* and *N* will be the same as the frequencies of these genes in the previous generation, and therefore the proportions of the various genotypes will be the same in the next generation as in the preceding one —hence the term *equilibrium*. In fact, the proportions of genes and genotypes will be constant from generation to generation. This means that, for alleles in Hardy-Weinberg equilibrium, any variability once gained by a population will be maintained on a constant level. One of the consequences of this situation is that as medical science, in effect, changes previously deleterious genes (hemophilia, phenylketonuria, etc.) into neutral ones, there will be an increase in the frequencies of these previously deleterious genes in the population. It will also follow that these genes will remain in the population indefinitely, in accordance with the Hardy-Weinberg equilibrium.

In our first discussion of the Hardy-Weinberg equilibrium, we chose the simplest situation—an autosomal trait in which the alleles show codominance. If we want to study a trait in which one allele shows complete dominance over the other so that the

homozygous dominant and the heterozygote are indistinguishable, we are very strictly limited to the proportion of the homozygous recessives for our information about gene frequency. We are in the further dilemma of not being able to check whether zygotic frequencies conform to those expected. However, with our increasing ability to detect the heterozygous state of so-called dominant genes, this problem should gradually fade away.

We want now to examine the form of the Hardy-Weinberg equilibrium for *X-linked genes*. Since one sex is haploid for genes in the X chromosome and the other sex is diploid for these same genes, we must consider the sexes separately. Let us assume that, as before:

p = frequency of *A*
q = frequency of *a*
p + q = 1

The equilibrium distribution of genotypes will be as follows:

In the XY sex: p *(A)* + q *(a)*
In the XX sex: p² *(AA)* + 2pq *(Aa)* + q² *(aa)*

It should be noted that the allelic frequencies of X-linked genes can be obtained directly from their frequencies in the sex having a single X chromosome. However, if there is strict dominance between the alleles, the correctness of these estimates can be checked only by noting the frequency of the homozygous recessives in the XX sex. Under these conditions, the frequency of an X-linked recessive trait among men (q) should be the same as the square root of its frequency among women (q²). A consequence of this relationship is that X-linked characteristics are more common in men than in women. As an example, if the frequency of color-blind men in a population is 8%, the expected frequency of color-blind women is 0.64%.

The last form of the Hardy-Weinberg equilibrium that we want to consider is that used for multiple alleles. If there is codominance among the alleles, the problem of estimating gene frequency is relatively simple, since this can be done directly from the

Table 13-2. Equilibrium frequencies of ABO genotypes

Phenotype	Genotype	Frequency
A	$I^A I^A$	p^2
	$I^A i$	$2pr$
B	$I^B I^B$	q^2
	$I^B i$	$2qr$
AB	$I^A I^B$	$2pq$
O	ii	r^2

frequency data of the various genotypes. However, it is far more difficult to obtain estimates of gene frequency if there is dominance among some of the alleles. We shall take as our example the alleles of the *ABO* locus, whose modes of inheritance we reviewed in Chapter 3. For the purposes of gene frequency analysis, we can let:

p = frequency of *A* (also designated as I^A)
q = frequency of *B* (also designated as I^B)
r = frequency of *O* (also designated as *i*)

Since these three alleles are all the alleles of this locus, $p + q + r = 1$. The equilibrium distribution of genotypes under a system of random mating is represented by the expansion of the trinomial $(p + q + r)$.[2] These values are shown in Table 13-2.

Obtaining the value of r presents no problem since it is the square root of the homozygous recessive O class. However, due to the dominance relationship among some of the alleles, the values for p and q must be obtained indirectly. Letting A, B, and O represent the proportions of these *phenotypes* in the population, we proceed as follows. From Table 13-2, we know the following relationship:

$$A + O = p^2 + 2pr + r^2$$

Since $p^2 + 2pr + r^2$ is the expansion of the binomial $(p + r)^2$:

$$A + O = (p + r)^2$$

We can now take the square root of both sides of the equation:

$$\sqrt{A + O} = \sqrt{(p + r)^2} = p + r$$

Since $p + q + r = 1$, we can also say that $p + r = 1 - q$. Therefore:

$$\sqrt{A + O} = p + r = 1 - q$$

If $1 - q = \sqrt{A + O}$, then $q = 1 - \sqrt{A + O}$.

Once we have obtained r from the square root of the homozygous recessive O class and q by the method shown above, the value of p can be gotten from the equation $p + q + r = 1$. In this manner, we can obtain allele frequencies from phenotype proportions.

We can then proceed to calculate the expected proportions of the various genotypes, estimate the number of individuals in a population that ought to be of each phenotype, and compare this estimate, by chi-square (again using one degree of freedom), with the numbers actually found by blood-typing.

Our discussion of Hardy-Weinberg equilibrium has indicated that as long as genotype frequencies remain constant in a population, no evolutionary changes will occur. However, evolutionary forces exist that can modify the gene frequencies of a population. These forces are (1) mutation, (2) selection, (3) migration, and (4) genetic drift. Taken together, they account for the origin and structuring of genetic diversity. We shall now consider each of these important evolutionary forces and indicate the kinds of effects they can have on gene frequencies in populations.

MUTATION

We reviewed the various methods of producing chromosomal rearrangements in Chapter 7 and point mutations in Chapter 9. For our present purposes, these two types of inherited genetic modification can be considered together, since they follow the same rules of population dynamics. Our interest is to discuss the possible fates of mutations after they have been produced. We shall begin by considering the fate of a mutant gene where there has been (1) a single occurrence of the mutation, (2) recurrent mutation, or (3) compensation by reverse mutation. In this section, the gene is assumed to be neither beneficial nor deleterious to the survival and reproduction of its possessor. Situations in which the gene is not neutral will be considered in the discussion on selection.

Single mutation

Let us begin with a population consisting of all AA individuals and consider the situation in which one of the A genes mutates to the allele a. There will then be only one individual in the population with the genetic constitution Aa. This individual must mate with an AA individual. If this mating leaves no offspring, the new allele a will be lost. If the mating yields just one offspring, the probability that this offspring will be AA and that the a will be lost is 0.5. Even if the offspring is Aa, there will still be only one a gene in the next generation. If the mating yields two offspring, the probability of a's being lost is 0.25. On the other hand, the chance that the number of a will be increased from one to two is also 0.25. This type of probability extends to any number of offspring that may be produced. The important point is that the new allele, purely by chance, may increase or decrease in frequency or be lost from the population. Since the allele is exposed to chance elimination in each generation, the most probable fate of a singly occurring neutral mutation is its eventual elimination from the population.

Recurrent mutation

The loss of an allele by chance, as just outlined, can occur whenever a mutation produces the allele. This will apply regardless of whether the mutation occurs singly or is recurrent. However, although many alleles are lost by chance soon after arising, in most cases identical alleles are formed anew in every generation. The rate of mutation per generation varies with different genes. For a particular gene, however, the rate seems to remain reasonably constant

from generation to generation if environmental conditions and genome are kept relatively unchanged. Let us, as before, take the normal or prevalent gene as A and the mutant gene as a. We shall, in addition, take p_n as the frequency of gene A in the nth generation and q_n as the frequency of gene a in the nth generation. Finally, the symbol μ will represent the rate of mutation of A to a. Under these conditions, the frequency of a in the next $(n + 1)$ generation will be increased by the amount μp_n, and the frequency of a in the $n + 1$ generation will be as shown below:

$$q_{n+1} = q_n + \mu p_n$$

From this formula we can see that the relative increase in a per generation will be larger when A genes are abundant (p_n is large) than when A genes are rare (p_n is small) in the population. This formula also indicates that if the mutation from A to a is unopposed by some counterforce, all the genes in the population will eventually be a.

Reverse mutation

Normally, all alleles mutate. When an allele is rare, its production of other alleles is hardly detectable because mutation rates tend to be small. However, as an allele becomes fairly frequent in a population, the mutations both to and from it have to be taken into consideration. Let μ be the mutation rate from A to a and let ν be the mutation rate from a to A. As before, let p be the frequency of A and let q be the frequency of a. In a situation involving reverse mutation, the net amount of change in the frequency of a per generation (Δq) is found through this equation:

$$\Delta q = \mu p \ (gain) - \nu q \ (loss)$$

The size and direction of Δq will depend on the relative magnitude of the gain or loss of q per generation. If, at a given stage, the gain in q is greater than its loss, then as q increases so does its loss, νq, increase. Eventually, the amount of gain per generation will balance the loss. When this equilibrium is reached, no further change in q will occur in subsequent generations. Under a condition of equilibrium $\Delta q = 0$, or $\mu p = \nu q$. Solving for the equilibrium values of p and q in the above formula (i.e., $p = 1 - q$ and $q = 1 - p$), we get the following:

$$\hat{q} = \frac{\mu}{\mu + \nu} ; \ \hat{p} = \frac{\nu}{\mu + \nu}$$

The "hat" ($\wedge$) denotes the equilibrium value. A graphic representation of this situation is given in Fig. 13-1 for the theoretical values of $\mu = 0.00006$ and $\nu = 0.00004$. Since μ and ν are constants, the above equilibrium is stable. Should the value of q depart from $\hat{q}$ for any reason, it will come back gradually to $\hat{q}$ when the cause for the departure ceases to operate. It is interesting to note that the equilibrium value of q is independent of the initial frequencies of the alleles in the population and is solely determined by the two opposing mutation rates. Also note that because mutation rates are so small, mutational equilibrium is achieved very slowly. What has been said for point mutations in this section will also apply to chromosomal rearrangements, provided they arise and change at some finite rate.

The consequence of the equilibrium just described is quite important to the population. It means that, given reverse mutation, all neutral alleles of a gene will remain permanently in a population, thus guaranteeing a certain amount of heterogeneity at all times. As will be discussed, some of this heterogeneity may become important if a change of environment makes a previously neutral gene either beneficial or deleterious to its possessor. Should the gene remain neutral, mutational equilibrium will result in some of the variability that we see in the population.

Spontaneous mutation rates

We have considered the theoretical consequences of mutation under three sets of conditions. We shall now look at some of the information that has been gathered on spontaneous mutation rates and the factors that can affect them. In Table 13-3 are listed the spontaneous mutation rates per generation for certain genes in various organisms. Three facts seem to stand out in the data. One is that different genes in the same organism can have quite different mutation rates. The second fact is that the different alleles of the same gene have different mutation rates toward each other. The third fact is that, in general, the genes of higher animals and plants mutate more frequently per generation than those of microorganisms. The greater mutation rates of higher organisms may be associated with the diploid state that permits the presence of a large number of potentially deleterious recessive genes whose detrimental effect can be neutralized by dominant alleles in the homologous chromosome. Haploid microorganisms have no such protective device against deleterious mutants and in the course of evolution have apparently developed a lower overall mutation rate.

Genetic control of mutation rates

A great deal of evidence exists to the effect that the mutation rate of a gene is itself under genetic control. An example of

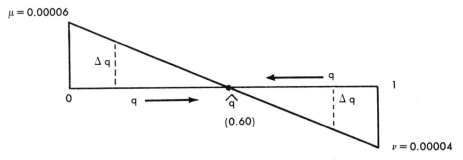

Fig. 13-1. Equilibrium involving reverse mutation.

this is found in the case of the mutation rate of wild type to yellow body color in the fruit fly. In males, the mutation rate (1×10^{-4}) is significantly higher than in the females (1×10^{-5}). Apparently the many differences associated with the two sexes of the fruit fly include differences in the mutation rate of this gene. Similar examples of the genetic control over spontaneous mutation rates exist, in which the strain of the organism rather than the sex is responsible. Laboratory stocks of *Drosophila melanogaster* collected from different sources have been found to show different rates of mutation. A number of such stocks were analyzed for the occurrence of X-linked recessive lethals. These represent the sum total of the lethals that were produced along the entire X chromosome. The Oregon R strain exhibited a mutation rate of 0.07%, which was significantly lower than the 1.09% of the Florida No. 10 stock. In 1937 M. Demerec published the results of genetic studies he had carried out on the Florida stock. These investigations revealed that the high mutation rate was due to the stock's being homozygous for a recessive gene on the second chromosome. When, by breeding experiments, that gene was replaced by its normal allele, the stock's mutation rate for X-linked lethal recessives dropped to 0.074%. Genes that increase the mutation rates of other genes are called *mutator genes*.

Another fine example of a mutator gene was discovered in maize (corn) by M. M. Rhoades in 1938. In maize, the dominant allele of a gene, A_1, located on the third chromosome, controls the production of anthocyanin, which imparts a purple coloration to various parts of the plant. The recessive allele a_1, when homozygous, produces green plants and colorless seeds. Under normal conditions, both alleles mutate to one another only infrequently. However, in the presence of the dominant gene *Dt* on the ninth chromosome, the gene a_1 mutates very frequently to A_1. The gene *dotted* has two alleles, *Dt* and *dt*, which by themselves have no visible effects. The recessive allele *dt* does not affect the mutation rate of the a_1 or any other gene, and the dominant allele *Dt* does not affect the mutation rate of any gene other than a_1. The mutation of a_1 to A_1 occurs in both the germ and the somatic cells of the plant. Somatic mutations show up as purple streaks in otherwise green leaves and

Table 13-3. Spontaneous mutation rates per generation for certain genes in various organisms

Organism	Character	Rate
Bacteriophage *(T2)*	Lysis inhibition, $r \rightarrow r^+$	1×10^{-8}
	Host range, $h^+ \rightarrow h$	3×10^{-9}
Bacterium *(Escherichia coli)*	Lactose fermentation, $lac^- \rightarrow lac^+$	2×10^{-7}
	Histidine requirement, $his^- \rightarrow his^+$	4×10^{-8}
	$his^+ \rightarrow his^-$	2×10^{-6}
Corn *(Zea mays)*	Shrunken seeds, $Sh \rightarrow sh$	1×10^{-5}
	Purple, $P \rightarrow p$	1×10^{-6}
Fruit fly *(Drosophila melanogaster)*	White eye, $W \rightarrow w$	4×10^{-5}
	Brown eye, $Bw \rightarrow bw$	3×10^{-5}
Man *(Homo sapiens)*	Normal $\rightarrow$ hemophilic	3×10^{-5}
	Normal $\rightarrow$ albino	3×10^{-5}

From Sager, R., and Ryan, F. J. 1961. Cell heredity. John Wiley & Sons, Inc., New York.

as purple dots on an otherwise white seed. The dots on the seeds are all small and of approximately equal size, indicating that the mutations of a_1 to A_1, induced by the Dt gene, all occurred late in development and at about the same time.

The occurrence of mutations in the triploid tissue of the seed's endosperm provides an excellent opportunity for the study of any dosage effects these genes may have. In endosperm tissue, by appropriate crosses, one can obtain various phenotypically distinguishable combinations of the a_1 gene. As an example, each of the following genotypes results in a unique phenotype: $a_1a_1a_1$ (colorless), $a_1a_1a_2$ (slightly pigmented), and $a_1a_2a_2$ (more heavily pigmented). In every case a mutation from a_1 to A_1 will result in an easily observed purple dot. Seeds of the above genotypes were obtained in combination with the Dt dt dt genotype. The numbers of colored dots formed on such seeds approach the ratio of 3:2:1, indicating that the frequency of mutation is proportional to the number of a_1 genes. Similarly, seeds may be obtained with one, two, or three doses of Dt in combination with a single a_1 gene. The number of purple dots observed per seed is as follows:

dt	dt	Dt	gives 7.2 dots per seed
dt	Dt	Dt	gives 22.2 dots per seed
Dt	Dt	Dt	gives 121.9 dots per seed

These results indicate that the number of mutations increases exponentially as the number of *Dotted* genes increases.

Mutator genes have also been discovered in microorganisms. In 1955 Goldstein and Smoot described a wild strain of *Escherichia coli* in which there was a very high frequency of mutants (up to 15% of the total number of viable cells) in every generation. Pathways for the biosynthesis of purines, pyrimidines, amino acids, vitamins, and for the fermentation of sugars were all affected. The mutants usually occurred singly. However, the unusual genetic instability of the strain was such that double mutants arose fairly frequently. This high mutability was itself observed to be a stable hereditary prop-

erty. The locus responsible for the high mutability was designated *astasia* and given the symbol *ast;* ast^+ cells were highly mutable, but ast^- cells were not. This locus was transferred to various strains through conjugation. In *E. coli* K12, it was found that the mutation from "azide-sensitive" to "azide-resistant" occurred in ast^+ cells 622 times more frequently than in ast^- cells. Comparable data for the mutation from "proline-requiring" to "proline-nonrequiring" gave the figure of 169, and that for "histidine-requiring" to "histidine-nonrequiring" was 11. The range of increased mutability per gene is highly variable, but in virtually every case the difference in mutation rates between ast^+ and ast^- cells is very significant.

In addition to mutator genes, other genetic factors may be involved in increasing mutation rates. In 1960 P. E. Thompson reported on the effects of chromosomal inversions on mutation rates in *D. melanogaster*. He found that inversions in one chromosome of an inversion-heterozygote caused an increase in the spontaneous lethal-mutation rate of the normal chromosome. This was found to be the case for the second and third chromosomes of the fly. At the present time, we can only speculate about the possible mechanism involved in this or any other previously mentioned examples of genetic control over spontaneous mutation rates.

Viral control of mutation rates

We have, in various discussions throughout this book, noted the large number of effects that viruses produce in host cells. There would appear to be no aspect of the host cells' metabolism that remains unaffected. We should not, therefore, be surprised at the reporting of evidence on the control that viruses have over mutation rates of the host's genes. One example of this type of virus-host interaction was reported by Sprague and colleagues in 1963. They took maize plants that were homozygous for the dominant genes A_1 (pigmented endosperm), *Su* (starchy endosperm), and *Pr*

Table 13-4. Frequencies of endosperm deficiencies in F_1 progeny of healthy and virus-infected male parental stocks of maize

| | Endosperm deficiencies per 1000 seeds | | | | | |
| | A_1 | | Su | | Pr | |
Type of plant	Entire	Fractional	Entire	Fractional	Entire	Fractional
Virus-infected	3.1	3.8	2.3	0.8	0.0	3.1
Control	0.2	1.4	0.3	0.4	0.4	0.1

From Sprague, G. F., McKinney, H. H., and Greeley, L. 1963. Virus as a mutagenic agent in maize. Science **141**:1052-1053. (Copyright 1963 by the American Association for the Advancement of Science.)

(purple endosperm) and inoculated them, in the seedling stage, with a strain of barley-stripe–mosaic virus. Pollen from the infected plants was applied to plants homozygous for the respective recessives of the above genes: a_1 (colorless endosperm), *su* (sugary endosperm), and *pr* (red endosperm). The F_1 seeds were then scored for entire and sectorial (fractional) losses of the dominant marker genes. The data obtained are shown in Table 13-4. The findings can be summed over all loci and thus provide a simple and direct measure of the effect of the virus. With this procedure, the frequency of effects is 1:108 seeds in the treated series, which is significantly greater than the 1:556 seeds in the controls. F_2 progenies were produced from both treated and control plants. Numerous mutants, as expected, were discovered among the treated plants but not among the control plants. A most interesting observation was that in none of the F_1, etc. plants could the virus be found. This would argue against the observed effects' being caused by the continued presence of the virus and would also imply that the genetic effects of the virus infection occurred prior to, or during, gamete formation in the treated plants.

Another example of increased mutation rates after virus infection was provided by R. C. Baumiller in 1967. The organism involved was *D. melanogaster,* and the virus used was Sigma, an organism that causes the fly it infects to be sensitive to carbon dioxide. Two lines of flies were set up, one a control line consisting of Sigma virus–free flies and the other an experimental line consisting of Sigma virus–containing flies. These two lines were tested for their frequencies of X-linked lethal mutants. The control line was found to have significantly fewer (0.125%) mutants than the experimental line (0.242%). The discovery that viruses increase mutation rates brings into question the concept of spontaneous mutation rates, since in none of the studies to determine these rates is the presence of viruses taken into account. No information is presently available as to how the virus causes an increase in the mutation rate of the host's cells.

Environmental influences on mutation rates

Three major environmental factors affect mutation rates: temperature, certain radiations, and certain chemicals. We shall consider the factors in this order and see to what extent they affect mutation rates.

Effects of temperature on mutation rates. A good deal of research on the effects of temperature on mutation rates has been performed on *D. melanogaster*. One such experiment on the occurrence of X-linked recessive lethals showed at 14° C., 0.087% lethals; at 22° C., 0.188% lethals; at 28° C., 0.325% lethals. It seemed clear that the mutation rate doubles or trebles with a 10°

C. rise in temperature. This twofold or threefold increase of effect caused by a 10° C. rise in temperature is what would be expected for most biological processes and chemical reactions. When this type of experiment is performed on bacteria, there does not appear to be any temperature effect comparable to the one observed in *Drosophila*. For the present, no general statement on the effect of temperature on mutation rates can be made that would apply to all organisms.

Instead of using temperatures within the normal range for the organism, some experimenters have used extreme temperatures for short periods of time. In one investigation, 3-day-old larvae of *D. melanogaster* were subjected to temperatures of 36° to 38° C. (heat shock) for 12 to 24 hours. This treatment resulted in a doubling of the X chromosome lethal mutation rate. Exposure of the animals to –6° C. (cold shock) for 25 to 40 minutes caused a tripling of the lethal mutation rate of the X and second chromosomes. That both heat and cold shocks should increase the mutation rate may at first seem contradictory. It must be borne in mind that temperature shocks do not merely increase or decrease biological processes but, rather, deal these processes a

severe jolt. The method (heat or cold) by which the shock is administered may be of little importance, and the net effect may very well be the same in both cases.

In a very interesting study of the effects of temperature on mutation rates, conducted on the mutator gene *Dt* in maize, it was found that the a_1 allele, in the presence of *Dt,* mutated to A_1 at a rate four to five times higher at 15° C. than at 27° C. This inverse relation between mutation rate and temperature is quite different from what is found when studying mutations from wild type genes to lethals in *D. melanogaster*. In studies conducted on genes with high mutation rates in different *Drosophila* species the results have not been consistent, and no generalization is possible at this time about the effects of temperature on the mutation rates of highly mutable genes.

Effects of radiation on mutation rates. We discussed the various types of radiation and the probable mechanisms by which radiations cause mutations in Chapter 9. Our interest in this chapter is to examine the relationships between ionizing radiations and mutation rates, as well as the factors that affect the relationships. This field of study was opened by H. J. Muller, who in 1927 announced that mutations are more

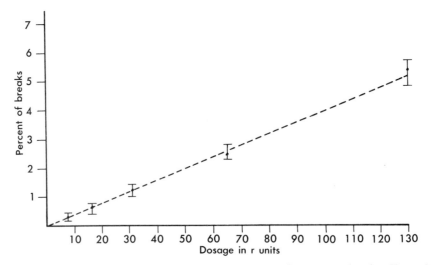

Fig. 13-2. Relation between x-ray dosage and frequency of chromosome breaks. (From Carlson, J. G. 1941. Proc. Nat. Acad. Sci. U. S. A. **27:**42-47.)

frequent in the progeny of *Drosophila* treated with x rays than in the offspring of those without such treatment. The basic unit of measurement of ionizing radiations is the *roentgen,* or *r,* unit. It is defined as the amount of radiation that will produce 1.8×10^9 ion pairs per cubic centimeter of air. Another unit, the *rad,* measures the amount of radiation absorbed by the organism's tissues.

One of the effects of ionizing radiations is the production of *single breaks* in chromosomes. The percentage of such single breaks occurring after irradiation can be estimated by examining the tissues microscopically for acentric fragments. The results of such a study of embryonic nerve tissue of the grasshopper are shown in Fig. 13-2. It can be seen that the number of breaks is directly proportional to the amount of dosage within the range tested, with 1% of breaks obtained for about every 20 r of radiation. Apparently there is no threshold amount of radiation, below which breaks are not produced.

Another effect of ionizing radiations is the production of *double breaks* in chromosomes. Should these occur in the same chromosome, they can result in deficiencies and inversions. If double breaks occur in nonhomologous chromosomes, they can result in translocations (Chapter 7). Investigations have been made of the occurrence of chromosomal rearrangements after exposure to various dosages of radiation. The results of such a study, using *Drosophila* spermatozoa, on the relationship of dosage to reciprocal translocations between chromosomes II and III are shown in Fig. 13-3. The data can be fitted to a curve that represents the dose raised to the power of 1.5. Actually, we would have expected a curve corresponding to the dose raised to the power of 2, since the probability of obtaining two breaks should be the square of the probability of obtaining a single break. This deviation from the expected is due to (1) high doses of radiation, frequently producing two or more ion clusters, each of which can cause a chromosome break independently of the other;

and (2) some single radiation "hits" that caused two or more chromosome breaks in the chromosomes of the same cell. Note that, as in the case of single breaks, there is no threshold amount of radiation below which there are no effects.

The last effect of ionizing radiations that we shall consider is the production of *point mutations*. Point mutations are changes confined to regions of the chromosomes so small that no loss, addition, or change in arrangement of genes can be demonstrated by microscopic examination or genetic tests. An example of the relationship of dosage to

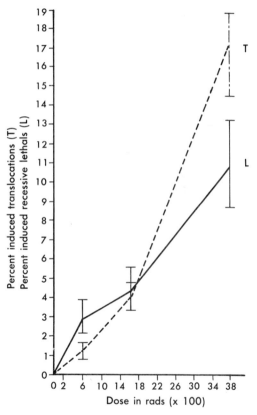

Fig. 13-3. Percentage of mutations, ± twice the standard error, recovered from *Drosophila* sperm exposed to different dosages. The X-linked recessive lethal frequencies (**L**) are joined by solid lines, and reciprocal translocation frequencies (**T**) between chromosomes II and III are connected by broken lines. (From Herskowitz, I. H., Muller, H. J., and Laughlin, J. S. 1959. Genetics **44:** 321-327.)

Table 13-5. Percentages of second-chromosome recessive lethals in *D. melanogaster* obtained under acute versus chronic irradiation

Total dose	Dose-rate		Percent lethals
3000 r gamma rays	2000	r/min.	4.5
	2.0	r/min.	4.7
3000 r x rays	600	r/min.	5.4
	2.0	r/min.	5.0
200 r gamma rays	2.0	r/min.	1.9
	0.01	r/min.	1.6

From Purdom, C. E., and McSheehy, T. W. 1961. Radiation intensity and the induction of mutation in *Drosophila*. Int. J. Radiat. Biol. 3:579-586.

X-linked lethals in *Drosophila* spermatozoa is shown in Fig. 13-3. Here, as in the case of single chromosome breaks, there is a direct proportion between dosage and point mutation: roughly 1% induced point mutations per 380 r.

One other aspect of the relationship between ionizing radiations and mutation rates warrants our attention. It is the question of whether the rate of exposure to radiations is a factor in the damage they cause. This question appears to have two answers, one obtained from experiments on *Drosophila* and the other from studies on mice. In the investigations utilizing *Drosophila,* the total dose of radiation has been consistently found to be the significant factor, and any fractionation of the dose, of no consequence. As an example, we can examine the findings of Purdom and McSheehy as reported in 1961. In their experiments *D. melanogaster* males were irradiated at various dose-rates and mated to females immediately thereafter. Fresh females were supplied every 3 days to produce a sequence of broods. Successive broods represented sperm derived from progressively earlier germ-cell stages, which were present in the testis at the time of irradiation.

The mutations scored were second-chromosome recessive lethals derived from cells that were in the spermatogonial stage at the time of irradiation. The data are shown in Table 13-5. There were no significant differences observed in the percentage of mutants obtained under acute versus chronic radiation conditions or between radiation with x rays and that with gamma rays.

Study of the effect of fractionated doses of radiation on mutation frequency in mice has yielded different results. As our example, we can consider the experiments reported by Russell and co-workers in 1958. Young mature male mice were exposed to various dosages of either acute x rays or chronic gamma rays. The males were then mated to a succession of test females who were homozygous for seven autosomal recessive mutant genes, each of which had a different visible effect. The offspring, resulting from fertilizations involving sperm that were in the spermatogonial stage at the time of irradiation, were examined for mutations at the seven loci. Results from the acute x-ray and chronic gamma-ray experiments are shown in Fig. 13-4. All the points for the acute x rays are considerably above the curve for the chronic gamma rays. It was concluded that acute x-ray radiation is significantly more effective than chronic gamma radiation in inducing specific locus mutations in mouse spermatogonia. This finding is in sharp contrast to the findings for *Drosophila* spermatogonia.

Effects of chemicals on mutation rates. In Chapter 9 we discussed the roles of various chemicals (nitrous acid, 5-bromouracil, proflavine, etc.) in producing different types of mutations. Now we want to consider the relationship between chemicals and mutation rates. This field of investigation was opened in 1947 when Auerbach and Robson reported on their findings that chemical treatment could be as effective as radiations in inducing mutations. In the original experiments *D. melanogaster* males were exposed to *mustard gas* and then tested for X-linked lethals. It was found that it was possible to increase the incidence of X-linked lethals from about 0.2% in the controls to about 24% in the treated flies. Related compounds, the nitrogen and sulfur mustards, mustard oil, and chloracetone, also proved mutagenic, although not to the same degree as mustard gas. A number of investigators soon found that these compounds were also mutagenic for other organisms. Further research disclosed an ever-increasing number of chemicals that had mutagenic effects. In addition, it was found that chemical mutagens could produce chromosomal rearrangements as well as point mutations.

When the dose-mutagenic response of a chemical like formaldehyde is determined, it is found to be linear, much as observed for point mutations and ionizing radiations. However, a different result is observed when

diethyl sulfate is used as a mutagenic agent. Diethyl sulfate is one of the alkylating agents (e.g., ethyl ethane sulfonate) whose effect is thought to be the result of a direct chemical reaction with the bases of DNA. A study of the mutagenic effects of diethyl sulfate was reported by Pelecanos and Alderson in 1964. The chemical, in various concentrations, was placed in the food on which either batches of male larvae were raised for their entire larval life or adult males were placed for 48 hours after hatching. In both cases, the first brood fathered by these males was studied for the incidence of X-linked recessive lethals. The data obtained are shown in Table 13-6.

A significantly greater percentage of lethals was obtained from those treated as adults than from those treated as larvae. This is considered to be due to the fact that in the treated adult males, the cells affected were all spermatozoa that had completed their meiosis at the time of treatment. Mature gametes that have been affected by mutagenic agents can still function in fertilization even if they contain chromosomal rearrangements, and the subsequent mitoses of the zygote and embryonic cells will preserve the total genome. In the larvae, however, spermatogonial and spermatocyte cells are affected. Any chromosomal rearrangements that result from the chemical treatment are exposed to elimination in the subsequent meiotic process. As a result, there remain mainly the point mutations and minor deletions. Thus, the effects of mutagenic agents on spermatogonia appear to be less than their effects on spermatozoa. However, much of this difference is due to the intervening meiotic process.

The data in Table 13-6 would also imply a threshold level of response, occurring at roughly the same percentage of diethyl sulfate for both treated larvae and adults. Here, too, the results have to be analyzed further. Since the mutagen is placed in the food, there is opportunity for the mutagen to react chemically with the food and lose its mutagenic power. Such reaction would occur until a high enough concentration of the muta-

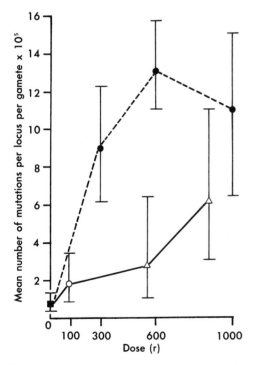

Fig. 13-4. Mutation rates at seven specific loci in the mouse, with 90% confidence intervals. Solid circles represent results with acute x rays (80 to 90 r per minute). Open points represent chronic gamma ray results (triangles, 90 r per week and the circle, 10 r per week). The point for zero dose represents the sum of all controls. (From Russell, W. L., Russell, L. B., and Kelly, E. M. 1958. Science **128**:1546-1550.)

Table 13-6. Dose-mutagenic response to diethyl sulfate of *Drosophila melanogaster* (first brood only)

Concentration of diethyl sulfate (%)	Percent of X-linked recessive lethals	
	Treated as larvae	Treated as adults
Control	–	0.3
0.3	1.8	4.6
0.35	2.9	–
0.375	10.3	4.7
0.4	9.7	19.8
0.45	–	19.0
0.5	11.6	27.1
0.55	–	26.6
0.575	–	27.9
0.7	10.7	–

From Pelecanos, M., and Alderson, T. 1964. The mutagenic activity of diethyl sulphate in *Drosophila melanogaster*. I. The dose-mutagenic response to larval and adult feeding. Mutat. Res. **1**:173-181.

gen is placed in the food so that the free compound is present. This possible explanation of the apparent threshold response was checked by repeating the experiment without yeast in the food. The yeast was believed to be the most likely food component to react with the diethyl sulfate. The results of this experiment clearly showed a percentage of X-linked recessive lethals that was directly proportional to an increasing concentration of mutagen. There was no sign of a threshold effect. Thus the dose-mutagenic response of diethyl sulfate was shown to be linear, as had been found for other mutagenic agents.

LSD and genetic damage

Our modern society has witnessed an increasingly widespread use of drugs as a mechanism either of achieving pleasure or of escape from problems. One of the more widely used drugs is the hallucinogen LSD (lysergic acid diethylamide). As the use of LSD increased, questions were raised about the possible genetic damage that might result from the ingestion of this drug. A large number of studies (68 from 1967 to 1970)

have been made that have dealt with this question. The various investigations have taken different experimental approaches.

One group of studies has centered on LSD as a mutagenic agent. A number of experiments have been performed on the induction of X-linked lethals in *Drosophila;* others have involved the induction of mutations from prototrophy to auxotrophy in *E. coli.* In both types of investigations, extremely high doses of LSD were found to be very mutagenic. However, when using doses comparable to those found in the blood of an "average" user of the drug, there was no observed increase in mutation rate.

Another experimental approach to the question of the possible production of genetic damage by LSD has been the use of in vitro experiments. These have involved the addition of LSD to cultures of different types of human cells. After varying periods of time, the chromosomes of the cells that had been exposed to LSD were examined, and any breaks occurring in the chromosomes were noted. These experiments have shown a significant increase in the number of chromosome breaks in cells treated with unusually high doses of LSD. However, when using amounts that more nearly approach the expected concentration of LSD in the blood of a person who has ingested a "usual" dose of the drug, there was no increase in chromosome breakage.

The last type of investigation we shall consider involves the incidence of chromosome breaks in the cells, usually lymphocytes, of individuals who have actually ingested the drug. Interestingly enough, the results obtained from these studies are quite fearsome. Two types of LSD-users were studied. One group consisted of people who received known quantities of pure LSD in medically supervised settings. The other group consisted of users of illicit LSD, which is of uncertain composition and potency. Both groups showed an increased frequency of chromosomal breaks. Of the 126 people treated with pure LSD, 18 (14.3%) showed a higher frequency of chromosome aberrations than a control group. Of the 184 individuals who ingested illicit LSD, 90 (48.9%) were observed to have an increased occurrence of chromosomal aberrations. The incidence of individuals with chromosomal damage among illicit drug users is more than triple that for persons given pharmacologically pure LSD. This last point deserves further consideration. The fact was just mentioned that illicit LSD is of uncertain composition and potency. Investigations of all types of illicit drugs show that they are generally contaminated with other chemicals that are extremely toxic to the human body and are known to cause genetic damage. These toxic chemicals include, among others, benzene, ether, chloroform, mercury, phenyl-2-propanone, methylamine, etc. Another aspect of this problem to be considered in evaluating the association of chromosome breakage and illicit LSD is the well-established fact that nearly all users of illicit LSD also use other drugs. These may include alcohol, amphetamines, barbiturates, marijuana, heroin, etc. The tremendously increased frequency of chromosome breaks found in the cells of users of illicit LSD may reflect a more general

deleterious genetic effect of drug usage and points up the need to study the actual conditions involving drug abuse when one attempts to evaluate the possible genetic damage from the use of any one drug.

• • •

We shall for the present leave our consideration of mutation. However, we shall return to it later in this chapter and also in Chapter 14. For the present, let us turn our attention to the topic of selection.

SELECTION

In our study of genes in populations, we have first considered the origin of new alleles through mutation. We reviewed the theoretical consequences of various patterns of mutation and then examined some of the ways in which mutation rates can be changed. Our discussion up to now has, in effect, dealt with the origin of genetic diversity in populations. However, once a new allele has arisen, the fate of that allele is determined by many factors. We have already noted certain possible fates that may befall a newly arisen allele, based on mutation rates and on chance. Now we wish to examine the survival or elimination of the gene due to the forces of the environment. The sum total of environmental forces acting to determine the fate of an allele in a population is called *selection*. It is clear that selection does not act on individual genes but rather on the organism possessing them. As a result, certain genes that greatly benefit an organism may ensure the survival into the next generation of other genes that are actually deleterious to their possessor. We shall also find that a compound, formed of genes that are individually homozygous lethal, which proves to be a superior genotype, will preserve even lethal genes in the population. A further point needs to be mentioned. In haploid organisms, each gene is expressed and is subject to the forces of selection. In diploid organisms, the fate of an allele will be determined in part by the type of interaction (complete dominance, incomplete dominance, codominance) it exhibits with other alleles at that locus. In most of our discussions, we shall be concerned with populations of diploid organisms and the relative survival of various genotypes.

The net effect of selection is to determine the relative contribution that carriers of different genetic constitutions in a population will make as parents of the next generation. The relative measure of the survival and reproductive efficiency of a particular genome, when compared to its competitors in a given environment, is called its *adaptive value* (symbolized by "W"). Adaptive value is a continuous quantity having values from zero to one. Genetic constitutions that are lethal or sterile to their possessors have an adaptive value of $W = 0$. For those genetic constitutions that have the highest adaptive value, relative to their competitors, $W = 1$. Those that have on the average only half as many offspring as the most efficient members of the population are said to be semilethals, and their $W = 0.5$. The origin of a superior allele will reduce the adaptive value of the previously most efficient gene and may even lead to its elimination from the population.

Another approach to achieving a quantitative estimate of the efficiency of a genotype in a population is to consider the magnitude of the forces acting to prevent a particular genetic constitution from surviving and reproducing. The measure of these forces is called the *selection coefficient* ("s"), which represents the converse of the adaptive value. The relationship between selection coefficient and adaptive value is as follows:

$$s = 1 - W \ or \ W = 1 - s$$

For a lethal or sterile genetic constitution, $s = 1$. For the genotype with the highest reproductive efficiency, $s = 0$, etc. We shall now consider the various types of selection that can operate on the different genotypes in a population.

Selection against recessive genes

Our first example of the consequences of the action of selection will be the case where the homozygous recessive individuals *(aa)* of a large population are prevented, to some degree, from reproducing. We need not specify, at this time, the magnitude of the selection coefficient (s), but we must realize that the adaptive value (W) of the *aa* individual will be $1 - s$. In our example we shall assume that there is complete dominance and that, as a result, the homozygous dominant and the heterozygous individuals produce, on the average, equal numbers of viable and fertile offspring. Under these conditions, the adaptive value (W) of *AA*, like that of *Aa* individuals will be 1. In Table 13-7 are shown the proportions of the three genotypes before selection, their adaptive values, and the proportions of the three genotypes after selection.

From Table 13-7, by the same method we used to get the frequencies of the *M* and *N* genes in the Hardy-Weinberg equilibrium, we can obtain an equation that yields the value of the *a* gene after a generation of selection. Following is the equation:

$$q_{n+1} = \frac{q_n(1 - sq_n)}{1 - sq_n^2}$$

The frequency of the *a* gene in a particular generation is indicated as q_n, and q_{n+1} is its frequency in the following generation. The selection coefficient, s, reflects the reproductive efficiency of the *aa* individuals compared to the other genotypes. The most interesting case is that in which $s = 1$, which means that the *aa* genotype is lethal or sterile. When this is the situation, the equation is expressed as follows:

$$q_{n+1} = \frac{q_n}{1 + q_n}$$

Using the above formulae, we can derive curves showing the effects of different strengths of selection for the various gene frequencies. Such curves are shown in Fig. 13-5. Curve a shows the effects of selection against a recessive lethal ($s = 1$); curve b shows the slower effects of selection against a recessive semilethal ($s = 0.5$). From curve a, we can see that, with this type of selection, the value of q de-

Table 13-7. Selection against recessive genes

	AA	Aa	aa	Total
Initial proportions	p^2	$2pq$	q^2	1
Adaptive values	1	1	$1 - s$	–
Proportions after selection	p^2	$2pq$	$q^2(1 - s)$	$1 - sq^2$

Table 13-8. Relative percentages of *Aa* and *aa* individuals under various frequencies of *a*

Frequency of a (q)	Percentage of Aa individuals (2pq)	Percentage of aa individuals (q²)	Ratio of Aa to aa individuals (2pq/q²)
0.5	50	25	2
0.2	32	4	8
0.1	18	1	18
0.01	1.98	0.01	198
0.001	0.1998	0.0001	1998

creases very rapidly when q is large and the rate of decrease lessens as q becomes small. This change occurs because as the *a* gene frequency is diminished, the gene is more likely to be found in the *Aa* combination where it is protected from selection than in the *aa* combination. The extent to which this factor serves to protect the *a* gene as it becomes rare can be seen in Table 13-8.

Effectiveness of selection in a eugenics program

The implications of Table 13-8 are extremely important for some of the programs that are designed to control man's heredity (eugenics) through the control of man's reproduction. There are two types of eugenics programs: (1) *positive eugenics,* which aims to have persons regarded as carriers of desirable gene combinations become parents, and (2) *negative eugenics,* which seeks the elimination of undesirable genes by discouraging individuals with undesirable traits from becoming parents. Table 13-8 indicates the limitations of a negative eugenics program designed to eliminate an undesirable trait that results from a recessive gene in the homozygous state (e.g., phenylketonuria). The direct effects of a negative eugenics program on such a trait are shown in Table 13-9, which starts with a theoretical frequency of the trait (q^2) of 25%.

The first two generations of the eugenics program would reduce the number of defectives by half in each generation. However, to go from 6% to 3% would require two generations, and to go from 3% to 1% would require five. Since a human generation is about 30 years, it would take about a century to make such a trait much less frequent. After that, the efficiency of the program is so greatly reduced as to make it virtually ineffective. Unfortunately,

Table 13-9. Effects of a negative eugenics program on a recessive gene

Generation	Percent of defectives (q^2)
Initial	25
1	11
2	6
3	4
4	3
10	0.8
20	0.2
30	0.1
50	0.04
100	0.01

since deleterious genes are usually in low frequency in the population, most negative eugenics programs are quite ineffective even when first begun. Even if the heterozygotes were detectable and also were discouraged from reproducing, the undesirable gene would not disappear from the population forever, since mutation would produce a certain number of them in each generation.

Selection against semidominant genes

Now we want to consider the situation in which the gene *a* does have a deleterious effect on the *Aa* genotype. To simplify the calculations, we shall assume that in adaptive value the heterozygote is exactly intermediate between the two homozygotes, which means that there is no true dominance. The effect of selection on the proportions of

Table 13-10. Selection against semidominant genes

	AA	Aa	aa	Total
Initial proportions	p^2	$2pq$	q^2	1
Adaptive values	1	$1 - s$	$1 - 2s$	—
Proportions after selection	p^2	$2pq(1 - s)$	$q^2(1 - 2s)$	$1 - 2sq$

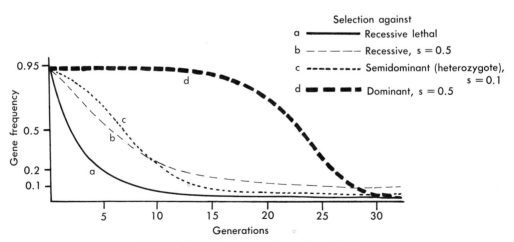

Fig. 13-5. Selection curves for various situations.

Table 13-11. Selection against dominant genes

	AA	**Aa**	**aa**	**Total**
Initial proportions	p^2	$2pq$	q^2	1
Adaptive values	$1 - s$	$1 - s$	1	–
Proportions after selection	$p^2(1 - s)$	$2pq(1 - s)$	q^2	$1 - 2sp + sp^2$

the *Aa* and *aa* genotypes is shown in Table 13-10.

From Table 13-10, we can obtain the following equation that yields the value of the *a* gene after a generation of selection:

$$q_{n+1} = \frac{q_n(1 - s - sq_n)}{1 - 2sq_n}$$

Genes subjected to this type of selection show a rapid decrease in frequency and, barring mutations, can be eliminated from the population by continued selection. In Fig. 13-5, curve "c" shows the effect of selection against a semidominant gene when the selection coefficient against the heterozygote is 0.10 and that against the recessive homozygote is 0.20. From the curves, we can see that this rather mild level of selection is actually more effective than the selection against a semilethal recessive.

Selection against dominant genes

Our consideration of the situation in which selection acts against the dominant gene will vary from our preceding discussions only insofar as the equation considers what will happen to the *A* gene rather than to *a*. Therefore, we shall be using the values of p rather than q. Where selection is against the dominant gene, the *aa* genotype will have an adaptive value of 1. We shall examine the case in which there is complete dominance and the adaptive value of both homozygous dominant and heterozygote is $1 - s$. The effect of selection on the proportions of the *AA* and *Aa* genotypes is shown in Table 13-11.

From Table 13-11, we can obtain an equation that yields the value of the *A* gene after a generation of selection. The equation is given below:

$$p_{n+1} = \frac{(1 - s)p_n}{1 - 2sp_n + sp_n^2}$$

The simplest case is that of a dominant lethal which takes effect before the age of reproduction. Anyone having such a gene has an adaptive value of $W = 0$. In the equation just given, $s = 1$ and $p_{n+1} = 0$. In this situation, the dominant gene is eliminated in one generation of selection and can only arise in the next generation through mutation.

When a dominant gene is not lethal before sexual maturity but takes its effect sometime during the reproductive period of the individual's life, some of the genes will be transferred to successive generations. The size of the fraction of genes that survive selection will depend on the value of s. In general, selection against dominant genes is quite sharp, and the gene can be eliminated, barring mutation, from the population. In Fig. 13-5, curve "d" shows the effect of selection against a dominant semilethal gene whose selection coefficient, s, is 0.5. The curve may appear to be in error, for when the frequency of the dominant gene is very high, selection is not very effective against it. The situation is easily explained. When gene *A* is very high in frequency, most of the *a* gene is in the heterozygous combination *Aa* ($2pq$), and very little of *a* is in the homozygous combination *aa* (q^2). As a result, the complete survival and reproduction of *aa* individuals has little effect on the gene frequency of *A* and *a* in the next generation. However, once the *a* genes become more frequent in the population (about 15%), selection begins to alter the subsequent gene frequencies by appreciable amounts.

The question may arise as to how genes that are so deleterious, as are lethals and semilethals, can ever achieve such high frequencies in a population. The answer is that these are theoretical curves, any part of which can be used as the situation requires. It is conceivable that a change in environment might rather suddenly make the previously normal allele a lethal or semilethal and bring into action the entire curve. In terms of present-day frequencies of most deleterious genes, the parts of the curve dealing with gene frequencies up to 50% are the more important.

Although a detailed discussion of selection against X-linked genes, dominant or recessive, is not included, such selection resembles that which operates against dominant alleles. This stringent selection against X-linked genes follows from the fact that X-linked recessive genes have the same effect in the XY individual as if they were dominant.

FACTORS MODIFYING SELECTION AGAINST A GENE

Other than mutations, there are two phenomena that tend to counter the effect of selection acting against a deleterious gene. One is a form of selection in which the heterozygote is favored and both homozygotes are selected against. The other phenomenon is an overcompensation in family size, found in individuals with a deleterious gene. We shall examine these two modifying factors of selection and see how they actually serve to maintain deleterious genes in a population.

Selection favoring heterozygotes

Up to now we have considered only the type of situation in which selection has acted against one of the homozygotes or against one of the homozygotes and the heterozygote. In these situations, the frequencies of the genes involved have approached either 0 or 1 as a limit. Now we shall examine the case of selection against the *AA* and the *aa* individuals but not against the *Aa* genotype. The severity of selection against the two homozygotes need not be the same. We shall, therefore, take as the adaptive value of *AA*, $W_1 = 1 - s_1$, and as the adaptive value of *aa*, $W_2 = 1 - s_2$. The adaptive value of the *Aa* genotype will be 1. The effect of selection on the proportions of the *AA* and *aa* genotypes is shown in Table 13-12.

From Table 13-12 we can obtain an equation that yields the value of the *a* gene after a generation of selection. The equation is as follows:

$$q_{n+1} = q_n + \frac{p_n q_n (s_1 p_n - s_2 q_n)}{1 - s_1 p_n^2 - s_2 q_n^2}$$

The value of q increases or decreases according to whether the factor $s_1 p$ is greater or less than $s_2 q$. There will be no change in

Table 13-12. Selection favoring heterozygotes

	AA	Aa	aa	Total
Initial proportions	p^2	$2pq$	q^2	1
Adaptive values	$1 - s_1$	1	$1 - s_2$	–
Proportions after selection	$p^2(1 - s_1)$	$2pq$	$q^2(1 - s_2)$	$1 - s_1 p^2 - s_2 q^2$

the value of q when $s_1p = s_2q$, and a stable equilibrium will result, in which both alleles will remain in the population indefinitely. The equilibrium values of A and a are given by the following equations:

$$\hat{p} = \frac{s_2}{s_1 + s_2}; \quad \hat{q} = \frac{s_1}{s_1 + s_2}$$

These equilibrium values are independent of the initial gene frequencies of the population and are entirely determined by the values of the selection coefficients against the homozygotes. The superiority of the heterozygote over the two homogyzotes is known as *heterosis*. One of the more interesting examples of selection favoring heterozygotes occurs when both homozygotes are lethal. Under such conditions, only the heterozygote survives in every generation. This is called a *balanced lethal system*. We discussed an example of this system in Chapter 11 when we considered the T locus in mice. It will be recalled that T/T homozygotes are lethal and t^0/t^0 are lethal, while T/t^0 are both viable and tailless. A cross of two tailless mice yields the following:

Parents	T/t^0	T/t^0		
Offspring	T/T	T/t^0	t^0/T	t^0/t^0
	(die)	(viable)		(die)

In this case, both s_1 against T/T and s_2 against t^0/t^0 have values of 1. The equilibrium frequency of T and t is 0.5, which is the frequency of all genes or chromosomal rearrangements involved in balanced lethal systems.

An example in humans of selection favoring heterozygotes is well illustrated by the gene that in double dose causes the hereditary disease known as sickle cell anemia (Chapter 10). The disease is usually fatal while the individual is quite young. Investigators have found, however, that persons who have a single sickle cell gene are more resistant to a lethal type of malaria (falciparum) than those in the population who do not have the sickle cell gene. As a result, individuals who are heterozygotes for the sickle cell gene are better able to sur-

vive and have larger families in malaria-infested areas than the rest of the population. Thus this gene, which is lethal in double dose, is maintained in some populations in quite high frequency by the superiority it confers to its carriers in heterozygous condition. In environments free of malaria, as expected, the gene is rare or, as in the case of American Negroes, seems to be rapidly decreasing in frequency.

The condition in which selection favors heterozygotes permits the different alleles of a gene to remain in stable equilibrium and retain substantial frequencies in a given environment. This results in what is known as *balanced polymorphism,* which is the maintenance of deleterious homozygotes in the population through the agency of heterozygote superiority. Balanced polymorphism furnishes genetic plasticity to a species and may be very important to the survival of the species under changing environments.

Overcompensation

A second phenomenon that tends to undo the effects of selection against a gene is an overcompensation in family size. An example of this occurs in the instance of persons possessing the dominant gene that, in heterozygous condition, causes the disease called Huntington's chorea. This is a disease of the nervous system that begins to show its effects when the individual is about 35 years old. The symptoms begin as occasional involuntary muscle spasms that progress until the individual is completely helpless, being unable to speak, walk, write, dress, or feed himself because of the uncontrolled twitching movements of the arms, legs, body, head, and face. A mental deterioration also occurs, and the person must eventually be committed to a mental institution. The disease is uniformly fatal. Persons coming from families in which this disease has occurred are acutely aware of its symptoms and consequences. They cease becoming parents once the disease has manifested itself. However, since the gene acts long after the reproductive period of the individual has begun, investigators have been able to compare family sizes of affected and unaffected branches of a family having the gene. This has been done for a pedigree extending over 120 years. It was found that the mean number of children of affected persons in this pedigree was 6.07, whereas that of unaffected persons was 3.33. The difference is statistically significant. The factors involved in this difference of reproductive rate are unknown. However, the consequences are clear: the higher reproductive rate serves to maintain the gene in the population despite its detrimental effects on its carriers. It should be realized that the phenomenon of overcompensation can occur only in situations where birth control is practiced. If family size is not limited by any of the genotypes involved, one would expect that the carriers of the detrimental gene would have fewer offspring than their normal counterparts.

• • •

We have discussed some of the factors that modify the effects of selection against a gene. We shall now consider the results of the interaction of mutation and selection.

INTERACTION OF MUTATION AND SELECTION

In preceding sections, we learned that mutation and selection, if unopposed, would change gene frequencies over varying periods of time. If mutation and selection exert their pressures in the same direction, the change in gene frequencies will be faster than that indicated in our earlier discussion. If these two forces oppose each other, their independent effects may be cancelled, and a stable equilibrium of the various alleles can result. Some examples of equilibria are examined in this section. We shall first look at situations in which the gene is rare in the population. When this is the case, we can ignore the factor of reverse mutation, which will simplify our equations a good deal. Then we shall examine the case of a gene that is frequent in the population. In this situation the factor of reverse mutation has to be considered.

Mutation and selection involving rare recessives

Let us assume that selection is against the recessive homozygote *aa,* with an intensity s, and that mutations from *A* to *a* occur at the rate of μ per generation. Selection will diminish the frequency of *a* (q), but mutation will increase it. The net change in q per generation will be as indicated below:

$$\Delta q = \mu p - spq^2$$

The second term, μp, represents the gain in q through incoming mutations; the third term, spq^2, represents the loss of q due to selection. If these two opposing pressures cancel each other, there will no longer be any change in gene frequencies and $\Delta q = 0$. The equilibrium value of q is shown as follows:

$$\hat{q} = \sqrt{\frac{\mu}{s}}$$

When the *aa* genotype is lethal (s = 1), the equilibrium value of q becomes

$$\hat{q} = \sqrt{\mu}$$

and the proportion of *aa* individuals in the population will be equal to the mutation rate:

$$q^2 = \mu$$

Mutation and selection involving dominants

If selection is against the homozygous dominants only, *AA,* a stable equilibrium may be established by mutations from the recessive, *a,* to the dominant, *A,* gene. The equation for equilibrium is the same as that for recessives, with the substitution of p for q and the substitution of the mutation rate of *a* to *A* (*v*) for μ. The equation for equilibrium follows.

$$\hat{p} = \sqrt{\frac{v}{s}}$$

Should the dominant gene be homozygous lethal (s = 1), the frequency of the dominant gene is the following:

$$\hat{p} = \sqrt{v}$$

If the heterozygote is also nonviable, the

frequency of the dominant lethal gene is equal to the mutation rate to that gene.

$$\hat{p} = \nu$$

In this last example, the dominant genes present in the population have newly arisen by mutation and are immediately eliminated.

Mutation and selection including reverse mutation

When the frequency of a gene in a population is small, the effect of reverse mutation is negligible and need not be considered. If q is of intermediate magnitude, then mutations in both directions should be taken into account unless mutations in one direction are known to be much lower than in the other. The total effects of mutation in both directions and of selection against recessive homozygotes are indicated below:

$$\Delta q = \mu p - \nu q - spq^2$$

At equilibrium $\Delta q = 0$ and the equation can be solved. The results show that there may be one or two stable equilibrium values, depending on the values of μ, ν, and s in the specific case.

OUR LOAD OF MUTATIONS

The title of this section owes its origin to a stimulating discussion by H. J. Muller in 1950 on the genetic penalty of increasing mutation rates through the widespread exposure of populations to atomic radiations. The overall argument goes as follows. Mutations occur spontaneously and will, in most cases, result in the individual's being heterozygous for the mutant gene. These mutants, like those produced by radiation, are usually deleterious to their carriers. This would be the expected effect of producing random changes in a highly integrated system. Since most of these mutations are not lethal, they will persist for a number of generations. These mutants will then constitute a *mutation load* that the population will carry. This mutation load will decrease the probability that the affected individuals

will survive and reproduce. The decreased probability of survival and reproduction will be manifested not only in a reduction of the average life-span in the population but also in an increase of its level of physical and mental illnesses.

Muller's argument rests on the assumption that all mutations, regardless of degree of severity, have equal consequences for the population. Let us examine this premise. The fully lethal mutation that unconditionally kills the first individual in whom it finds expression would appear to be the worst mutation that can occur. However, consider, for comparison, a slightly detrimental mutation that reduces the chances of its carrier's surviving and reproducing by only 1% (s = 0.01). This gives the average individual having it only 1/100 greater chance of death or failure of reproduction than the average individual not carrying the mutation. With the population at equilibrium, the carrier would be expected to survive and pass along this gene in 99 out of 100 cases. This gene would then tend to continue on down through some 100 individuals of later generations, on the average, before it had piled up enough of these 1/100 risks of dying out to be eliminated. Therefore the gene with so minor an effect does as much damage in the population, in the end, and gives rise to as much mutational load as the fully lethal mutation. Each results in a single death. The same is true of detrimental mutations of every grade. They differ only in the extent to which they divide up and distribute their load and their risk of death or sterility. This mutational damage results in what have been called *genetic deaths*. A genetic death is the extinction of a gene lineage through the premature death or reduced fertility of some individual bearing the gene. Consider a large population whose size remains constant from generation to generation. If it starts out with a certain number of harmful mutant genes, this number will decrease in each generation because the carriers of the mutants have a smaller chance of surviving and reproducing than individuals without them. This process will continue until all the mutant genes are eliminated.

Under the force of selection described above, the persistence of deleterious mutations in a population depends on recurrent mutation. An equilibrium is then achieved in which the elimination of deleterious mutants, by selection, is balanced by the production of new mutants.

The relationship between mutation rate and mutation load is very important. Assume, as above, a detrimental gene with the selection coefficient s = 0.01. Such a mutant gene will persist in the population an average of 100 or $1/s$ generations. If the normal gene has a probability of μ of mutating each generation, then in each generation the proportion of individuals carrying newly mutated genes at this point on the chromosome will be 2μ. The factor "2" is due to the two sets of genes in everyone. If the mutant is rare, as is usually true, reverse mutations can be safely ignored. Since each mutant persists an average of $1/s$ generations, the population will eventually contain an average of $2\mu/s$ mutant genes of this type. However, the effect of each mutant gene is to decrease reproductive fitness by a fraction, s. Hence the average reduction in fitness of the population is $(2\mu/s)(s)$, or simply 2μ. The interesting result is that the s factors cancel out. Thus the effect of mutation on a population is proportional to the mutation rate and is independent of the effect of the mutant on the individual. When the mutants of all genes are considered (if they act independently and are rare, relative to normal genes), their total effect is approximately the sum of the individual effects. Therefore the fitness of an equilibrium population is decreased, because of mutation, by a fraction approximately equal to the total mutation rate per individual or twice the gametic mutation rate. Under these conditions, it matters little what the severity of the mutant may be. All new deleterious mutants add to the mutational load of the population and result eventually in a genetic death. Such factors as heterosis and overcompensation add to the genetic load

by increasing the frequency of the deleterious genes and thereby increasing the number of gene lineages that must terminate in the death of some individual possessing the gene.

MIGRATION

Another method by which the frequencies of alleles in a population can be changed is through *migration*. The effect of migration on gene frequency can be exceedingly small. This is especially true if two populations are located close to one another. Such populations tend to be nearly alike in their gene frequencies. Although a great deal of migration may take place between these populations, it will have little effect on the frequencies of the various alleles. A situation of greater consequence occurs when two widely separated populations exchange migrants. Such populations may differ considerably in their gene frequencies, and migrants between them may alter these frequencies to a great extent. The most obvious effect of continued intergroup migration is to make the gene frequencies of the various groups more nearly alike and thus, in the absence of any countermeasures, render the total population more homogeneous.

In considering the equation for the effects of migration on gene frequency, let us designate Q as the frequency of a gene in one population that we shall call "migrant," q as the frequency of the gene in a second population that we shall call "native," and m as the number of migrants from the first to the second population, expressed as a fraction of the size of the second population. It seems clear that, after migration, the second population will consist of mQ genes acquired from the first population (migrants) plus (1 − m) q genes remaining from itself (natives). Therefore the frequency of the gene being studied after migration will be

$$q' = mQ + (1 - m)q$$

Equations have been derived that include the interactions of mutation and selection with migration, but these are beyond the scope of our discussion.

A most interesting use of the above equation was reported by Glass and Li in 1953. Surveys of white and Negro populations have shown that the frequencies of the allele Rh^0 (a member of the rh series) are American Negro, 0.45; American white, 0.03; and African Negro, 0.63. The assumption was made that American Negroes carry a mixture of genes obtained from their African Negro ancestors and from white Americans. It was further decided, for the purpose of this study, to equate the ancestral African Negroes with the modern African Negroes. Under these assumptions the equation was written as follows:

$$0.45 = 0.03m + 0.63(1 - m)$$

Solving the equation, we find that $m = 0.30$. This means, assuming the findings for the Rh^0 allele apply to all other genes, that 30% of the genes of the American Negroes have been acquired from the American white population, while 70% have been retained from the ancestral African Negroes. If one further assumes that the American Negroes have been in the United States for some 300 years and that an average generation time can be taken as 30 years, then the average rate of migration of genes between the two groups has been 3% per generation. This estimate has to be considered with some reservation, since the amount of gene migration, m, has undoubtedly varied from generation to generation. We can take this analysis one step further. If the ultimate effect of intergroup migration, as stated earlier, is to render the total population more homogeneous, we can estimate how long it will take for the American Negro to disappear as a distinguishable segment of the American population. Assuming that the gene migration m will remain at 3% per generation, we can expect that some 700 more years will be required for this to occur.

GENETIC DRIFT

The last method that we shall consider by which gene frequencies in a population can be changed is *genetic drift*. This factor represents the chance events that may result in the elimination of an allele from the population ($q = 0$) or in the fixation of that allele in the population ($q = 1$). The basis for genetic drift lies in the fact that the frequencies of alleles among the gametes that give rise to the individuals of the next generation may not represent exactly the gene frequencies of the population from which these gametes were formed. This type of "sampling accident" that can occur in a random-breeding population has more chance of taking place in a small population than in a large one. The association of genetic drift with small populations can easily be seen when we compare the amount of chance variability that is expected with two different-sized populations.

Let us consider two populations, one of 500,000 individuals and the other of 50. In addition, let us assume that in both populations a gene is represented by two alleles, A and a, which are of equal frequency, so that $p = q = 0.5$. The larger population has been formed from 1,000,000 gametes, but the smaller came from only 100. Gametes in both populations will reflect the gene frequencies of the original populations within certain limits. The expected limits of deviation of the gametic frequencies from the populational frequencies is given by the standard error. The expected numbers of A and a alleles in a sample of 1,000,000 gametes with the standard error of the number is indicated here:

$$500,000 \pm \sqrt{\frac{500,000 \times 500,000}{1,000,000}} = 500,000 \pm 500$$

(i.e., $q_{n+1} = 0.5 \pm 0.001$)

For the smaller population, the expected numbers of A and a alleles in a sample of 100 gametes, with the standard error of the number, are calculated as below:

$$50 \pm \sqrt{\frac{50 \times 50}{100}} = 50 \pm 5$$

(i.e., $q_{n+1} = 0.5 \pm 0.1$)

In the case of the large population, the standard error of 500 represents only 0.1% of the gametes. On the other hand, the standard error of 5 in the small population represents 10% of the gametes. Under these conditions, the proportions of the alleles *A* and *a* will remain quite constant from generation to generation in the large population but will be much more variable in the small one. Should a series of deviations in allele frequency occur in the same direction, by chance, the elimination of one of the alleles and the concomitant fixation of the other could be expected.

Laboratory experiments designed to test the occurrence of genetic drift were reported by Kerr and Wright in 1954. They set up a large number of breeding lines of *D. melanogaster* in which the progenitors of each line and the parents of each generation consisted of 4 randomly chosen males and 4 randomly chosen females. At the beginning of the experiment, each line contained a mutant gene whose frequency was 50%. The lines were continued for 16 generations unless fixation of the mutant or its wild type allele occurred sooner. In most of the experiments, the action of selection against one or both of the homozygotes did not permit genetic drift to take place. However, in the case of the sex-linked gene *forked*, selection against the mutant was sufficiently small to permit random fluctuations in gene frequency to occur. Ninety-six lines of flies were started, and by the sixteenth generation 29 lines had become fixed for *forked*, 41 lines had become fixed for the wild type allele, and 26 lines remained unfixed. These results show somewhat less fixation of the mutant gene than expected, but this is probably due to the small amount of selection against the gene. Despite the action of selection, the experiment did demonstrate the occurrence of genetic drift in small laboratory populations.

Good examples of the action of genetic drift in natural populations are few in number, due to the great problems involved in obtaining organisms from nearly identical environments that have been derived from a common ancestral stock. A report on the possible occurrence of genetic drift in a natural population was made by H. Grüneberg

in 1961. He studied the frequencies of eight skeletal variants in five populations of Delhi (India) rats, taken from grain shops in one part of the city. He found that the five groups differed statistically from one another in a number of skeletal variants. This implied that the five groups differed also in their genetic constitutions, since the grain shops appeared to be extremely similar and provided approximately the same range of foodstuffs. It was assumed that the rat population of each grain shop had its origin in a small number of foundation animals from some ancestral population and that the unique skeletal characteristics of each group of rats reflected the occurrence of genetic drift.

Genetic drift in a human isolate

It has been most difficult to positively identify a human population in which genetic drift has occurred. The population has to consist of a very small group of individuals who have, for a number of generations, married only among themselves. A small population that has been reproductively restricted from receiving genes from surrounding groups is called an *isolate*. The rare instances in which such small human groups exist usually involve religious sects that, as a result of their marriage customs, have been reproductively isolated from the surrounding population. To test such a group for the occurrence of genetic drift, one has to know and be able to examine the descendants of the parent population from which the small group originated, and in addition, one must be able to also examine the population that now surrounds the isolate.

An isolate that lends itself to this type of study is to be found in the "Dunker" community in Pennsylvania. It is a Baptist group that migrated from Germany about 250 years ago. The present community consists of some 298 individuals. The results of a genetic study of this group was reported by Glass and co-workers in 1952. Data were gathered on seven different genetic traits of which we shall consider two in some detail:

ABO and MN blood groups. The blood types of the members of the Dunker community were ascertained, and the frequencies of the blood group alleles were calculated. From other studies, the frequencies of the *ABO* and *MN* alleles were obtained both for the German population now living in the region from which the Dunkers migrated and for the present surrounding American population in Pennsylvania. The calculated allelic frequencies were as follows:

Population	A	B	O	M	N
U. S. A.	26%	4%	70%	54%	46%
German	29%	7%	64%	55%	45%
Dunker	38%	3%	59%	65%	35%

A statistical test of the data indicated no significant difference in allelic frequencies between the surrounding American population in Pennsylvania and the German population. However, the Dunker population did differ significantly from both. It seems quite clear that genetic drift has occurred in the Dunker population, resulting in distinctive allelic frequencies for the blood group genes listed above. However, it must not be supposed that all gene frequencies will change whenever conditions permit genetic drift to occur. In the present case, for example, the various *Rh* alleles in the Dunker group showed a frequency pattern that was not statistically different from that of the other populations. The same was true of the number of left-handed versus right-handed individuals. The existence of some gene frequencies that were the same in the Dunker and in the other populations must be recognized as actually a support for the hypothesis that genetic drift has occurred in this human isolate, since by its very nature, genetic drift is a random process and will not affect all genes in the same way.

• • •

In our discussion of genetic drift we have used the terms *large* and *small* in describing populations without attempting to set any limits on their sizes. This raises the more general issue of the *effective size* of a population. In a natural population, the total number of individuals of all ages may be very large, but not every individual reaches sexual maturity and mates. Even those that mate do not necessarily leave offspring that survive to maturity in the next generation. Thus the effective size of a population may be much less than its absolute number. Unfortunately, in most cases direct information on the effective sizes of natural populations is difficult to obtain. As a result, geneticists have not been able to decide on a clear-cut line between "large" and "small" populations.

SUMMARY

In this chapter we have studied (1) the origin of genetic diversity and (2) the structuring of genetic diversity. The origin of genetic diversity is in essence mutation, which for this purpose we consider as any inheritable change. We have examined three different patterns of mutation (single mutation, recurrent mutation and reverse mutation) and the consequences of each for the frequencies of different alleles in the population. We then reviewed the evidence for genetic and viral control of mutation rates and concluded the section with an analysis of the effects of various environmental factors (temperature, radiation, and chemicals) on mutation rates.

The structuring of genetic diversity is governed by such forces as selection, migration, and genetic drift. We first considered selection against different types of genes (recessive, semidominant, and dominant), as well as the effectiveness of selection in a eugenics program. Factors such as heterosis and overcompensation that modify the effects of selection against a gene were then reviewed. This section of the chapter continued with an examination of the interaction of mutation and selection in establishing equilibria for recessive and dominant genes and the population's load of mutations. The chapter closed with a review of migration and genetic drift and the various ways in which these factors affect gene frequencies. Having studied the factors involved in the origin and structuring of genetic diversity, we shall now consider the segregation of genetic diversity as seen in race and species formation.

Questions and problems

1. Discuss genetic and viral control of mutation.
2. Assess the ramifications on the genetic structure of human populations, resulting from medical advances that allow individuals harboring deleterious genes to survive and reproduce.
3. Discuss the effect of migration on gene frequency.
4. Discuss how the phenomenon of heterosis serves to modify the process of selection.
5. What exactly is the portent of the Hardy-Weinberg equilibrium? What factors are responsible for deviations from the expected equilibrium?
6. Discuss the effect of genetic drift on gene frequency.
7. Assume that gene B mutates to b at a rate of $1(10)^{-5}$. If BB and Bb are equally fertile, and 0.1% of the population is homozygous recessive, what selection pressure exists against bb?
8. The frequency of a gene a in a certain native population of $D.$ $pseudoobscura$ is 0.6, and the frequency of that same gene in a migrant population is 0.8. If the frequency of new immigrants is 1/100, what will be the frequency of gene a in the native population after one generation?
9. Assume the mutation rate, μ, of A to a is 0.00003, and the mutation rate, ν, of a to A is 0.00002. With all other forces disregarded, what will be the equilibria values of A and a?
10. Assume that the Hardy-Weinberg equilibrium is operating. What is the frequency of gene X in each of the populations listed below if its allele, x, is homozygous in the following percentages of individuals in population:

    a. 48%   c. 39%
    b. 3%   d. 23%

11. In a randomly mating population in which the Hardy-Weinberg equilibrium is operative, the frequencies of gene X and its allele, x, are 0.2 and 0.8, respectively.
    a. What proportion of this population is heterozygous with respect to these genes?
    b. What proportion of this population will be homozygotes?
    c. After two generations of mating hybrids with hybrids only, what percent of this population will be homozygotes?
    d. What ramifications would the situation described in c have on the composition of the population's gene pool?
12. The frequency of children homozygous for the recessive lethal gene causing infantile amaurotic idiocy is approximately 1 in 25,000. What is the frequency of heterozygotes in the population?
13. The ability to taste PTC is due to the dominant allele T. Among a group of 215 students, 150 could detect the taste of PTC, and 65 could not. Calculate the gene frequency for t and T.
14. Among a group of 890 students, 70.5% could taste PTC.
    a. What percentages of the students were respectively heterozygous, homozygous recessive, and homozygous dominant?
    b. What proportion of the tasters marrying nontasters might expect some offspring unable to taste PTC?
    c. What proportion of the tasters marrying nontasters could expect only offspring able to taste PTC?
15. Among a population of 11,535 people, the following distribution of blood types was noted: type A—4841; type O—5200; type B—1082; type AB—412. Calculate the gene frequencies of the three alleles, I^A, I^B, and i.
16. Discuss how the phenomenon of overcompensation serves to modify the process of selection.
17. Compare recessive and dominant genes with respect to the ease with which they can be eliminated from a population.
18. When the relative frequency of alleles in a population is changed, what forces operate to establish a new equilibrium?
19. Among a population of 300 Africans, the frequencies of I^A, I^B, and i alleles are 0.03, 0.01, and 0.96, respectively. What percent of the population has type A blood? type AB? type O? type B?
20. Discuss the following facets of mutation: frequency, direction, and type (spontaneous versus induced), in terms of their ramifications in population genetics.

References

Crow, J. F., and Kimura, M. 1970. An introduction to population genetics theory. Harper & Row, Publishers, New York. (A highly detailed consideration of the mathematics involved in population genetics theory.)

Dahlberg, G. 1948. Genetics of human populations. Advances Genet. 2:69-98. (A discussion of the factors influencing the genetic structure of human populations.)

Dishotsky, N. I., Loughman, W. D., Mogar, R. E., and Lipscomb, W. R. 1971. LSD and genetic damage. Science **172**:431-440. (A thorough review of the literature on this subject from 1967 to 1970.)

Falconer, D. S. 1960. Introduction to quantitative genetics. The Ronald Press Co., New York. (A fine discussion of population genetics and selection is included.)

Fisher, R. A. 1930. The genetial theory of natural selection. Clarendon Press, Oxford. (A pioneer work in the quantitative treatment of the effects of selection.)

Glass, B. 1954. Genetic changes in human populations, especially those due to gene flow and genetic drift. Advances Genet. **6**:95-139. (An examination of the forces of genetic drift and gene flow on human population genetics.)

Gowen, J. W. 1952. Heterosis. Iowa State College Press, Ames. (A symposium dealing with discussions of the problems of heterosis.)

Hardy, G. H. 1908. Mendelian proportions in a mixed population. Science **28**:49-50. (One of the classic statements of the binomial distribution of genotypes.)

Lerner, I. M. 1954. Genetic homeostasis. John Wiley & Sons, Inc., New York. (An interesting discussion of the problem of heterosis and its evolutionary ramifications.)

Lerner, I. M. 1958. The genetic basis of selection. John Wiley & Sons, Inc., New York. (A very good textbook dealing with material presented in this chapter.)

Li, C. C. 1955. Population genetics. University of Chicago Press, Chicago. (A clear and excellent coverage of the mathematics of population genetics.)

Population genetics. 1955. Cold Spring Harbor Symp. Quant. Biol. **20**:1-346. (A series of excellent articles dealing with various aspects of population genetics.)

Stern, C. 1943. The Hardy-Weinberg law. Science **97**:137-138. (A discussion of the Hardy-Weinberg law of genetic equilibrium, including a translation of part of the original statement set forth by Weinberg in 1908.)

Wallace, B. 1968. Topics in population genetics. W. W. Norton & Co., New York. (A well-written, nonmathematical analysis of selected areas of population genetics research.)

Wright, S. 1968. Evolution and genetics of populations. Vol. 1, Genetic and biometric foundations. University of Chicago Press, Chicago. (The first of a three-volume series that will stand as a classic analysis of the field of population genetics.)

Wright, S. 1969. Evolution and genetics of populations. Vol. 2, The theory of gene frequencies. University of Chicago Press, Chicago. (The second volume in the series described above.)

14 Race and species formation

In this chapter we shall continue our study of population genetics. We have considered the forces that are involved in the origin of genetic diversity (mutation) and those involved in the structuring of genetic diversity (selection, migration, and genetic drift). We now want to examine some natural populations of organisms to see how genetically diverse they are from one another. We also want to discover the factors that result in one particular genotype's being favored over another. This may lead us to laboratory experiments where the various factors can be studied one at a time. Finally, we want to analyze the mechanisms that keep populations of organisms that are adapted to one environment from exchanging genes with organisms adapted to a different environment. This last topic constitutes the study of the segregation of genetic diversity *(species formation)*.

GENETIC DIVERSITY IN DROSOPHILA POPULATIONS

Most species have a reasonably wide distribution over at least a portion of the earth. Within its range, each species is broken up into a number of populations that vary in the amount of contact they have with one another. It is of interest to establish whether these populations differ from one another in their frequencies of various genotypes and, if so, to what extent. A species whose populations have been well studied for their genetic diversity is *Drosophila pseudoobscura*. Most of the investigations on this species have been conducted by T. Dob-

zhansky, his students, and co-workers, beginning in 1935 and continuing to the present. *D. pseudoobscura* is a forest-dwelling form whose habitat has been left relatively undisturbed by man. Its range lies in the western part of North America from British Columbia southward to Guatemala and from the Pacific Ocean eastward to the Rocky Mountains.

This species of fly has five pairs of chromosomes in its karyotype. These chromosomes can be easily studied in late larval salivary gland preparations. (See Fig. 11-3.) The bands of each chromosome are of different characteristic widths, and the banding pattern of any given chromosome is easily identified. The third chromosome pair of *D. pseudoobscura* attracted the attention of investigators when the sequence of bands along this chromosome was found to vary, depending on the population from which the fly was obtained. Some fifteen different band sequences were found, variously distributed, within the range of the species. These fifteen band patterns were analyzed and found to be inversions of one another. As we discussed in Chapter 7, an inversion represents the result of (1) two breaks occurring in a chromosome, (2) a twisting of the center segment 180 degrees, and (3) a healing of the three segments, reconstituting the chromosome with presumably no loss of genetic material. The change in gene sequence, as seen in a change of banding pattern, is illustrated below:

I. A B C D E F G H

II. A E D C B F G H

The underlined portion represents the inverted section of genes in the chromosome. The gene arrangement of the chromosome can be further altered by the occurrence of a second inversion that overlaps the first, thus:

III. A E D G F B C H

Overlapping inversions permit us to describe the possible courses of the evolution of these inversions. We can postulate from the above illustrations that the descent relationships can be any of the following

I → II → III or III → II → I or III ← II → I

but that there is no other possibility. In other words, the relationship I ⇌ III is impossible except with II as an intermediate link. An extensive analysis of the various banding patterns showed that the fifteen gene arrangements form an interlocking series of overlapping inversions. These have been arranged in a phylogenetic series as shown in Fig. 14-1. The reliability of this method of analysis is attested to by the fact that in some of the early investigations the existence of certain inversions and the sections of the chromosome involved were predicted and later discovered. Note in Fig. 14-1 that one arrangement (hypothetical) remains undiscovered.

Geographical distribution of various gene arrangements

All fifteen gene arrangements mentioned above are not distributed throughout the range of the species. In any one locality, seven chromosome types has been the maximum number observed; in some places only one gene arrangement is present, all the flies of that locality being homozygous for one of the chromosome types. It was also found that each gene arrangement differs in frequency from locality to locality. Collections of flies were made along various transects across the distribution area of the species. The chromosomes of the flies' salivary gland cells were analyzed, and the frequencies of the various gene arrangements were recorded. One such series of collections was made at various localities along the United States–Mexican border, which forms a transect in a generally west-east direction. The frequencies of the five most common gene arrangements in this region are shown in Fig. 14-2. It can be seen that the frequency of the Standard gene arrangement decreases from west to east but does not completely disappear. Arrowhead reaches very high frequencies in Arizona and New Mexico and declines both to the east and to the west. Pikes Peak shows a very smooth gradient. It does not occur in California, is rare in Arizona, be-

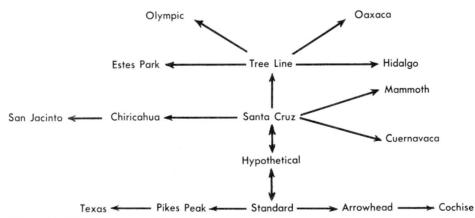

Fig. 14-1. The phylogeny of the gene arrangements in the third chromosome of *Drosophila pseudoobscura.* (From Dobzhansky, T., and Epling, C. 1944. Carnegie Inst. Wash. Publ. 554, pp. 1-183.)

comes common in New Mexico, and is the predominant arrangement in Texas. Chiricahua has a low frequency along the Pacific coast, is more common in the mountains of California, and then declines in frequency from California eastward, being present only in the western part of Texas. Tree Line shows gradients decreasing in frequency eastward from California, being rare or absent in the middle of the transect, and increasing eastward from Arizona.

Clearly, each of the above common gene arrangements has a distinctive frequency pattern along the west-east transect. It is also apparent that some gene arrangements (Arrowhead and Standard) are more frequent than others (Chiricahua, Santa Cruz, and Tree Line) in virtually every locality. Finally, we note that no chromosome type has the same frequency throughout the transect. The fact that populations living in different habitats often differ in the relative frequencies of their gene arrangements means that carriers of different gene ar-

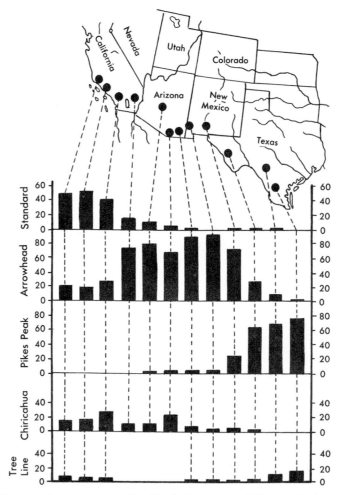

Fig. 14-2. West-east transect across the distribution area of *D. pseudoobscura,* along the United States–Mexican border. The heights of the black columns symbolize the relative frequencies of the third-chromosome gene arrangements, the names of which appear on the left. The numerals indicate percentages of the respective gene arrangements in the populations. Black spots on the map show the geographical origin of the population samples. (From Dobzhansky, T., and Epling, C. 1944. Carnegie Inst. Wash. Publ. 554, pp. 1-183.)

rangements may be favored or discriminated against by different environments. Not enough is known about the environmental characteristics of each locality to indicate the factors that favor one gene arrangement over the other. However, temperature and rainfall differences and the biological ramifications of such differences immediately present themselves as subjects for further research.

Collections of flies from transects that run from north to south across the distribu-

tion area of *D. pseudoobscura* were also made, and the frequencies of the various gene arrangements were determined. The data obtained from a transect along the Pacific coast are shown in Fig. 14-3. It will be noted that the Pikes Peak arrangement is absent from this region, while the Santa Cruz arrangement is present and relatively frequent. An examination of the data shows a very different frequency pattern than that seen in the west-east transect. Standard appears to be relatively constant in frequency

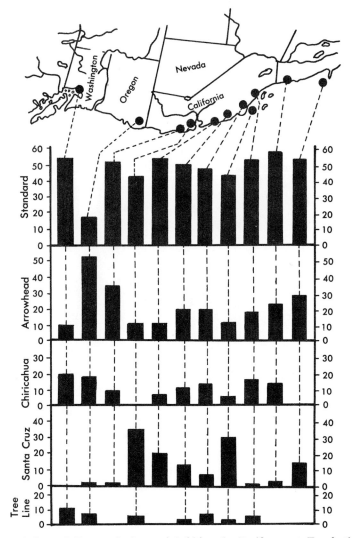

Fig. 14-3. Populations of *D. pseudoobscura* inhabiting the Pacific coast. For further explanation see Fig. 14-2. (From Dobzhansky, T., and Epling, C. 1944. Carnegie Inst. Wash. Publ. 554, pp. 1-183.)

throughout the region except in Oregon. Arrowhead varies irregularly along the coast, being highest in Oregon where it supercedes Standard as the most frequent gene arrangement. The distributions of Chiricahua, Santa Cruz, and Tree Line show no regularity in frequencies along the transect. One fact stands out. Standard has the overall highest frequency along the entire coast. This is in sharp contrast with its extremely limited frequency and distribution away from the coast, along the west-east transect.

Another north-south transect was examined for inversion frequencies. This transect was inland and was made at higher altitudes along the general course of the Cascade Range, the Sierra Nevada, and the mountains of southern California. The data are shown in Fig. 14-4. No geographical gradients of inversion frequencies are apparent on this transect either. However, it is easily seen that Arrowhead has supplanted Standard as the overall most frequent chromosome type. Unfortunately, not enough is

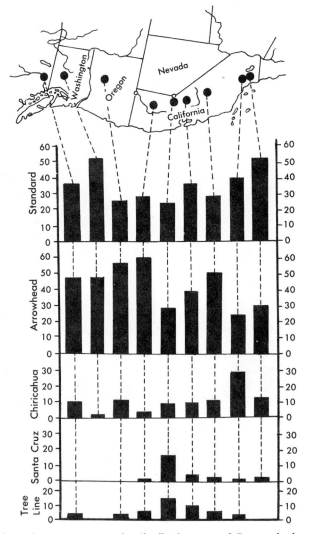

Fig. 14-4. North-south transect across the distribution area of *D. pseudoobscura,* in the region of the Cascade and Sierra Nevada ranges. For further explanation see Fig. 14-2. (From Dobzhansky, T., and Epling, C. 1944. Carnegie Inst. Wash. Publ. 554, pp. 1-183.)

known about the environmental differences between localities at the two elevations to explain the variations in chromosome frequencies that were observed.

Races of D. pseudoobscura

Each of the localities on the three transsects just discussed contains a genetically distinguishable population, based on the frequencies of its different gene arrangements. The population of *D. pseudoobscura* in each locality may then be considered as a separate *race* of the species. Members of one population can migrate and become incorporated into another population. Should a great deal of migration occur between the populations, they will eventually become very similar in the frequencies of their chromosome types. The populations will then be genetically indistinguishable from one another and will be considered members of the same race. Frequencies of gene arrangements are not the only criterion for distinguishing *Drosophila* populations. Various populations may actually have the same frequencies of their chromosome types and yet be genetically distinguishable on the basis of frequencies of different alleles at various loci. One might argue that, if he chose small enough groups of flies, he could establish that each group was genetically distinguishable and hence ought to be considered a

separate race. This problem is obviated by the definition of "population." A population, regardless of how genetically diverse its subgroups, is defined as a reproductive community of sexual and cross-fertilizing individuals among whom matings occur regularly. Races refer only to populations, not to any of their subgroups (clans, families, colonies, etc.) and certainly not to their individual members.

Altitudinal gradients of gene arrangements

In an effort to gain understanding of the kinds of selective forces acting on gene arrangements, studies were made of the frequencies of the various chromosome types at different altitudes in the same region. Three localities on Mount San Jacinto, California, were sampled over the entire year, repeatedly, between 1939 and 1946. Three gene arrangements (Standard [ST], Arrowhead [AR], and Chiricahua [CH]) comprise about 95% of the chromosome types in the region. The frequencies of these three gene arrangements, averaged over the entire year, and the elevations of the three localities are shown in Table 14-1. The table shows that Standard chromosomes are most frequent at the lowest altitude and least frequent at the highest. Chiricahua chromosomes show the opposite relationship, while Arrowhead shows no significant differences

Table 14-1. Percentages of gene arrangements in populations of localities on Mount San Jacinto (California)

Elevation of locality	Gene arrangements			
	ST	AR	CH	Others
4500 feet (Keen Camp)	34	24	38	4
4000 feet (Piñon Flats)	41	25	29	5
800 feet (Andreas Canyon)	58	24	15	3

From Dobzhansky, T. 1947. Adaptive changes induced by natural selection in wild populations of *Drosophila*. Evolution 1:1-16.

in the three localities. These findings revealed the occurrence of altitudinal gradients for gene arrangements and reinforced the idea that the various gene arrangements are important contributors to the adaptive values of their carriers' genotypes.

Seasonal cycles of gene arrangements

The discovery of an altitudinal gradient in the frequencies of gene arrangements led to investigations of possible seasonal cycles in these frequencies. Repeated samplings, month by month, were taken in the three localities on Mount San Jacinto. It was found that the composition of a population may change quite significantly from month to month. These changes are regular and follow the annual cycle of seasons. The data obtained at Piñon Flats are shown in Fig. 14-5. It can be seen that, at the beginning of spring (March), the frequency of Standard is highest in the population. It then declines during the spring only to rise again during the summer and to remain at its highest frequency through the winter. Chir-

icahua has an inverse relationship to Standard, while Arrowhead appears to follow Chiricahua in the spring and early summer, but not thereafter.

We can correlate the results obtained from the altitudinal studies with those obtained from the investigations on seasonal cycles. It will be recalled that at an elevation of 800 feet (Andreas Canyon) the average annual frequency of Standard was highest (58%) whereas at an elevation of 4000 feet (Piñon Flats) its average annual frequency was 41%. Yet at Piñon Flats the frequency of Standard rises to 50% during the summer months. These findings complement each other in that, during the warm season, the populations at higher elevations come to resemble in chromosomal composition the populations of lower altitudes.

Role of gene arrangements in particular environments

The discovery of an altitudinal gradient at Mount San Jacinto, California, raised the question of whether such altitudinal

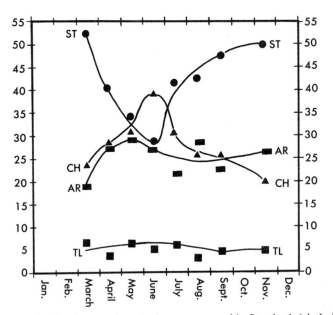

Fig. 14-5. Changes in the frequencies of chromosomes with Standard (circles), Chiricahua (triangles), Arrowhead (horizontal rectangles), and Tree Line (squares) gene arrangements in the population of Piñon Flats, California. Ordinate—frequencies in percent; abscissa—months. Combined data for six years of observation. (From Dobzhansky, T. 1947. Evolution **1**:1-16.)

clines existed elsewhere in the distribution of the species. The discovery also raised the question of whether a particular gene arrangement invariably bestows upon its carrier an advantage or disadvantage in a particular type of environment. Fortuitously, both questions were answered as a result of a single investigation. Populations at different elevations in the region of the Yosemite National Park in the Sierra Nevada in California were studied. This region lies some 300 miles north of Mount San Jacinto. The data obtained are listed in Table 14-2. Here, as was true at Mount San Jacinto, Standard is highest at low altitudes and decreases in frequency as the elevation of the locality increases. However, in the Sierra Nevada, Standard and Arrowhead, and not Standard and Chiricahua, vary in frequency in an inverse relationship. This investigation has demonstrated two points: (1) altitudinal gradients exist in more than a single region of the species distribution and (2) there is nothing inherent in a particular gene arrangement that adapts its carrier for a particular environment.

In Chapter 7 we learned that an inversion acts as a crossover suppressor when the homologous chromosome contains a different gene arrangement. This, then, limits crossing-over to homologous chromosomes carrying the same type of gene arrangement. Under these conditions, each gene arrangement, in fact, becomes a "supergene" whose component parts can be transferred only to other identical chromosomal types. This means that if a mutation adapting its carrier to high altitudes arises in a Chiricahua chromosome (e.g., at Mount San Jacinto), the newly arisen allele will always remain in that gene arrangement. However, since inversions vary only in their gene arrangements and not in their gene contents, this same mutation can arise in an Arrowhead chromosome in some other locality (e.g., Sierra Nevada). Adaptation to a particular environment is usually a function of many interacting genes. However, at each stage of the evolution of a highly adapted genotype, each improvement, no matter how small, will be favored and spread by natural selection throughout the population in succeeding generations. The gene arrangement that contains this beneficial allele will likewise be spread and will be the most likely place where other beneficial alleles could also arise. Thus any slight initial advantage presented to a particular gene arrangement in a given environment will tend to promote the incorporation of further genetic improvements for that environment in the same chromosome type.

Electrophoretic studies of D. pseudoobscura gene arrangements. In recent years a number of population geneticists have

---

Table 14-2. Percentages of gene arrangements in populations in the Sierra Nevada (California)

Elevation of locality	Gene arrangements			
	ST	AR	CH	Others
6800 feet (Aspen Valley)	26	44	16	14
4600 feet (Mather)	32	37	19	12
3000 feet (Lost Claim)	41	35	14	10
800 feet (Jacksonville)	46	25	16	13

From Dobzhansky, T. 1948. Genetics of natural populations. XVI. Genetics **33**:158-176.

Table 14-3. Frequencies of alleles in different gene arrangements of
D. pseudoobscura from Mather, California

Gene arrangement	Pt-10*		Pt-12†		α-Amylase*	
	1.04	1.06	1.18	1.20	0.84	1.00
Standard	1.00	—	0.20	0.80	0.05	0.85‡
Arrowhead	0.90	—§	0.95	0.05	0.05	0.95
Pikes Peak	1.00	—	1.00	—	—	1.00
Chiricahua	0.50	0.50	1.00	—	0.36	0.64
Tree Line	—	1.00	0.95	0.05	0.90	0.10

*From Prakash, S., and Lewontin, R. C. 1968. A molecular approach to the study of genic heterozygosity in natural populations. III. Direct evidence of coadaptation in gene arrangements of *Drosophila*. Proc. Nat. Acad. Sci. U. S. A. **59**:398-405.

†From Prakash, S., and Lewontin, R. C. 1971. A molecular approach to the study of genic heterozygosity in natural populations. V. Further direct evidence of coadaptation in inversions of *Drosophila*. Genetics **69**:405-408.

‡α-Amylase allele 0.92 is found in the remaining 10% of the Standard gene arrangements.

§Pt-10 allele 0.94 is found in the remaining 10% of the Arrowhead gene arrangements.

turned their attention to the proteins that characterize different *Drosophila* populations. In the case of *D. pseudoobscura* this research has focused on the protein variation to be found among the homozygous carriers of different gene arrangements. The experimental technique consists in subjecting flies to electrophoresis and determining the frequencies of different forms of specific proteins. As examples of this type of investigation, we can consider the data reported by Prakash and Lewontin in 1968 and 1971. These studies involved two larval proteins of thus far unknown function, designated as Pt-10 and Pt-12, and the glycogen-splitting enzyme, α-amylase. The data pertaining to the different alleles of each of these proteins, as found in the Mather population of the Sierra Nevada mountains of California, are shown in Table 14-3.

It is quite clear from the data in Table 14-3 that each gene arrangement in the Mather population produces a unique complex of protein types. Although we are presently limited to the alleles of these three proteins, we can nevertheless unequivocally characterize each gene arrangement of this population, based on the frequencies of its various alleles. In similar fashion we should eventually be able to distinguish the Standard, Arrowhead, Chiricahua, etc. gene arrangements of the Mather population from their corresponding inversion types in other *Drosophila* populations. Future investigations will undoubtedly also permit us to characterize each *Drosophila* population both as to its allelic as well as its inversion composition.

Demonstration of the action of selection. The presence of geographic and altitudinal gradients, as well as seasonal variations, indicates that the various inversions are not equally adapted to the different environments in which the species exist. This has raised the question as to the method by which two or more gene arrangements can be maintained in a population, since the adaptively superior chromosome type would be expected to eliminate the adaptively inferior type. One mechanism that would provide for the maintenance of the different gene arrangements is heterosis (superiority of the heterozygotes over the homozygotes). The operation of such a mechanism can be demonstrated through the use of the Hardy-Weinberg equilibrium.

We can analyze frequencies of gene arrangements much as we do the individual genes. Since we can distinguish cytologically both homozygotes and heterozygotes for the various types of chromosomes, the analysis of chromosomal frequency follows the pattern used for allele frequencies of the *MN* locus. Once the chromosome frequencies have been estimated, the expected distribution of zygotic genotypes can be calculated. The observed distribution of zygotic genotypes can then be compared with the expected, by chi-square tests. In 1948 Dobzhansky and Levene reported their findings on the action of selection in natural populations of *D. pseudoobscura.* Female flies that had mated with males in their natural habitats were collected and allowed to produce offspring in the laboratory. In these offspring, the inversion homozygotes and heterozygotes were found to have frequencies that agreed with those expected from the Hardy-Weinberg equilibrium. These investigators also collected male flies and had them mate individually in the laboratory with females that were homozygous for one of the gene arrangements. By examining the chromosomes of the salivary gland cells of the offspring, the gene arrangements of the males' chromosomes could be ascertained. When this was done with a great number of males, it was found that there was a small but significant excess of heterozygotes, with a corresponding deficiency of homozygotes, differing from results expected on the basis of the Hardy-Weinberg equilibrium. It was concluded that a differential mortality takes place during the development of the flies, favoring the heterozygotes over the homozygotes. The retention of the different gene arrangements in the population through the mechanism of heterosis demonstrated that selection was operating to produce and maintain a *balanced chromosomal polymorphism.*

In a study of natural populations of a different species of flies, *Drosophila paulistorum,* an example of heterosis involving an enzyme locus was reported by Richmond and Powell in 1970. They were investigating the X-linked gene that specifies the enzyme

tetrazolium oxidase, whose function in the animal's body is as yet unknown. They found that, in a given female fly, the enzyme will have one of three electrophoretic banding patterns (F, F/S, or S), depending on whether the female's genotype is, respectively. *F/F, F/S,* or *S/S.* An analysis of the frequencies of genotypes of females collected at a single locality showed a highly significant excess of heterozygotes. This indicated that selection was operating to produce and maintain a *balanced allelic polymorphism* in this *Drosophila* population.

Genetic diversity in experimental populations

Very little is known about the intensities of the selective forces that operate in natural populations. To obtain some indication of how severe these may be, we must rely on laboratory experiments. One approach to this problem has been the use of *population cages.* These cages consist of wooden or plastic boxes. The bottom of the cage has circular openings into which glass containers with food medium are placed at periodic intervals. Correspondingly, worked-out food cups are removed as cups with fresh food are added. A rate of cup replacement is chosen that will ensure each cup's remaining in the cage for more than the generation time of the flies. With this experimental arrangement, a population may be maintained indefinitely, generation after generation. The experimental population will resemble a natural population in that there will be present at all times members from overlapping generations, in the various stages of the life cycle. An example of a population cage is shown in Fig. 14-6.

Several hundred flies are introduced into the cage at the beginning of the experiment. These flies are a mixture of individuals with different gene arrangements in desired proportions. Within a single generation, the population of the cage increases to the maximum that the food will support. The competition for survival is intense, since the numbers of eggs deposited are tens to hundreds of times greater than the numbers of

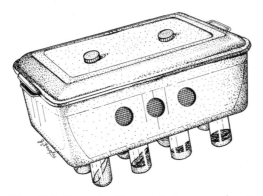

Fig. 14-6. Example of a population cage. A rate of cup replacement is chosen that will ensure each cup's remaining in the cage for more than the generation time of the flies.

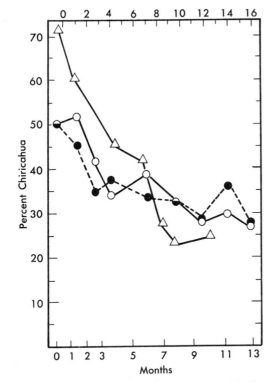

Fig. 14-7. Frequencies of chromosomes with the CH gene arrangement in populations of geographically uniform origin. Dobzhansky's 1948 data—white triangles; Levine's 1955 data—white and black circles. (From Dobzhansky, T. 1948. Genetics **33:**588-602; and Levine, L. 1955. Genetics **40:**832-849.)

adult flies that hatch. Egg samples are periodically taken from a cage, the larvae are dissected, and their salivary gland cells are examined for chromosome types. From these data, the frequencies of the gene arrangements are determined.

In one such series of experiments, flies derived from the Piñon Flats population and carrying the Arrowhead and Chiricahua gene arrangements were placed in cages. The data obtained in two such experiments are shown in Fig. 14-7. The frequency of Chiricahua comes to an equilibrium at about 25% and continues thereafter at that level. Some populations that have been maintained for more than eight years still retain this equilibrium level. The experiments show that the initial frequency of the gene arrangements does not determine its equilibrium level. From the amount of change in frequency of the two chromosome types per generation, we can calculate the adaptive values of the three genotypes. For the two experiments illustrated in Fig. 14-7, the adaptive values are given here:

Karyotype	Dobzhansky (1948)	Levine (1955)
AR/AR	0.81	0.86
AR/CH	1.00	1.00
CH/CH	0.53	0.48

The adaptive values, based on two different sets of data, are in good agreement. These values indicate that in these experimental populations, we have witnessed the establishment and maintenance of a balanced chromosomal polymorphism in which all the gene arrangements introduced into the cage at the beginning were preserved in certain proportions in the final population.

Analyses were made in the 1948 experiment of the Hardy-Weinberg proportions among larvae and among adults. As in natural populations, the larvae of the 1948 experiment showed karyotype frequencies that were close to the expected, but the adults showed an excess of heterozygotes and a deficiency of homozygotes. Here, too, there appears to be a differential mortality that takes place during development. Thus, selection appears to act in similar fashion

Table 14-4. Number of matings recorded in observation chambers with various ratios of *AR* and *CH* homozygotes

Pairs of flies	Number of groups	Having mated					
		AR ♀ ♀	CH ♀ ♀	AR ♂ ♂	CH ♂ ♂	♀ ♀ p ♂ ♂	
12*AR*, 12*CH*	15	137	128	131	134	>0.05	>0.05
20*AR*, 5*CH*	17	138	69	136	71	<0.02	<0.02
5*AR*, 20*CH*	19	69	140	105	104	<0.02	<0.02

From Ehrman, L., Spassky, B., Pavlovsky, O., and Dobzhansky, T. 1965. Sexual selection, geotaxis, and chromosomal polymorphism in experimental populations of *Drosophila pseudoobscura*. Evolution **19**:337-346.

both in natural and in experimental populations.

Selective advantage of the rare karyotype

An interesting problem arose in studies of some experimental populations. It was found that the adaptive values of the different karyotypes varied with the frequency of the karyotypes. In general, as a genotype became rare, its adaptive value increased, and it appeared to be favored to some extent in the population. This was best seen in populations that did not initially exhibit heterosis. Under these conditions, one of the homozygotes usually had a higher adaptive value than the heterozygote and other homozygote. This situation would be expected to lead to elimination of the gene arrangement forming the less-favored homozygote. However, in many such cases, as the gene arrangement became rare, the selection against it seemed to lessen, and it maintained itself at a low level. It did not seem reasonable to involve the sudden appearance of heterosis when a gene arrangement is rare, since heterosis should be an inherent property of the chromosome types and should not be frequency-dependent.

A resolution of this problem appears to have been found in observations of the frequencies of the different types of matings that occur when flies with rare and common genotypes are available as mates. Experiments dealing with this phenomenon were reported by Ehrman and co-workers in 1965. These investigators observed the frequencies of matings that occurred when flies homozygous for either Arrowhead or Chiricahua, in varying frequencies, were placed together in an observation chamber. The data are shown in Table 14-4. They indicate that when the two types of gene arrangements are equally frequent, mating is purely random, and all genotypes mate with equal frequency. However, when one of the karyotypes is rare, it mates proportionally much more frequently than expected, as seen by the "p" values obtained from chi-square tests. The most striking example of the mating superiority of the rare individual is seen in the case of the rare Arrowhead males. The physiological basis of the mating advantage of the rare forms is at present unknown. As we discussed in Chapter 12, in *Drosophila* the females are believed to be responsible for whether a mating will or will not be successful. Apparently, they are attracted in some fashion to the male who is "somehow different." This selective advantage of the rare karyotype has important implications for the maintenance of polymorphism in populations. It implies that polymorphism can persist in a population even if the heterozygous genotypes do not possess an advantage compared to the homozygotes. Here, we have an interaction of behavior and population genetics that can have important evolutionary consequences for the species.

• • •

We shall now turn our attention to the study of genes in human populations. We do not have available as sophisticated a series of observations from which to draw conclusions as we have with *Drosophila*. This is to be expected when one considers man's long generation time and the lack of controlled matings.

GENETIC DIVERSITY IN HUMAN POPULATIONS

On investigating the types of genetic differences that occur among human populations, we find that these differences are in frequencies of alleles at various loci, rather than in inversions or other genetic variations. We shall discuss the geographic distributions of the alleles of some genes and attempt, whenever possible, to indicate the forces that might account for the observed frequencies. All our considerations will be confined to genes whose effects are relatively large because they are the most easily studied and have been the most completely investigated. Unfortunately, so little is known of the inheritance of polygenically controlled traits that we cannot even begin to characterize genetically the populations that are known to differ in these traits. The blood groups and hemoglobins provide the best illustrations we have of the genetic diversity that exists in human populations.

ABO blood group genes

The blood group gene that shows the most regular pattern of distribution is *B*. Its highest frequency is observed in central Asia (25% to 30%), with the incidence decreasing in all directions as we move out from this center. An east-west transect across Asia and Europe shows western Asia with some 20% to 25% of this allele, Russia with 15% to 20%, central Europe with 10% to 15%, France and England with 5% to 10%. Specific localities may have frequencies of *B* that vary from that charac-

teristic of the region, and these localities are especially interesting to study. In the native populations of Australia and North and South America, the *B* gene is either completely absent or is found in rather low frequencies (5% to 10%). In Africa, the *B* gene has its highest frequency, among native populations, in the central Congo region (15% to 20%) and then decreases to 10% to 15% in both north and south directions, with isolated localities containing even less (5% to 10%) of this allele.

Blood group gene *A* has a more universal but less regular distribution than *B*. Gene *A* has its highest frequencies (35% to 45%) in western Europe, Australia, and certain areas of North America. Spreading out from these localities are gradients of decreasing frequency of *A*. In Europe the gradient moves toward Asia, with central Europe having 25% to 30% of this allele; Russia and most of Asia, 20% to 25%; and eastern Asia, 15% to 20%. Similar frequency gradients occur in Australia and North America. There are isolated localities with high frequencies of *A* among the Tibetans, the Bushmen in South Africa, Congo Pygmies, Negritos in the Philippine Islands, and some "hill tribes" of southern India.

Blood group gene *O* is the most frequent allele of the series and is found in high frequency over most of the world. Most native populations of North and South America have *O* gene frequencies of about 75%. Regions with similarly high frequencies of this allele are to be found in Africa, Australia, and the Middle East. Western Europe is characterized by a frequency of 60% to 65%, while the rest of Europe and most of Asia have a frequency of about 55% to 60%. Most African populations average about 65% to 70% for this allele. Some American Indian tribes (e.g., Kwakiutl of British Columbia, Utes of Montana, and Toba of Argentina) are almost 100% homozygous for the *O* gene. Other tribes have lesser amounts of *O*, with *A* as the other allele: the Blackfeet of Montana have 49% *O* and 51% *A;* the Navaho of New Mexico have 87% *O* and 13% *A*.

The great variation in the frequencies of the different blood group genes has raised the question as to whether these alleles bestow any selective advantage or disadvantage on their carriers in the different parts of the world. Thus far, no evidence has been found that any of these alleles make their carriers more fit for a given environment. This last statement does not mean that selective forces are not operating on these alleles. Geneticists know, for example, that in marriages of blood type O mothers and heterozygous blood type A fathers (as shown by the birth of a blood type O child) there is a very significant deficiency of blood type A children, amounting to almost 25%. When the mother is heterozygous blood type A and the father O, this deficiency does not occur. The most apparent explanation rests on the fact that type O persons normally produce. anti-A and anti-B antibodies. Mainly during the birth process, some of the blood type A cells formed by an A fetus enter the mother's bloodstream through breaks that occur in the placenta. The A cells act as antigens that stimulate the production of additional large amounts of anti-A antibodies. In subsequent pregnancies, these antibodies pass through the placenta into the fetus' bloodstream, destroy the blood cells of the fetus, and cause the death of the fetus. This situation can also arise if blood type B males or AB males marry type O females. Such matings are termed *ABO-incompatible*. Although the knowledge that selective forces are acting at this locus is very interesting for the study of population genetics and despite the fact that this information may be useful in genetic counseling of prospective parents, it does not help explain the observed frequencies of the various alleles in different parts of the world.

A striking example of the action of selection in relation to the ABO blood groups was reported by Osborne and de George in 1957. These investigators studied pairs of twins who were demonstrated to be dizygotic (i.e., each twin was formed from a separate fertilized egg). One group of such twins were juveniles when studied (4 to 8 years old), while the others were adults (18 to 55 years of age). The twins were examined for ABO blood groups and classified as to whether they had the same (concordant) or different (discordant) blood groups. It was found that 91% of the juvenile pairs and 70% of the adult pairs of twins were concordant for ABO blood groups. The excess of dizygotic pairs concordant for ABO blood groups among the juvenile twins was statistically significant, but the excess among the adults was not. The results from the juvenile twins strongly support a hypothesis of selective survival in dizygotic twins in relation to the ABO system. The results from the adult twins may indicate a further selective factor acting between the ages of 8 and 18 that adversely affects the survival of concordant twins. The mechanisms by which these selective forces are acting are unknown.

Another interesting discovery about the alleles of this locus is their relationship to the incidence of some diseases. There seems to be a positive correlation between the occurrence of duodenal ulcers and blood type O. The relative incidence of duodenal ulcer in people of blood type O, as compared to persons of other blood groups, is about 1.4:1. This means that persons of blood type O have a 40% greater chance of developing duodenal ulcers than people of other blood types. Again, this does not help explain the known distributions of these alleles.

If we cannot explain the observed gene frequencies by selection, we must consider other possibilities. The most plausible alternative to selection appears to be genetic drift. It is generally agreed that until rather recently in his history man was a hunter, lived in small groups, and traveled a good deal in search of food. Under such conditions, it is very likely that many of the modern-day populations originated from small migrant groups that became isolated from the parental population. This would explain the extremely high incidence of the *O* gene in American Indian populations despite their assumed origin from Asiatic populations that are high in *B*.

Rh blood group genes

The *Rh* locus is considered by some investigators to be a single locus with many alleles, whereas other investigators believe that three closely linked genes *(cde)*, each with a dominant and recessive allele, are responsible for the observed antigen-antibody reactions. There is only one Rh-negative genotype, which may be symbolized as *rh/rh* or *cde/cde*. The Rh-positive genotypes are all characterized by at least a single *Rh* or *D* gene. The *rh (cde)* gene is most frequent (55%) in the Basque population of Spain, is about 40% among the majority of the European populations, and drops to zero or near zero in the Chinese and Japanese populations. It is relatively low in frequency in Africa (23%) and somewhat higher in American Negroes (28%) due undoubtedly, as discussed in Chapter 13, to an admixture with American whites who reflect the European frequency of 40%. Among the native Indian populations of America and Australia, the *rh (cde)* gene is completely absent. Obviously, very significant differences exist in the frequencies of this gene throughout the world's populations. However, no gradual gradients in frequency, such as that found for the *ABO* genes, are found for the *rh (cde)* gene.

The Rh-positive genes are of a number of types and of irregular distribution. In Chapter 13, we discussed the allele *Rh⁰ (cDe)* with reference to the phenomenon of migration. This gene has a frequency of 5% or less in Europeans, Chinese, American Indians, and Australian aborigines but of 60% or more in African Negroes and Pygmies. Another Rh-positive gene is *Rh¹ (CDe)*. It has a high frequency among Europeans (55%), an even higher frequency among Asiatics (60%), a relatively low frequency among Africans and American Indians (25%), but a high frequency among Australian aborigines (58%). The other Rh-positive alleles show equally irregular distributions.

As in the case of the *ABO* alleles, it is of interest to ask whether the alleles of the *Rh* locus are subject to the action of selection in a given environment. Here, too, no evidence exists that the various alleles affect the adaptive values of their carriers differentially in any geographic area. However, in all localities selection is operating on the alleles of this locus. The selection force has its basis in the phenomenon that people with an Rh-negative phenotype can produce antibodies against Rh-positive blood cells, while the reverse does not occur. Under these conditions, an Rh-negative female married to an Rh-positive male *(RhRh or Rhrh)* will produce all or some Rh-positive children. The Rh-positive blood cells of the child will enter the mother's bloodstream, mainly through breaks that occur in the placental blood vessels at birth, and will stimulate the production of anti–Rh-positive antibodies. Since no antibodies of this type existed in the bloodstream of the mother until the seepage of these fetal blood cells occurred, the antibodies produced cannot affect the fetus of the first pregnancy. However, starting with the second or later Rh-positive babies, the antibodies may cause a hemolytic anemia of the fetus or newborn child. The disease is called *erythroblastosis fetalis,* and the marriage is termed *Rh-incompatible.* Rh-incompatibility has serious consequences in about 1 birth in 200 or 300 in the United States. Through the development of a transfusion technique for replacing the Rh-positive blood of the infant with Rh-negative blood, most of the affected children that are born alive can be saved. It is now also possible to prevent the production of anti–Rh-positive antibodies by Rh-negative females by giving the mother, immediately after the birth of an Rh-positive child, an injection of anti–Rh-positive antibodies. This will serve to destroy the fetus' incompatible Rh-positive cells before the mother's antibody-producing machinery has an opportunity to begin functioning.

The loss of those Rh-positive children that are not born alive acts as a selective force against the alleles at this locus. Each

zygote eliminated is a heterozygote, and there is therefore the loss of one Rh-positive and one Rh-negative allele. The effect of this kind of selection on a population depends on the relative frequencies of the alleles. If they are equally frequent in the population, equal proportions of the two alleles will be lost from the population through the death of the heterozygous fetus. If, however, one allele is more common than the other, the elimination of equal numbers of the two alleles results in the loss of a *greater proportion* of the less frequent allele. This will be reflected in a change in allele frequencies in the next generation. If we look at the frequencies of the two alleles in different populations, we find that in the Basque population of Spain Rh-incompatibility should be operating to eliminate the Rh-positive genes from the population whereas in the rest of the world the reverse should be true. Disregarding other evolutionary forces, this process should eventually lead to one human population (Basque) that is completely Rh-negative, while the rest of the world's populations are completely Rh-positive. The selection process has been complicated, at least in the United States, where marriage of Rh-negative females to heterozygous Rh-positive males (i.e., as shown by the birth of an Rh-negative child) results in a larger number of offspring than the average. The children born alive later in these marriages are all Rh-negative and represent an *overcompensation* for the loss of earlier Rh-positive fetuses.

As in the ABO situation, the selective forces at work on the *Rh* alleles are not correlated, in any known fashion, with environmental factors. Again the question arises as to what explanation can be offered for the observed distributions of the various alleles. Once more, genetic drift appears to be the most plausible answer. This explanation is supported by other evidence that the Basques, living in the Spanish Pyrenees, are descended from an ancient European population that has been largely displaced elsewhere on that continent by later immigrants. The Basque population thus represents an

isolate from a population and could have been subject to the action of genetic drift.

Hemoglobin genes

Our discussion of genetic diversity in human populations has thus far been restricted to genes that determine the antigenic properties of red blood cells. We now want to consider the distribution of some of the genes that affect the hemoglobin molecules of red blood cells. As our first illustration we shall take the sickle cell gene, which we discussed in Chapter 10 as an example of a gene whose alleles have been shown to differ from one another in causing the substitution of one amino acid for another in the final gene product. The distribution of the sickle cell gene has been best studied in Africa, where it reaches its highest frequencies, although the gene is also found in the Mediterranean region, the Middle East, southern India, and Indochina. The data are generally presented as the frequencies of the "sickling phenomenon," which includes the total proportions of heterozygotes and homozygotes. In the Congo region of Africa, the sickling phenomenon reaches its highest known incidence (40%) in some localities, although the average for the region is about 20%. Much lower frequencies of this trait are observed in both North and South Africa (0% to 5%). There are striking fluctuations in the incidence of the trait from tribe to tribe in central Africa. In 1949 Lehmann and Raper reported on their studies in the Upper Nile region. They found that in four different tribes speaking a Hamitic type language the frequency of individuals whose blood cells sickled varied between 0.8% and 3.9%; in seven different Nilotic-language tribes, the incidence varied between 21.0% and 28.0%; in tribes speaking a Bantu type language, the frequency of sicklers varied all the way from 2.0% to 40.0%. These tribes are located relatively close to one another, and the gene frequencies may represent the action of genetic drift in this region, since the members from different tribes do not normally intermarry. The frequencies of the sickling phenomenon are somewhat less (5% to 15%) in other areas of their distribution than they are in Africa.

The high incidence of a homozygous lethal gene, such as the sickle cell gene, immediately raised the hypothesis that the heterozygotes must in some way be adaptively superior to the normal homozygote. Since the sickle cell gene has a somewhat limited distribution over the earth, it seemed plausible that the observed heterosis was a response to an environmental factor. In 1954 A. C. Allison presented evidence of an association between the sickle cell gene and malaria. This was later confirmed by others who demonstrated conclusively that sickle cell trait carriers (heterozygotes) have a selective advantage over other genotypes in areas with endemic malaria due primarily to *Plasmodium falciparum*. The concurrence of the sickle cell gene and falciparum malaria over the areas of their distributions is shown in Fig. 14-8. This illustration shows only the distribution of sickle cell genes, not their frequencies.

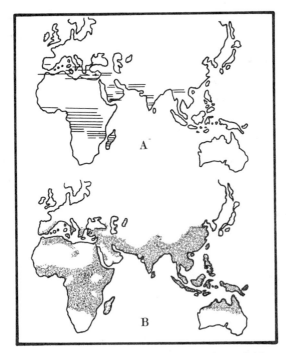

Fig. 14-8. Distribution of sickle cell hemoglobin (bars in map **A**) and falciparum malaria (shaded areas, map **B**). (From Motulsky, A. G. 1960. Hum. Biol. **32**:28-62.)

If we assume that the sickle cell gene in a particular area is at equilibrium with its normal allele, we can calculate the selective disadvantage of the "+/+" genotype in that region.

Table 13-12 gives the form for selection that favors heterozygotes. The equilibrium equation is shown below:

$$\hat{q} = \frac{s_1}{s_1 + s_2}$$

In our case, $\hat{q}$ is the frequency of the sickle cell gene at equilibrium, s_1 is the selection coefficient against the normal homozygote, and s_2 is the selection coefficient against the sickle cell homozygote. If we consider all sickle cell homoyzgotes as lethals, which is probably the case in most of the medically deprived areas of its distribution, $s_2 = 1$. The equation then becomes

$$\hat{q} = \frac{s_1}{s_1 + 1}$$

Taking an area where 40% of the population are carriers (2pq) of the gene and allowing for the complete absence of the homozygous lethal class (q^2), we arrive at the value of q = 0.25. Substituting this number in the equation and solving the equation, we obtain $s_1 = 0.33$. This means that, on the average, normal homozygotes produce only 67% as many offspring as do sickle cell heterozygotes. We have taken as our example of the equilibrium frequency for this gene the highest frequency yet recorded. Areas with lower frequencies of sickle cell gene may represent regions where malaria is less common. It should be kept in mind that if falciparum malaria is eradicated from a given area, the selection process will change and will follow the pattern for the elimination of a recessive lethal. This, in effect, is what has happened in the case of the American Negro whose forced transfer to the United States resulted in his removal from a falciparum malarial region. The frequency of the sickle cell gene in the American Negro population is 4% to 5%, as compared to the 15% to 25% that is characteristic of the area of Africa

from which the population was derived. Some of the reduced frequency of this gene in American Negroes is undoubtedly due to the admixture which has occurred with the American white population. However, the extreme reduction in frequency of the sickle cell gene in the American Negro population must, in large part, be the result of selection operating against the gene.

Another hemoglobin gene that we can consider is that for thalassemia, which, as will be recalled, is normally homozygous lethal. Interestingly enough, a map of the distribution of thalassemia would in general coincide with that showing the distribution of malaria. However, it does not appear that it is the falciparum malaria which is correlated with the thalassemia but, rather, the quartan and benign tertian types that are caused, respectively, by *Plasmodium malariae* and *Plasmodium vivax*. Thalassemia does not occur in as high frequencies as the sickling phenomenon. The largest recorded frequency of thalassemia heterozygotes is 25%. However, this is much too high a frequency for a recessive lethal gene unless there is some selective advantage that the heterozygote has over the normal homozygote. The thalassemia heterozygote is believed to have some resistance to the two forms of malaria mentioned above. Much more work remains to be done on the distributions of thalassemia and other hemoglobins before their frequencies can be satisfactorily explained. However, unlike the dilemma we faced with the blood groups, the distributions of the various hemoglobins can be correlated with specific environmental factors.

Human races

As stated earlier, the concept of *race* is a population concept. To have a race, one must have a population of organisms among whom matings occur regularly and which is genetically distinguishable from other populations of the same species. Applying the criterion of "random mating" to a human population, we find a very confusing situation. Races of sexually reproducing species

usually occur in different territories. This is to be expected, for if more than one population lived in the same locality, the members of the two populations would interbreed, and the exchange of genes between the populations would fuse them into a single population. However, in human populations this postulated fusion of adjacent populations need not occur. The development of different cultures has allowed human races, which in the past were restricted to different territories, to exist today side by side in the same locality without immediate fusion. An explanation is that intermarriage of members of different human populations is often prevented by social, religious, economic, and other cultural differences. On the other hand, modern transportation has permitted very widespread travel of people, with the inevitable consequence of marriage between persons from different populations. Thus we have two forces in operation, each tending to undo the effects of the other. The balance of victory seems to be on the side of the force that is tending to break down the previous isolation of human populations. These are now being fused into larger, more variable groups.

Although the subdivision of the human species into distinct races may in time disappear, at present several genetically distinguishable populations can still be recognized. Most physical anthropologists agree on the following six racial types: (1) Negroid, (2) Caucasoid, (3) Mongoloid, (4) Bushmen, (5) Australoid, and (6) Polynesian. The original criteria used by physical anthropologists to distinguish these racial groups included skin, hair, and eye colors, stature, hair form, conformation of nose and lips, and shape of head. Unfortunately, the modes of inheritance of these traits are not well understood, since they are determined mainly by polygenes (Chapter 4). As can be easily appreciated, the geographic distribution of polygenes is extremely difficult to determine. We should note that the criteria used by anthropologists for the classification of races do not result in an infallible system. The differences between the races are relative rather than absolute, and each race contains types that are considered typical of some other group.

If we attempt to use blood group data for purposes of racial classification, the results are no better than with the original anthropological criteria. We can characterize Asiatic populations (Mongoloids) by their frequencies of the *ABO* genes and find that these are different from European populations (Caucasoids). However, this type of analysis would result in placing the American Indians in the same race as the Europeans, although there is ample evidence that the American Indians are derived from Asiatic populations. In like fashion, we would have problems with using the *Rh* genes even though the *Rh⁰* allele is found, to any appreciable extent, solely in African Negroes and Pygmies. The use of the *Rh⁰* allele as a criterion would then classify all the other human populations into a single race. It should also be apparent that the hemoglobin genes cannot be used for racial classification, since malaria is not restricted to any racial group. It follows from even this brief discussion that human populations differ mainly in frequencies of certain genes rather than in the absence in some populations of certain genes that are present in every individual in other populations. The genetic fusion of human populations, as discussed above, will make such classifications much less meaningful in the future except for historical purposes.

SPECIES FORMATION

A discussion of races leads quite naturally to a consideration of species formation. This follows from the fact that races are usually genetically distinguishable populations living in different territories. The genetic differences between races are reflections of the environmental differences that exist between territories. It should be clear that the same mutations arise in all populations of a species. Due to environmental demands, some mutations will be favored in one locality but selected against in another. From such differences in selection will develop populations with characteristic frequencies of certain alleles or chromosome arrangements, hence races. These genetic differences will result in structural, physiological, and behavioral differences in the members of one race as compared to another. When these differences become so great that they prevent the members of two races from breeding with one another, we say that we have two species. Under such a sequence of events, race formation represents one stage of speciation; races can then be considered *subspecies*. As we have seen in the case of human beings, race formation does not guarantee species formation. Race formation merely provides the opportunity for species formation. However, while some races may remain apart and diverge sufficiently to form species, other races may come together, fuse, and thus eliminate the possibility of species formation. From what we have said, it must follow that speciation is usually a relatively slow process, and we should be able to find examples of races whose members have varying degrees of success in mating with members of different related races. When the process of speciation is complete, the *genetic diversity* that existed between the particular races will have become *segregated,* since it will no longer be possible for the members of the two species to breed. It should then be possible for the two species to come together and live in the same locality but still maintain their separate identities. We may, then, define a species as the largest assemblage of populations whose members are *reproductively isolated* from members of other populations. Under this definition each species is a "closed genetic system." As in the case of race, species is also a population concept.

Reproductive isolating mechanisms

We have stated that structural, physiological, and behavioral differences between members of different populations can prevent their breeding with one another. As might be anticipated, there are a large number of mechanisms that can achieve this effect. No one should be surprised to learn that different species, even within the same

genus, are kept reproductively isolated from one another by different mechanisms. We shall now consider some of the types of reproductive isolating mechanisms that can effect the *segregation of genetic diversity.*

To prevent the meeting of potential mates

Any factors that prevent the meeting of potential mates will act to reproductively isolate members of different populations. Two kinds of barriers can serve this end: a difference of habitat and a difference in the season of sexual maturity. An example of *habitat,* or *ecological isolation* can be seen in two populations of the stickleback fish *Gasterosteus.* One population lives the year round in fresh waters, mainly in small creeks. The other population lives in the sea in the winter, but migrates into river estuaries in spring and in summer, where it breeds. The members of the two populations are reproductively isolated from one another because of their adaptations to different salt concentrations. Since, under these conditions, there is virtually no opportunity for interbreeding between members of the two populations, we may consider these populations as separate species, kept apart by differences in habitat.

A fine illustration of the action of *seasonal,* or *temporal, isolation* has been discovered in the toads *Bufo americanus* and *Bufo fowleri,* which inhabit eastern and central United States. Members of these species can be crossed in the laboratory, and they produce hybrids that are fully viable and fertile. However, in nature they do not produce hybrids despite the fact that their distributions overlap. The reproductive isolation between these species is effected through differences in breeding season. *B. americanus* breeds early in the season, whereas *B. fowleri* breeds late. These species are prevented from exchanging genes by differences in mating seasons.

To prevent the formation of zygotes

Included in this category are the barriers that prevent gene flow between populations whose members do meet and attempt to mate. These barriers may be mechanical, physiological, or behavioral in nature. Examples of *mechanical isolation* are best found among the insects. Many species of insects have complex genitalia. In some cases, copulation with members of another species is impossible because of noncorrespondence of their genitalia. In other cases, copulation may be effected, but the genitalia of one or both of the participants are destroyed, leading to their deaths or making egg deposition impossible. An interesting discovery has been made concerning the importance, for successful copulation, of differences in body size. Variations in body size *within* insect species do not seem to hinder copulation. In *Drosophila,* as an example, giant and dwarf mutants cross easily and produce viable and fertile offspring. Thus, mechanical isolation would appear to be restricted to the structural arrangement of the genitalia.

Physiological barriers to the exchange of genes between populations involve factors that prevent fertilization from occurring should copulation take place. These factors are quite variable from species to species and have been grouped under the rubric of *gametic isolation.* This type of reproductive isolation has been studied in many organisms, including two species of sea urchins *Strongylocentrotus purpuratus* and *Strongylocentrotus franciscanus.* Eggs of each species were placed in seawater containing different concentrations of spermatozoa of its own, or the other, species. The concentration of *S. franciscanus* sperm necessary to produce 100% fertilization of the eggs of the same species was determined. This same concentration of *S. franciscanus* sperm produced from 0% to 1.5% fertilization of *S. purpuratus* eggs. Attempts at achieving fertilization of *S. franciscanus* eggs by *S. purpuratus* sperm gave similar results. There seems to be no doubt that some as yet unknown physiological differences between the gametes from members of the two species prevent fertilization from occurring.

Two interesting phenomena involving physiological isolation have been observed in

Table 14-5. Results of multiple-choice experiments involving two closely related *Drosophila* species

	D. pseudoobscura ♀♀		D. persimilis ♀♀	
	Inseminated	Virgin	Inseminated	Virgin
D. pseudoobscura ♂♂	88%	12%	10%	90%
D. persimilis ♂♂	23%	77%	64%	36%

From Mayr, E., and Dobzhansky, T. 1945. Experiments on sexual isolation in *Drosophila*. IV. Proc. Nat. Acad. Sci. U. S. A. **31**:75-82.

various species of the Jimson weed, *Datura*. In the Jimson weed, as in all advanced seed-bearing plants, pollen grains are deposited on the stigma of the pistil of the flower. The pollen grains must then develop into pollen tubes which grow through the style and eventualy reach the ovary. Different species of *Datura* have styles of different lengths. The crosses in which the species with a short style is used as the female parent and that with a long style as the male parent are more likely to succeed than the reciprocal crosses. This appears to be due to differences in rates of pollen tube growth. Pollen from species with long styles develop pollen tubes that tend to have more rapid rates of growth than pollen tubes in species with short styles. The slow-growing pollen tubes are not able to reach the ovary before the flower withers. Another phenomenon was observed involving pollen tubes growing in the styles of a foreign species. For reasons to date unknown, these pollen tubes have a very high frequency of bursting. Thus we have speed of pollen tube growth and frequency of pollen tube burst acting to prevent crosses between different species of this genus.

Behavioral barriers may exist that prevent matings from occurring between members of different species. This type of reproductive isolating mechanism has been variously called *sexual, psychological,* or *ethological isolation*. It exists when the mutual attraction between males and females of the same species is greater than the attraction between males and females of different species. In Chapter 12, the basic courtship pattern of *Drosophila* was described. Direct observations of courtship of many species were reported in 1952 by H. Spieth. He found that every species has its own courtship pattern and mating techniques and confirmed the observation that the females decide whether a mating will occur. The strength of the sexual isolation existing between two species can be measured experimentally in what is called a "multiple choice experiment." In this procedure, equal numbers of virgin females of two species are placed together with males of one of the species. After a particular time interval, the females are dissected and their sperm receptacles are examined for the presence or absence of sperm. If the species involved are very different from one another, only females of the same species to which the males belong are inseminated. If the species are closely related, some females of both species may be inseminated, but the proportion of inseminated females of the same species as the male will be higher than that of the females of the foreign species. An example of the data obtained from experiments involving the closely related species *D. pseudoobscura* and *Drosophila persimilis* is shown in Table 14-5. It is evident that each type of male inseminated more females from his own than from the other species.

Table 14-6. Percentages of inviable hybrids in crosses of species of frogs *(Rana)*

Males / Females	sylvatica	pipiens	palustris	clamitans	catesbeiana
sylvatica	—	100	100	100	100
pipiens	100	—	0	100	100
palustris	100	0	—	100	100
clamitans	100	100	100	—	100
catesbeiana	100	100	100	100	—

From Moore, J. A. 1949. *In* Jepsen, G. L., Mayr, E., and Simpson, G. G. (ed.). Genetics, paleontology, and evolution. Princeton University Press, Princeton, N. J.

To handicap species hybrids

The last group of factors that we shall discuss are those that act to prevent gene flow between species by rendering species hybrids either inviable or sterile. A number of excellent examples of *hybrid inviability* are known. One such case involves hybrids between species of flax. In the cross of *Linum perenne* and *Linum austriacum,* the hybrid seeds will fail to germinate regardless of which species acts as male or which acts as female. The cause of the lack of germination was discovered to be due to an inability of the embryos to break the seed coats that surround them. If the embryos are freed from their seed coats by an experimenter, they are perfectly viable and produce fertile plants.

Another type of hybrid inviability was reported for a group of North American species of frogs *(Rana)* by J. A. Moore in 1949. The distribution areas of all the species studied overlap, although some are, to varying extents, reproductively isolated by habitat and seasonal isolation. In the laboratory, the eggs of each species were placed in separate batches and each batch was fertilized by the sperm of one of the species. The results, shown in Table 14-6, indicate that the percentage of the various species hybrids which developed to the adult stage was very low and that most of the hybrids died at various stages of development. It is clear that even if the habitat and seasonal isolation referred to above is not effective in preventing breeding between members of these species, hybrid inviability will serve to reproductively isolate them.

Should all previously mentioned barriers to species hybridization fail to isolate two species, there is still the possibility that *hybrid sterility* will do so. The classic examples of this type of reproductive isolating mechanism are the mule and the hinny. The mule is the result of a cross that uses the horse as the female parent and the donkey as the male parent. The hinny is the result of using the donkey as the female parent and the horse as the male parent. Both mules and hinnies are either males or females. These species hybrids can mate, but no offspring result regardless of whether the mate is a horse, donkey, mule, or hinny. A study of the chromosomes of the four different animals involved shows the following diploid numbers: horse, 64; donkey, 62; mule, 63; hinny, 63. Not only are there differences in chromosome numbers between horse and donkey, there are also structural differences—most noticeable in the greater numbers of metacentric chromosomes in the donkey than in the horse. It has not yet been demonstrated whether the sterility of the mule and hinny is due to a lack of pairing of horse and donkey chromosomes during meiosis, with the resultant formation of nonfunctional sperm.

Hybrid sterility has been well studied in

Drosophila. In some species crosses (e.g., *D. melanogaster* × *D. simulans*), the hybrids of both sexes are always sterile. In other species crosses (e.g., *D. pseudoobscura* × *D. persimilis*), the males are sterile, but the females are fertile. The sterility of the male hybrids has been analyzed genetically. This was done by using flies of the two species whose chromosomes contained mutant genes. With such an arrangement, the distribution of the chromosomes in the hybrid progenies can be followed by observing the phenotypes of the flies. On repeated backcrosses of hybrid females to males of either species, one obtains some males whose chromosomes are all of the same species. These males are fertile. The other males contain mixed chromosomal sets and are sterile. The conclusion drawn from these experiments is that each species has in its chromosomes complementary genes, all of which must be simultaneously present to produce fertile sons.

• • •

The various reproductive isolating mechanisms can be viewed as a series of hurdles that prevent members of different species from breeding with one another. However, not all these factors need be operating in any one case. Furthermore, if a number of these barriers are involved in a particular situation, none of them need be 100% effective by itself in order to ensure the reproductive isolation of the two species. Those mechanisms that serve to prevent matings (e.g., habitat, seasonal differences, or some forms of mechanical and sexual isolation) are beneficial to the species concerned because they do not result in a loss of gametes. The mechanisms that act after mating (e.g., gametic isolation, hybrid inviability, and hybrid sterility) are deleterious to the species involved because they result in a loss of gametes and, hence, potential progeny. It is interesting to note that in nature the most common isolating mechanisms are those that prevent wastage of gametes.

Selection for reproductive isolation

We stated earlier in this discussion that reproductive isolation originates as a by-product of the genetic divergence of races. Since the genetic characteristics of a race serve to adapt it to a particular environment, one may assume that racial crosses would result in offspring that are less well adapted for any available habitat than either parental type. Under such conditions, any offspring produced from racial crosses would be relatively poorly fit for survival and reproduction and would tend to be eliminated (i.e., hybrid inviability). The elimination of these offspring would tend to lead to the extinction of the parental genotypes involved in the interracial matings. Thus there would be a selective advantage to those genotypes that participated in intraracial crosses. In effect, natural selection would be promoting an increase in the degree of sexual isolation between races through the elimination of unfit racial hybrids. A long-term consequence of this type of selection is a continued increase in the degree of reproductive isolation between the races until each becomes a totally "closed genetic system," hence a full species.

Experiments designed to test the hypothesis that selection can increase the reproductive isolation between populations were reported by K. F. Koopman in 1950. In these experiments the species *D. pseudoobscura* and *D. persimilis* were used. When flies of these species are kept together at 16° C., roughly one third of the matings are interspecific. The investigator took equal numbers of males and females from both species and placed them in population cages. Each species had been made homozygous for a different recessive mutation so that the proportions in the progeny of both species and of the species hybrids could be determined by inspecting the offspring. Although the matings and egg deposition took place in the population cages, the cups were removed, and the offspring produced by each cup were scored as to whether they were the result of an interspecific or an intraspecific mating. The species hybrids were then destroyed, and equal numbers of males and females from both species were collected from the cups for use as parents of the next

Table 14-7. Selection for reproductive isolation between *D. pseudoobscura* and *D. persimilis*

Generation	Percent hybrids		
	Experiment 1	Experiment 2	Experiment 3
1	37	23	49
2	24	6.0	17.6
3	40	—	3.3
4	15	—	1.0
5	54	5.1	1.4
6	5	31.4	3.4
7	1.5	2.7	1.8
10	6.3	3.2	0.6
11	2.9	5.8	0.7
12	1.2	4.9	1.7

From Koopman, K. F. 1950. Natural selection for reproductive isolation between *Drosophila pseudoobscura* and *Drosophila persimilis*. Evolution **4**:135-148.

generation. Since the hybrids were destroyed in every generation, the flies that mated solely with members of their own species left more surviving offspring than did flies that mated solely or even occasionally with individuals of the foreign species. The data obtained are shown in Table 14-7.

It is quite clear that selection against the species hybrids was very effective in increasing the reproductive isolation between these species. From the fifth generation on, the proportions of the hybrids were, with some exceptions, below 5%. Multiple-choice experiments were conducted for various generations of the experimentals as well as for flies obtained from the stock cultures (controls), to determine whether the sexual isolation between these species had in fact been increased. It was found that there was a significant increase in the amount of sexual isolation between these species after selection had been conducted for a number of generations. Thus the experimental results confirm what had been hypothesized—that selection acts to strengthen the reproductive isolation of genetically divergent populations if the hybrid formed by their members is less fit than either parental type.

Certain expectations follow from the above findings. If natural selection strengthens the degree of reproductive isolation between species by acting against species hybrids, it should follow that contiguous populations of two species should exhibit greater sexual isolation from each other than widely spaced populations of these same two species. This has been found to be the case. The sexual isolation between *Drosophila miranda* and *D. pseudoobscura* is stronger or weaker depending on the geographic origin of the flies used. The flies of *D. pseudoobscura* from regions in which *D. miranda* also occurs exhibit much greater sexual isolation than do flies from distant regions. Similar results have been obtained by using many other related species belonging to various genera. Another expectation is that species that are kept apart by seasonal or habitat isolation should show relatively little sexual isolation if brought together. This has also been found to be the case. An example of this is seen in the toad species *B. americanus* and *B. fowleri,* whose members are normally kept reproductively isolated from one another by differences in breeding periods (seasonal isolation). In some years when the spring weather is very erratic, both species emerge simultaneously to breed. When this occurs, members of the two species will breed with one another, and a large number of species hybrids will be produced. This observation indirectly confirms the hypothesis that natural selection can act to promote species formation when members of genetically divergent populations are able to breed and produce offspring.

Protein differences in Drosophila species

Although all species are reproductively isolated from one another, not all species are morphologically distinct from one another. In fact, a fair number of instances are known in which two or more species are morphologically very similar. Species that resemble one another very closely in structure and appearance are called *sibling species*. However, it is also found that within the same taxonomic species group or subgroup that contains the sibling species, there are other species that are morphologically distinct from the sibling species. An interesting question has been raised as to whether the degree of morphological resemblance among taxonomically related species is paralleled by a similar degree of genetic resemblance among these same species. An investigation of this question was conducted by Hubby and Throckmorton, whose results were reported in 1968. These investigators studied the electrophoretic banding patterns of 18 different proteins in twenty-seven species of *Drosophila*. The twenty-seven species were chosen so as to form nine similar groups. Each group consisted of two sibling species and one species that belonged to the same taxonomic species group or subgroup as the sibling pair, yet was morphologically distinct from the other two.

Comparisons were made of the electrophoretic mobility of the 18 chosen proteins in each group. A great deal of variation was noted in the results. In one extreme case, 85% of the proteins of the sibling species were found to have identical mobilities, but only 21% of the proteins of either one or both of these sibling species had mobilities

identical to the corresponding proteins of the morphologically distinct species. In contrast to this case, one pair of sibling species had only 23% of their proteins in common, and when considered either together or separately, the pair had a total of 27% of their proteins in common with the morphologically distinct species. When one considers all nine groups of species together, one finds that the sibling species pairs, on the average, shared proteins with identical mobility 50% of the time and that a sibling and a morphologically distinct species shared proteins with identical mobility only 18% of the time. These results indicate that, on the average, a high degree of similarity in morphology is paralleled by a high degree of similarity at the protein, and hence on the genetic, level. Obviously much more work has to be done before we can speak with confidence on the genetic compositions of different species and, as a result, are able to achieve a better understanding of the speciation process.

Speciation in man

Our discussion of races in man indicated that these populations are in the process of genetic fusion due to widespread travel and marriages of people derived from different geographic areas. One might very well ask whether the racial hybrids formed are less fit than the parental types and, if so, why natural selection does not act to promote reproductive isolation among the races. The answer is to be found in the tremendous control that man has over his environment. The relative fitness of any genotype is a function of the environment in which the genotype must survive and reproduce. Those environmental factors that channeled the early evolution of man toward race formation are no longer effective. Through modern medicine and social welfare programs, man is steadily moving to the point where he will be able to give each genotype an adaptive value of 1 and thus establish a Hardy-Weinberg equilibrium for all human alleles. Under such environmental conditions, all offspring, regardless of genotype, will have equal chance of survival. Without a differential survival of the various genotypes, there can be no change in gene frequency and, hence, no further divergence of human populations. The existing genetic differences between human populations will then be eroded away by the action of migration, which appears to be increasing with time.

It is quite apparent that the early evolution of man did *not* proceed to the point of species formation of any of the racial groups, since under modern environmental conditions, all racial crosses result in viable and fertile offspring. Also evident is the fact that the evolutionary process of speciation has been reversed in man and that future populations of human beings will be larger and genetically more variable than those we find today.

SUMMARY

In this chapter we have considered the process of segregation of genetic diversity. This has included a study of race and species formation. We first examined an example of genetic diversity in *Drosophila* populations—the inversion system found in *D. pseudoobscura*. We found that the frequencies of the different inversions exhibit geographic, altitudinal, and seasonal variations. We also found that the population of each locality can be characterized as a genetically distinguishable unit, which is called a race. We then reviewed studies of natural and experimental populations that demonstrated the action of selection in determining the genetic composition of a particular population. There followed a discussion of the selective advantage of the rare karyotype and its effect on the maintenance of polymorphism in a population. We then turned our attention to human populations and considered (1) their genetic diversity as seen in blood group genes and hemoglobin genes, (2) the possible role of selection in determining this diversity, and (3) race formation. After our discussion of race formation, we considered the process of species formation. This process involves the development of reproductive isolating mechanisms that prevent gene exchange between members of different populations. Such mechanisms may include habitat and seasonal, mechanical, gametic, and sexual isolation as well as hybrid inviability and hybrid sterility. Experiments showing that selection could increase the reproductive isolation between species were discussed. The chapter was concluded with an examination of the speciation process in man, including an analysis of the role played by modern man in his own evolution.

Questions and problems

1. How has the study of chromosomal banding patterns provided information regarding karyotype phylogeny in the species *Drosophila pseudoobscura?*
2. What is the significance of the finding that populations living in different habitats often differ in the relative frequencies of their gene arrangements?
3. Distinguish between species, populations, and race.
4. Discuss the statement: "Any slight initial advantage presented to a particular gene arrangement in a given environment will tend to promote the occurrence of further genetic improvements for that environment in the same chromosome type."
5. Define the following terms:
   a. Balanced chromosomal polymorphism
   b. Heterosis
   c. Subspecies
   d. Adaptive value
   e. Erythroblastosis fetalis
   f. ABO incompatibility
6. Describe an experiment designed to study the establishment and maintenance of balanced polymorphism.
7. What ramifications does the advantage of the rare karyotype have in the maintenance of polymorphism in populations?
8. a. How is the "favored" inheritance of type O blood—from a marriage between a homozygous type O female and a heterozygous type A male—explained?
   b. Why is the deficiency in type A offspring not evident in marriages of the reciprocal types (i.e., type O male and heterozygous type A female)?
9. What explanation is offered for the observed distribution of the various alleles of the ABO system and of the Rh system? Why is this the most plausible explanation?
10. a. Why is the sickle cell gene present in such high frequencies among certain populations if the trait is known to result in a

slight anemia and the homozygous condition is known to be fatal?

b. Under what conditions would the sickle cell gene be selected against?

11. How is genetic diversity reflected among human populations, as compared to populations of *Drosophila?*

12. Discuss briefly, the major steps involved in species formation.

13. According to information presented in this chapter, what do you think is the most important factor that transforms races into species? Why?

14. Describe and illustrate the various kinds of reproductive isolating mechanisms discussed in this chapter.

15. What evolutionary purpose is served by the formation of sterile or inviable hybrids?

16. *Drosophila* brought to the laboratory from the wild are often found to be heterozygous for recessive lethal genes. Why has natural selection not acted to eliminate these lethals?

17. *D. pseudoobscura* and *D. persimilis* exhibit a great deal of hybridization at 16° C. Experiments have been conducted wherein *pseudoobscura* and *persimilis* were introduced into a population cage at 16° C. After a series of generations it was noted that the species exhibited an increase in the amount of their sexual isolation from each other. How do you account for these findings?

18. Discuss the feasibility of the following situations:

a. Selection operating before fertilization

b. Speciation occurring without previous geographic isolation.

19. A hereditary neurological disease called kuru is found almost exclusively and in very high frequency among a group of natives of New Guinea. Females and males homozygous for the *Ku* gene contract the disease early in childhood and die. Females heterozygous for the *Ku* gene contract the disease during adulthood and die from the disease, usually in the postreproductive period of their lives. Males heterozygous for the *Ku* gene are unaffected. How do you explain the widespread nature of such a deleterious gene in a human population?

20. Assume that a heterotic gene, which is homozygous lethal, has an equilibrium value of $q = 0.16$. Give the adaptive value of each of the following:

a. Normal homozygote

b. Heterozygote

References

Allison, A. C. 1954. Protection afforded by sickle-cell trait against subtertian malarial infection. Brit. Med. J. **1**:290-294. (Original announcement of the discovery of a selective advantage for sickle cell heterozygotes, in regions where malaria is endemic.)

Darwin, C. 1859. On the origin of species by means of natural selection, or the preservation of favoured races in the struggle for life. Appleton-Century-Crofts, New York. (The classic statement of the effects of the forces of selection on evolution.)

Dobzhansky, T. (ed.). 1942. Biological symposia. Vol. 6. The Jacques Cattell Press, Lancaster, Pa. (A group of papers dealing with the various mechanisms of reproductive isolation and their roles in speciation.)

Dobzhansky, T. 1970. Genetics of the evolutionary process. Columbia University Press, New York. (An excellent book dealing, in general terms, with the biological theory of evolution as discussed in this chapter.)

Dobzhansky, T., and Epling, C. 1944. Contributions to the genetics, taxonomy, and ecology of *Drosophila pseudoobscura* and its relatives. Carnegie Inst. Wash. Publ. 554. (An examination and discussion of the genetic diversity seen in *Drosophila pseudoobscura*.)

Ehrman, L., Spassky, B., Pavlovsky, O., and Dobzhansky, T. 1965. Sexual selection, geotaxis, and chromosomal polymorphism in experimental populations of *Drosophila pseudoobscura*. Evolution **19**:337-346. (An interesting paper dealing with karyotype rarity in *Drosophila* and its advantage in the face of evolutionary forces.)

Harris, H. 1969. Enzyme and protein polymorphism in human populations. Brit. Med. Bull. **25**:5-13. (An excellent review of allelic polymorphisms in human populations.)

Hubby, J. L., and Throckmorton, L. H. 1968. Protein differences in *Drosophila*. IV. A study of sibling species. Amer. Natur. **102**:193-205. (An investigation of the electrophoretic-mobility characteristics of proteins from taxonomically related species.)

Origin and evolution of man. 1950. Cold Spring Harbor Symp. Quant. Biol. **15**:1-415. (A collection of articles dealing with various aspects of race formation in humans.)

Patterson, J. T., and Stone, W. S. 1952. Evolution in the genus *Drosophila*. The Macmillan Co., New York. (A detailed discussion of the geographical distribution, gene variation, chromosomal variation, and reproductive isolating mechanisms in various *Drosophila* species.)

Population genetics. 1955. Cold Spring Harbor Symp. Quant. Biol. **20**:1-346. (A group of 31 papers written by specialists in the field of population genetics.)

Prakash, S., and Lewontin, R. C. 1971. A molecular approach to the study of genic heterozygosity in natural populations. V. Further direct evidence of coadaptation in inversions of *Drosophila*. Genetics **69**:405-408. (A continuing study of the electrophoretic-mobility characteristics of the proteins from flies carrying different gene arrangements.)

Race, R. R., and Sanger, R. 1968. Blood groups in man. 5th ed. Blackwell Scientific Publications, Oxford. (An authoritative review of our knowledge on human blood group alleles.)

Richmond, R. C., and Powell, J. R. 1970. Evidence of heterosis associated with an enzyme locus in a natural population of *Drosophila*. Proc. Nat. Acad. Sci. U. S. A. **67**:1264-1267. (Probably the first reported case of heterosis associated with an enzyme locus in a natural population of *Drosophila*.)

Spencer, W. P. 1947. On Rh gene frequencies. Amer. Natur. **81**:237-240. (An article discussing selection in the Rh system of humans.)

Author index

Subject index

A

A (amino-acyl) site, 24
ABO blood groups, genes for, 324-326
ABO incompatibility, 325
Abortion, chromosome abnormality and, 125-126
Acenaphthene, 122
Acetabularia
  cap formation in, 239-240
  crenulata, 239, 240
  mediterranea, 239-240
  nucleus-cytoplasm interaction, 239-240
  plastids research, 164
  protein synthesis regulation in, 239
Acetylcholine, 268
Acetyltransferase, 223
Acridine dyes, 107, 166, 191, 192
Actinomycin D, 207
  –brain RNA synthesis and memory relation, 278-280
  in mRNA half-life research, 28
  in recombination frequency research, 148
Adaptive value, 298
Additions
  in mutations, 191-192
  proflavin induction, 191-192
Adenine, 6
  tautomerization and, 186-187
Adenosine diphosphate (ADP), 8
Adenosine triphosphate (ATP), 8
  in amino acid activation, 24
ADP; *see* Adenosine diphosphate
Aegilops squarrosa, 127
Aggression, human, chromosomal complement and, 260-261
Alanine, 19, 34
Albinism, 59-61, 73, 221-222
  probability analysis in, 62-64
Alcaptonuria, 220-221
Alcohol preference in mice, 261-263
Algae; *see* Blue-green algae; Green algae
Alkaline phosphatase in *D. melanogaster* research, 76
Alleles, 59, 199; *see also* Nonallelic gene interactions
  codominant interacting, 76
  interactions, 59-66
  multiple, 66
  transformer, 106
Allopolyploid(s), 123
  laboratory production, 127-128
  wheat as, 126-127
Allotetraploid(s), 123, 125
Alpha four, 87

Alpha three, 86-87
Altitudinal gradients of gene arrangements in *D. pseudoobscura*, 317-318, 319
Amber mutant, 209
  suppressor, 209
Ambystoma jeffersonianum, 124
American foulbrood (AFB), 255-256
Amino acid(s), 16-17; *see also* specific compounds
  abbreviations, 31
  activation by ATP, 24
  attachment to tRNA, 24
  –nucleotide sequence relation, 28-29
  in translation, 208
Amino acid adenylate, 24
Amino acid sequence
  of hemoglobin, 214, 215
  of tryptophan synthetase A protein of *E. coli*, 216
Amino-acyl synthetase, 24
Amoeba, 123
Amphibians
  nuclear transplantation in, 240-241
  polyploidy in, 123, 124-125
Amphidiploid(s), 123, 125
Anaphase
  chromosome in, 48
  in mitosis, 48
Anaphase I, 52
Anaphase II, 54
Anemia, 211-212; *see also* specific diagnoses
  hemoglobin and, 211-215
Aneuploidy, 111-121
  addition vs. loss of chromosomes, 120-121
  autosomal, 113, 115
  dermatoglyphics and, 117-120
  height and, 120, 261
  human aggression and, 260-261
  in man, 112, 113, 115-116
  maternal age and, 120
  origin of, 120-121
  sex chromosome, 115-116
Angiosperms, 128
Animals; *see also* specific species
  meiosis in, 47, 51-54
  polyploidy in, 123, 124-126
  totipotency in, 241
  tumor-inducing viruses in, 242-244
Anthocyanin, 72
Antibiotics, 211
  in protein synthesis inhibition, 168
Anticodon, 33-34
Arithmetic mean, 85
Artemia salina, 124